CLIMATE CHANGE IN THE SOUTH PACIFIC: IMPACTS AND RESPONSES IN AUSTRALIA, NEW ZEALAND, AND SMALL ISLAND STATES

ADVANCES IN GLOBAL CHANGE RESEARCH

VOLUME 2

The titles in this series are listed at the end of this volume.

CLIMATE CHANGE IN THE SOUTH PACIFIC: IMPACTS AND RESPONSES IN AUSTRALIA, NEW ZEALAND, AND SMALL ISLAND STATES

Edited by

Alexander Gillespie
University of Waikato,
Hamilton, New Zealand

and

William C.G. Burns
Pacific Institute for Studies in Development,
Environment, and Security,
Oakland, CA, U.S.A.

KLUWER ACADEMIC PUBLISHERS
DORDRECHT / BOSTON / LONDON

Library of Congress Cataloging-in-Publication Data

ISBN 978-90-481-5365-7 ISBN 978-0-306-47981-6 (eBook)
DOI 10.1007/978-0-306-47981-6

Published by Kluwer Academic Publishers,
P.O. Box 17, 3300 AA Dordrecht, The Netherlands.

Sold and distributed in North, Central and South America
by Kluwer Academic Publishers,
101 Philip Drive, Norwell, MA 02061, U.S.A.

In all other countries, sold and distributed
by Kluwer Academic Publishers,
P.O. Box 322, 3300 AH Dordrecht, The Netherlands.

Printed on acid-free paper

This book is dedicated to Al's son, Lex
and
Wil's partner, Tamar

CONTENTS

INTRODUCTION

ALEXANDER GILLESPIE & WILLIAM C.G. BURNS

The idea for this book grew out of the Ecopolitics conference in Canberra, Australia in 1996. The conference captured the ferment of the climate change debate in the South Pacific, as well as some its potential implications for the region's inhabitants and ecosystems. At that conference, one of the editors (Gillespie) delivered a paper on climate change issues in the region, as did Ros Taplin and Mark Diesendorf, who are also contributors to this volume.

This book focuses on climate change issues in Australia, New Zealand, and the small island nations in the Pacific as the world struggles to cope with possible the impacts of environmental change and to formulate effective responses. While Australia and New Zealand's per capita emissions of greenhouse gases are among the highest in the world, their aggregate contributions are small. However, both nations may exert a disproportionate influence in the global greenhouse debate because their obstinate positions at recent conferences of the parties to the United Nations Framework Convention on Climate Change (FCCC) may provide justification for other developed nations, as well as developing countries, to refuse to make meaningful reductions in their greenhouse gas emissions.

Small island states in the Pacific, as well as other regions of the world, may be in the unenviable position of serving as climate change "canaries," manifesting some of the ramifications associated with global warming well in advance of other nations. Moreover, their limited ability to cope with the effects of climate change may further strain their economies and ecosystems, undermining their autonomy, and in some cases, even threatening their existence as nations.

The book is divided into three sections, assessing the impacts of climate change and policy responses in Australia, New Zealand, and small island states in the South Pacific.

The Australian section begins with an assessment of the potential impacts of climate change upon Australia by Jones, Pittock and Whetton. The second part of their chapter describes recent developments in the methodology of impact assessment. These aim to move beyond the identification of sensitivity, to the assessment of 'dangerous' levels of climate change and adaptation options. The basic structure of a common assessment framework incorporating comprehensive scenarios of climate change and variability, thresholds linking states of climate change to levels of impact, and risk assessment are also described.

The second chapter by Harriet Bulkeley focuses on the formation of Australian climate change policy between 1985-1995. Bulkeley argues that the Australian government's leadership at the Rio Conference subsequently devolved into a confrontation between

A. Gillespie and W. C. G. Burns (eds.), Climate Change in the South Pacific: Impacts and Responses in Australia, New Zealand, and Small Island States, 1–5.

those who contended that greenhouse gas emission reductions would imperil the Australian economy, and supporters of "ecological modernisation," who argued that "existing political, economic, and social institutions can internalise . . . care for the environment," and "that there need be no fundamental contradiction between environmental protection and economic growth." Ultimately, the government eschewed the adoption of a carbon tax in favour of voluntary measures that resulted in Australia's failure to meet its obligations under the FCCC.

In the third chapter, Clive Hamilton focuses on the development of the Australian negotiating position prior to the Third Conference of the Parties at Kyoto. He concludes that the Australian government exaggerated the costs of meeting emissions reduction targets and that the models used to justify Australia's position actually demonstrated that the costs of meeting reasonable emission targets would be low. Hamilton also questions the Australian government's trepidation about possible "carbon leakage" to developing nations, as well as the soundness of its advocacy of differentiation. He concluded that Australia's "victory" in the short-term might ultimately be a "poisoned chalice" that undermines global efforts to confront climate change and the need for Australia to move beyond a dependence on fossil fuels.

In the fourth chapter, Mark Diesendorf critiques the arguments proffered by the Australian government to justify its successful effort to obtain permission under the Protocol to *increase* its emissions above 1990 levels. Diesendorf focuses his analysis on the energy sector, the largest producer of greenhouse gas emissions in Australia. He argues that over the course of the last decade the government made virtually no effort to wean the country off fossil fuels and failed to encourage energy efficiency, undermining the legitimacy of positions at Kyoto. Moreover, he suggests that government made serious errors in its assumptions and methodologies in the economic models it used to justify its positions at Kyoto and to the Australian public.

The fifth and sixth chapters by Taplin and Yu also focus on the formation of climate change policy in Australia between 1995 to 1998, and the Australian position at the Kyoto conference. In Chapter 5, they note that Australia's emissions were already 11 percent above 1990 levels by 1995-1996, and slated to increase substantially without additional measures. Unfortunately, the authors contend, the government lacked the political will to meaningfully confront climate change during the latter part of this decade given its commitment to fossil fuel exports. Moreover, it has failed to implement energy efficiency and renewable energy programs. The authors also argue that increased public engagement in the climate change debate might ultimately increase pressure on the government to do more in the future. In Chapter 6, Taplin and Yu focus on the Australians' positions at the Third Conference of the Parties at Kyoto. Australia's "unvarnished emphasis on economic considerations" manifested itself in advocacy of differentiated targets for Annex I nations and reliance on flawed economic modelling to support its positions. This is unfortunate, the authors conclude, because the impact of emissions cutting on the Australian economy remains speculative. Moreover, a commitment to alternative energy sources might actually prove beneficial.

In Chapter 7, the first chapter in the second section of the book, Reid Basher examines the "paradox" of New Zealand. This develops with its "clean low-population environment, [and] its inhabitants predisposed to environmental smugness, but with very little motivation to take environmental action." This chapter opens with an assessment of the potential impacts of climate change upon New Zealand - "some good, many bad, and most with a high degree of uncertainty." While noting that the government has done a reasonably good job of assessing the possible impacts of climate change, the authors outline three shortcomings of government policy responses: its failure to develop an integrated climate change research program, a failure to engage industry in businesses in science impacts and research, and an absence of a national political consensus and will to undertake "no-regrets" responses.

Chapter 8 assesses the formation of New Zealand's climate change policy between 1990 and 1996. Kirsty Hamilton provides a detailed account of the government's reliance on free market responses to "deliver energy efficiency and CO_2 reductions" despite recognition of market failure in the context of energy pricing and investment. Moreover, the government avoided pressures to reduce emissions in the energy and transportation sectors by adopting a controversial "net approach," contending that overall net emissions would be reduced by 50-59% by 2000 as a consequence of tree planting. Hamilton concludes that New Zealand's "victory" in eschewing serious commitments to reducing emissions may ultimately prove to be "Pyrrhic" "if it ultimately fails to prepare itself for a world in which the pressure grows stronger to reduce emissions and the need for a diversified energy base becomes more critical."

Chapter 9, by Alexander Gillespie, builds on the preceding chapter by tracking the debate within New Zealand between 1995 to 1998. Characterizing New Zealand's climate response policy response during the 90s as moving between "the sublime and the ridiculous" Gillespie concludes that New Zealand has been "among the more notable failures in reducing its carbon dioxide emissions to 1990 levels." In commenting on New Zealand's advocacy of the net approach during the negotiations at Kyoto, Gillespie concludes that there remains great uncertainty as to the sources and potential of sinks. Moreover, there is the threat that developing countries may tilt their policies toward creating sinks that can be used in an emissions trading regime to the detriment of other environmental objectives. Gillespie also assesses New Zealand's failure to reduce methane emissions, currently the largest source of greenhouse gas emissions in the nation, and its refusal to use measures such as taxes and subsidies to encourage conservation and energy efficiency.

The role of carbon sequestration as a response strategy to global warming is examined from a scientific point of view in Chapter 10. Justin Ford-Robertson, Maclaren and Wakelin set out the critical methodological factors associated with the measurement of carbon sequestration in planted forests in New Zealand. Moreover, the paper explains the different ways sequestration can be measured, and the substantial differences that such approaches can make in assessing the potential of sinks in New Zealand's greenhouse gas reduction strategies.

The final chapter in the second section is by Peter Alsop. Chapter 11 focuses on the role of joint implementation in international climate change policy. Alsop outlines the history of joint implementation (JI) in the context of the FCCC, and its evolution in the Kyoto Protocol. While concluding that JI could be a "viable and cost-effective instrument for climate protection," he cautions that JI "is no panacea for climate change." Moreover, he contends there are a "myriad of methodological, institutional, financial and political barriers that need to be overcome if the full potential of JI is to be realised." Alsop also outlines JI projects launched over the past few years and discusses the prospects of JI for Australia, New Zealand and South Pacific island nations.

The third section of the book, focusing on small island nations in the South Pacific, begins with an analysis by William Burns of the possible impacts of climate change over the next century. While acknowledging that regional assessments of climate change remain highly speculative, Burns in Chapter 12 outlines the possible impacts of climate change for Pacific Island Developing Countries (PIDCs). These include threats posed by sea-level rise, temperature increases, and violent weather events. Burns concludes that climate change may result in devastation of critical ecosystems in PIDCs, severe damage to economic infrastructure, the loss of species, and a substantial increase in the incidence of serious diseases, e.g., dengue fever and malaria. Burns also assesses the prospects for the FCCC to prevent these possible impacts from occurring in the next century, concluding that industrialized states may find it economically rational to delay action well past the time when PIDCs begin to experience the serious effects of climate change.

In Chapter 13, Michael Edwards advances the provocative argument that "climate change discourse has been dominated by actors in rich and powerful nations, thus marginalising the concerns of people who have little or no control over the nature, magnitude and timing of the threat - the real victims of climate change." Edwards contends that climate change discourse has encouraged "worst case/doomsday scenarios" that contribute to destruction of the region's environment by encouraging short-term planning. The author outlines four areas of research that could help to reduce the security threats that climate change could engender in the South Pacific.

In Chapter 14, John Hay analyses composite records, based on instrumental, proxy, and anecdotal information, to assess the prospects for climate change in the South Pacific. Additionally, Hay outlines some of the present research programs in the region and assesses the methodologies and consistency of global climate models used to conduct regional assessments. Hay concludes that the dominant impacts on small island states may not be related to sea-level rise, but rather changes in rainfall regimes and soil moisture budgets, prevailing winds, and in short-term variations in regional and local sea levels and patterns of wave action.

In the book's final chapter, Joanna Ellison focuses on how mangroves in South Pacific island states may respond to rising sea levels. After outlining the extent of mangroves and their uses in Pacific island states, Ellison looks at some present case studies of mangrove diebacks in other parts of the world that may illuminate the potential consequences of climate change in the south Pacific. However, Ellison also concludes that climate change may have some beneficial effects for mangroves, increasing the productivity of some species and expanding their range. Given the current shortfalls in research in this context, and the likelihood of substantial variations in mangrove responses in the region, Ellison emphasises the need for a global monitoring system to facilitate informed management decisions in the future.

Climate change may prove to be the most serious environmental challenge in the South Pacific in the next century. The modest objective of this book is to frame the ecological and policy issues for those concerned about the fate of the region in the future.

1. THE POTENTIAL IMPACTS OF CLIMATE CHANGE

R.N. JONES, A.B. PITTOCK AND P.H. WHETTON
CSIRO Atmospheric Research
Private Bag No. 1, Aspendale 3195, Australia

1. Introduction

Concern over the potential impacts of future climate change on both natural and human systems drives research into the enhanced greenhouse effect. There are two fundamental questions that need to be addressed by this research: whether atmospheric pollution currently being emitted, and likely to be emitted in the coming decades, will result in dangerous climate-related impacts for future generations, and if so, what action should be taken to limit that danger.

The United Nations Framework Convention on Climate Change (FCCC) has been drafted to address this issue. Within the FCCC there are two complementary strategies that focus on mitigation and adaptation. The first (mitigation) aims to stabilize greenhouse gas concentrations in the atmosphere at levels preventing "dangerous" anthropogenic interference with the climate system. The second (adaptation) recognises that some climate change is inevitable, and that some systems sensitive to climate will prove to be vulnerable, requiring adaptation measures.

The FCCC also requires that strategies dealing with climate change must take place within a framework of sustainable development, by accounting for the ramifications of climate change for both current and future generations. The three key systems focused upon within this framework are natural ecosystems, food security and economic development. This approach has been incorporated into the Mission Statement of the 1997 National Greenhouse Response Strategy for Australia:

> Australia intends to play its part in the global effort to stabilise greenhouse gas concentrations in the atmosphere at a level that would prevent dangerous interference with the climate system and within a time frame sufficient to: allow ecosystems to adapt naturally to climate change; ensure that food production is not threatened; and enable economic development to proceed in a sustainable way.[1]

Impact assessment is an integrative activity requiring knowledge of the processes contributing to both climate variability and change, and within natural and human systems sensitive to climate change. This requires the integration of the biophysical and social sciences to diagnose dangerous impacts within these systems based on physical, cul-

1. Intergovernmental Committee on Ecologically Sustainable Development (ICESD), *Future Directions for Australia's National Greenhouse Strategy* (1997), at 7.

A. Gillespie and W. C. G. Burns (eds.), Climate Change in the South Pacific: Impacts and Responses in Australia, New Zealand, and Small Island States, 7–32.

tural and economic criteria, and to understand the links between systems.

Working Group I of the Intergovernmental Panel on Climate Change (IPCC) concluded that: "The balance of evidence suggests a discernible human influence on global climate and Climate is expected to change in the future."[2] Working Group II concluded that natural ecological systems, socio-economic systems, and human health are all sensitive to both the magnitude and rate of climate change.[3] As a result, the emphasis in climate change research is progressing beyond the modelling of climate change to modelling of the impacts of climate change.

Two recent reports on impact assessment have been released by the IPCC. The first, the Second Assessment Report of the IPCC Working Group II,[4] is a global synthesis of published impacts research to 1995. Although its aims were to address mitigation and adaptation as described above, the research it summarised could only identify sensitivity to climate change in a number of processes, systems and regions. The second report, one of a series of regional impact reports commissioned by the IPCC, describes this research in more detail for Australia and New Zealand.[5]

This chapter is divided into two parts. The first part summarises recent impact assessments for various Australian sectors. The second part describes recent developments in the methodology of impact assessment. These aim to move beyond the identification of sensitivity, to the assessment of "dangerous" levels of climate change and of adaptation options. The basic structure of a common assessment framework incorporating comprehensive scenarios of climate change and variability, thresholds linking states of climate change to levels of impacts, and risk assessment, are described.

---

2. IPCC, *Climate Change 1995: The Science of Climate Change*, in CONTRIBUTION OF WORKING GROUP I TO THE SECOND ASSESSMENT REPORT OF THE INTERGOVERNMENTAL PANEL ON CLIMATE CHANGE (J.T. Houghton et al. eds., 1996).
3. IPCC, *Climate Change 1995: Impacts, Adaptations and Mitigation of Climate Change: Scientific-Technical Analyses*, in CONTRIBUTION OF WORKING GROUP II TO THE SECOND ASSESSMENT REPORT OF THE INTERGOVERNMENTAL PANEL ON CLIMATE CHANGE (R.T. Watson, M.C. Zinyowera & R.H. Moss eds., 1996).
4. *Id.*
5. IPCC Working Group II, *The Regional Impacts of Climate Change: An Assessment of Vulnerability*, (R.E. Basher & A.B. Pittock eds., 1998), at 105-148.

2. Palaeoclimatic Impacts in Australia

Although palaeoclimatic analogues for patterns of regional climate change *per se* have been largely discarded in favour of output from global climate models (GCMs), understanding the *response* of Australian biological and physical systems to past climate changes is an important part of understanding future long-term responses. This is particularly important for understanding natural rates of adaptation by species and ecosystems to climate change.

As a large, isolated and tectonically stable continent, Australia's geological and evolutionary history has been dominated by climate and climate change for millions of years. Australian landforms have been shaped by wind and water movements, the geochemistry of surficial processes and interactions with the biota, all of which are influenced by climate. Most Australian ecosystems have also evolved to cope with seasonal to semi-permanent water deficits. Those ecosystems requiring constant moisture are restricted to Tasmania and the coastal and highland regions of eastern Australia.

During the past 2.5 million years, the earth has undergone a series of ice-ages, where relatively short warm and wet interglacial periods lasting about 10–15,000 years have alternated with longer, glacial periods containing both wet and dry phases. The final phase of an ice-age is the glacial maximum, which is cold and dry over most of Australia, followed by an interglacial period. The current interglacial, the Holocene period, is between 10,000 and 10,500 years old and is comparatively warm and wet.

Climate changes within glacial cycles can be substantial, involving large changes in temperature and/or rainfall. While many of these climate changes had widespread impacts on ecosystems, palaeoecological studies have not found many direct links between extinction and climate change.[6] For instance, during glacial maxima, many species intolerant to cold and drought retreated to refugia, re-establishing broader ranges as conditions became warmer and wetter during interglacial periods. Conversely, cold-adapted species that thrive during ice-ages persist in refugia during interglacials. Even after rapid climate change, most ecosystems had time to recover during ensuing periods of stability. The response time of biota to climate change ranged from rapid (e.g., bird species nesting in wetlands) to thousands of years (e.g., slowly migrating tree species).

Processes associated with land degradation have also occurred previously as a result of climate change. Two examples are salinity and erosion, which were both widespread during the last glacial period. Following a relatively wet phase from 34–25,000 years ago, salinity affected large parts of the Murray Basin as aridity accompanied the onset of the glacial maximum.[7] High evaporation, combined with

---

6. T.F. FLANNERY, THE FUTURE EATERS (1994).

7. J.M. Bowler & R.J. Wasson, *Glacial Age Environments of Inland Australia*, in LATE CAINOZOIC PALAEOCLIMATE OF THE SOUTHERN HEMISPHERE 183–208 (J.C. Vogel ed., 1984).

high water tables and reduced vegetation cover led to the concentration of salt in the water table. Over large parts of the Murray Basin, saline groundwater actively discharged into a number of lake basins, as shown by relict hypersaline groundwater and sub-surface gypsum deposits. Erosion was widespread, as arid conditions led to vegetation loss with both wind and water erosion mobilising large quantities of soil. Productivity in these areas increased naturally in response to the warmer and wetter conditions of the Holocene.

These earlier periods of erosion and salinity were caused by natural processes and sometimes took hundreds or even thousands of years to recover. Today, we face similar problems caused by land-use change in a little over one hundred years. These changes have already caused a loss of productivity in natural and human systems, and may compound problems associated with climate change over the next century. An understanding of the palaeoclimatic impacts of processes such as salinity and vegetation change can contribute to our ability to manage their recovery, particularly under a changing climate.

The dynamics of past climates can also shed light on how climate may change in the future. The IPCC states that future climate change might involve "surprises."[8] This conclusion was based on the understanding of past climate, where large and rapid changes occurred in the space of decades. One such example comes from southern Victoria, where water-level changes in crater lakes for the past 16,000 years have been linked to regional climate and changes in precipitation/evaporation (P/E) ratios.[9]

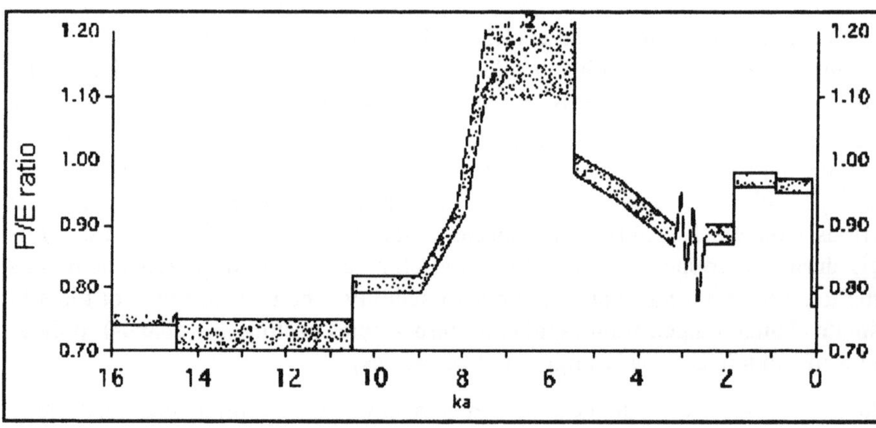

Figure 1. Estimated Range Of P/E Ratios for Lakes Keilambete, Bullenmerri and Gnotuk from 16,000 Years Ago To The Present.[10]
(Ka = Thousands of Years)

8. IPCC, *supra* note 2, at 7.

9. R.N. Jones, J.M. Bowler & T.A. McMahon, *A High Resolution Holocene Record Of P/E Ratio From Closed Lakes, Western Victoria*, 4 PALAEOCLIMATES: DATA & MODELLING 1-32 (1997).

10. *Id.*

Jones[11] cites at least eleven changes of P/E ratio >0.05 that have occurred during the Holocene at a frequency of about 1 every 1,000 years. As the cores that supplied the data on which this model was based were sampled at high-resolution (30–200 years), at least seven of these changes occurred relatively abruptly. Two periods of the Holocene (9,000–7,500 and 5,500–3,000 years ago) show gradual increases and decreases; however, these trends may also have consisted of a series of incremental changes. Jones[12] also demonstrated that the modern instrumental record was homogeneous across south-eastern Australia and oceans to the south, suggesting that these previous changes were at least regional.

Changes in solar radiation, the major forcing mechanism of climate during the Holocene, were gradual, whereas the record of P/E ratios in Figure 1 demonstrates that stable conditions can persist for hundreds of years before changing relatively abruptly. This suggests that gradual forcing can produce abrupt changes in regional circulation that affects the regional atmospheric moisture balance.[13] Similar dynamics could be a part of future climate change. It is possible that even if global climate changes produce only gradual trends as simulated by GCMs, regional changes may be more abrupt, similar to those observed in the palaeoclimatic record.

3. Climate Scenarios

Regional climate change scenarios for Australia are periodically prepared by the CSIRO to inform policymakers and guide research into climate change and its impacts. The latest scenarios take into account the global warming scenarios of the IPCC[14] and the cooling effect of sulfate aerosol pollution.[15] The IPCC has identified a broad range of future greenhouse gas and sulfate aerosol emission levels, based on differing assumptions of economic activity, population growth and fuel use.[16] Only one of these scenarios provides any allowance for mitigation.[17]

---

11. R.N. Jones, *Climate-Induced Changes In Lake Levels*, in AUSTRALIAN J. WATER RES. (in press).
12. R.N. Jones, *Modelling Hydrologic and Climatic Controls of Closed Lakes Western Victoria*. Ph.D. Thesis, University of Melbourne, Melbourne (unpublished, 1995).
13. B. Van Geel, J. Buurman & H.T. Waterbolk, *Archaeological And Palaeoecological Indications For An Abrupt Climate Change In The Netherlands, And Evidence For Climatological Teleconnections Around 2650 BP*, 11 J. QUATERNARY SCI. 451-460 (1996); P.A. Mayewski et al., *Climate Change During The Last Deglaciation In Antarctica*, 272 SCI. 1636-1638 (1996); F. Sirocko et al., *Teleconnections Between The Subtropical Monsoons And High-Latitude Climates During The Last Deglaciation*, 272 SCI. 526-29 (1996).
14. IPCC, *supra* note 2.
15. CSIRO, *Climate Change Scenarios for the Australian Region*, Climate Impact Group, CSIRO Division of Atmospheric Research, Melbourne 8 (1996).
16. *See* J Alcamo et al., *Climate Change 1994: An evaluation of the IPCC IS 92 Emission Scenarios*, in CLIMATE CHANGE 1994 (J.T Houghton et al. eds., 1994), at 251–304.
17. *Id.*

When combined with ranges of global climate sensitivity taken from a number of GCMs assessed by the IPCC, the IS92a–f scenarios provide a range of global warming from 1990–2100. These scenarios include the cooling effect of sulfate aerosols, where sulfate emissions increase in line with fossil fuel use until 2075, at which point the rates decline slightly. The inclusion of sulfate aerosol serves to reduce global warming by 20 to 30% relative to scenarios that do not incorporate sulfates.[18] Reductions in sulfate emissions earlier than this would lead to greater warming.

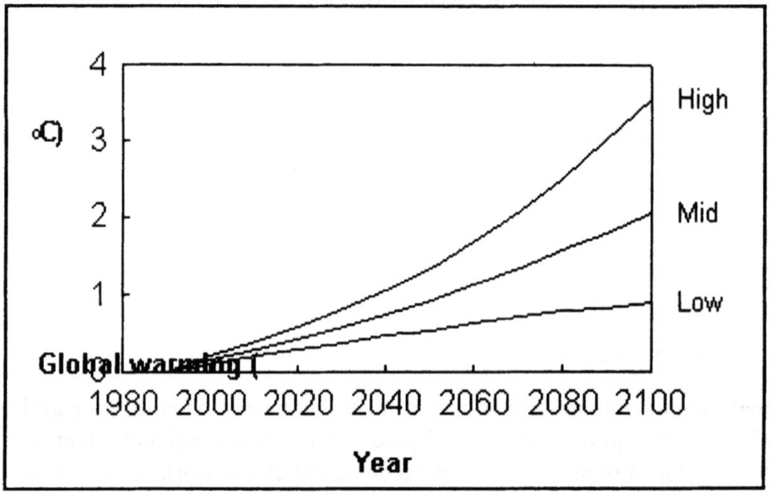

Figure 2. Low, Mid and High Case Scenarios Of Global Warming Assuming Variable Sulfate Emissions. (IPCC, 1996a)

Two types of GCMs are used in the scenarios. The first is the slab-ocean GCM, which consists of a layered atmosphere with a single-layer ocean that is flux-adjusted to simulate deeper water. The second is a coupled GCM, which consists of both a layered ocean and atmosphere. The coupled GCM is the more recent type of model and is considered to generate a more realistic representation of the ocean-atmosphere system.

A range of regional warmings is suggested by five coupled GCMs. Global warming from each GCM is averaged globally and the regional pattern of temperature response calculated for that GCM as change per degree of global warming. This effectively distinguishes regional deviations from global average warming. The range produced for each region shows the variation in patterns of warming between models. That range is further modified by an adjustment to allow for a regional response to sulfate aerosols, as the latter have a limited influence in the

18. CSIRO, *supra* note 15.

southern hemisphere.

The regional warming scenarios over Australia are shown in Table 1. Small changes in mean temperature can have large effects on extreme temperatures. For example, Hennessy and Pittock found that a 2°C warming could double the number of summer days over 35°C and halve the number of winter days below 0°C.[19]

TABLE 1. Warming Scenarios (_C) for Locations in the Australian Region

Region	Local warming per degree of global warming	Warming in 2030	Warming in 2070
Northern Coast (north of 25_S)	0.9 to 1.3	0.3 to 1.0	0.6 to 2.7
Southern Coast (south of 25_S)	0.8 to 1.6	0.3 to 1.3	0.6 to 3.4
Inland	1.0 to 1.8	0.4 to 1.4	0.7 to 3.8

Precipitation patterns were calculated in a similar way, but with no allowance made for sulfate aerosols due to high uncertainty. Regional scenarios from both slab- and coupled-ocean models were presented separately for the following reasons:

- The difference between the rainfall changes simulated by the two classes of GCMs is greater in Australia than it is elsewhere in the world. Over Australia, coupled models tend to simulate summer rainfall decreases and slab models simulate summer rainfall increases.
- The rainfall change in coupled models differs from that in slab models due to a strong delay in warming in the higher latitudes of the southern hemisphere in coupled models. There is considerable uncertainty regarding the reliability of the processes simulated in ocean models that lead to this result.[20]
- Some aspects of the pattern of warming in the southern hemisphere simulated by coupled models conflict with observed 20th century trends (although the latter may not be due to the enhanced greenhouse effect).[21]

Scenarios of precipitation change are shown in Figure 3. Changes in winter are broadly

19. K.J. Hennessy & A.B Pittock, *Greenhouse Warming and Threshold Temperature Events in Victoria, Australia*, 15 INT'L J. CLIMATOLOGY 591-612 (1995).
20. *See* P.H.Whetton et al., *Southern Hemisphere Climate: Comparing Models With Reality*, in CLIMATE CHANGE, PEOPLE AND POLICY: DEVELOPING SOUTHERN HEMISPHERE PERSPECTIVES 89-130 (A. Henderson-Sellers & T. Giambelluca eds., 1996).
21. A.B. Pittock et al., *Progress Towards Climate Change Scenarios For The Southwest Pacific*, 15(2) WEATHER & CLIMATE 21-46 (1995).

similar in the coupled and slab models. Both show a pattern of rainfall decreasing over most of Australia, increases over the oceans to the south, and a region of uncertain change in between. In the coupled models, the area of precipitation decline is stronger and more extensive and the bands of no change or increase are located a little further south. However, in summer, the slab models are dominated by rainfall increases, especially in the north-west, whereas the coupled models are dominated by decreases. This reflects a major systematic difference in regional response of coupled and slab models, and demonstrates the sensitivity of simulated rainfall change over Australia to oceanic processes.[22]

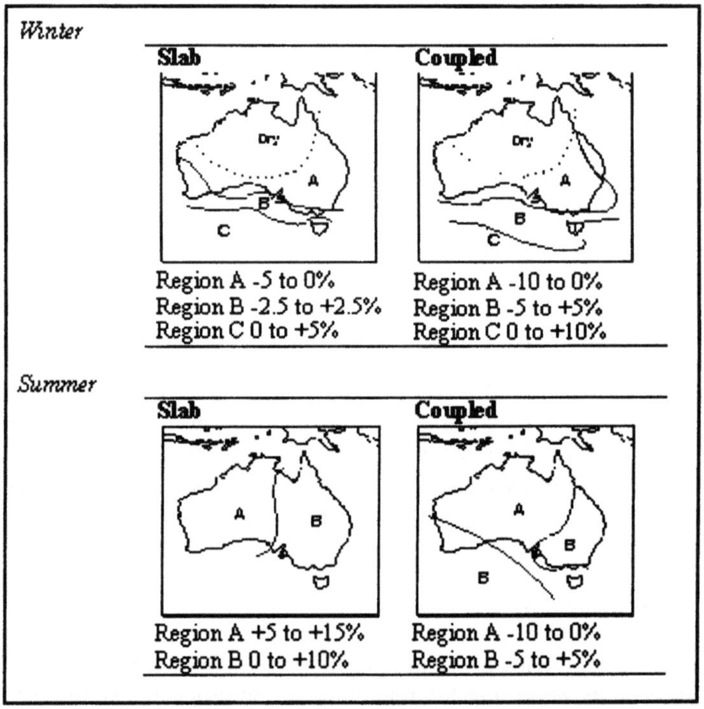

Figure 3. Scenarios of Precipitation Change for Australia from Slab and Coupled Models As Percentage Change Per Degree Of Global Warming.

4. Other Changes In Climate

Other variables such as evaporation, humidity and solar radiation are important to climate impact analysis, especially for use in impact models, e.g., rainfall-runoff, plant growth and species distribution models. The use of direct model output for these variables is preferred to the use of statistical regressions based on current climate, as the relationships between climatic variables are likely to change under cli-

22. P.H. Whetton et al., *Global Comparison Of The Regional Rainfall Results Of Enhanced Greenhouse Coupled And Mixed Layer Ocean Experiments: Implications For Climate Change Scenario Development*, 33 CLIMATIC CHANGE 497-519 (1996).

mate change. This requires an improved understanding of a whole host of potential climate changes rather than just changes to temperature and rainfall.

Changes to climate variability are also important. For instance, most significant impacts occur in response to climate extremes and thresholds, such as heat, cold, drought, flood, wind storms, or to phenomena that are climate-related such as fire. The gradual improvement of climate models is now facilitating some forms of climate variability to be assessed under climate change. For instance, changes in precipitation variability will have impacts on runoff, flood and drought frequency. The analysis of climate model output suggests that heavy rainfall events may become more frequent and/or intense under enhanced greenhouse conditions.[23]

Interannual variability of rainfall over Australia and many other parts of the Asia-Pacific region is strongly affected by the El Niño Southern Oscillation (ENSO). Recent model simulations contain ENSO-like phenomena that reproduce realistic oscillations under control (1′ CO_2) conditions.[24] There is now some evidence showing that ENSO may change under the enhanced greenhouse effect;[25] however, other studies suggest little change.[26] Accordingly, results may be too preliminary to draw firm conclusions.

Long model runs also show significant interdecadal variability.[27] A 600-year run with the CSIRO coupled GCM shows periods of very high and low rainfall in both the control and transient greenhouse run based on the IS92a emission scenario. This variability appears to have similar amplitude to variability observed in the short (100–120 year) historical record. This has implications for prediction of future climate change because the noise involved in regional signals may be as

23. P.H. Whetton et al., *Implications Of Climate Change Due To The Enhanced Greenhouse Effect On Floods And Droughts In Australia.* 25 CLIMATIC CHANGE (1993), at 289-317; A.M. Fowler & K.J. Hennessy, *Potential Impacts Of Global Warming On The Frequency And Magnitude Of Heavy Precipitation*, 11 NATURAL HAZARD 283-303 (1995); J.M. Gregory & J.F.B. Mitchell, *Simulation Of Daily Variability Of Surface Temperature and Precipitation Over Europe*, 121 Q. J. ROYAL METEOROLOGICAL SOC'Y 1451-1476 (1995); L.O. Mearns et al., *Analysis Of Daily Variability Of Precipitation In A Nested Regional Climate Model: Comparisons With Observations And Doubled CO_2 Results*, 10 GLOBAL & PLANETARY CHANGE 55-78 (1995); K.J Hennessy et al., *Fine Resolution Climate Change Scenarios For New South Wales*, (NSW Environment Protection Authority, 1997)

24. K.R. Knutson, S. Manabe & D. Gu, *Simulated ENSO In A Global Coupled Ocean-Atmosphere Model: Multidecadal Amplitude And CO_2 Sensitivity*, 10 J. CLIMATE 138-161 (1997); S.G. Wilson & B.G. Hunt, *Impact Of Greenhouse Warming On El Niño/Southern Oscillation Behaviour In A High Resolution Coupled Global Climatic Model*, Report To Environment Australia, CSIRO Division Of Atmospheric Research (1997).

25. *See* for example, D.Z. Sun, *El Niño, A Coupled Response To Radiative Heating?* 24 GEOPHYSICAL RES. LETTERS 2031-2034 (1997); Wilson & Hunt, *supra* note 24.

26. Knutson· *supra* note 24.

27. B.G. Hunt & H.L. Davies, *Mechanism Of Multi-Decadal Climatic Variability In A Global Climate Model*, 17 INT'L J. CLIMATOLOGY 565-580 (1997); R Suppiah et al., *Climate Change Under Enhanced Greenhouse Conditions In Northern Australia: Final Report*, CSIRO Division Of Atmospheric Research (1998); Knutson, *supra* note 24.

large, or larger, than the signal itself (perhaps out to 2′ CO_2).

If the trend of interdecadal variability supplements the trend of climate change, then impacts will be larger than those caused by climate change alone. Conversely, climate variability could offset the mean climatic trend, leading to reduced impacts. However, multi-decadal variability usually only persists for several decades before changing, whereas the pattern of warming from greenhouse gas emissions is expected to last for hundreds of years.[28]

Periods of very heavy rainfall, high winds and oceanic storm surge occur in association with tropical cyclones (TCs) around the northern Australian coastline, throughout much of the South Pacific and in parts of Southeast Asia. With implications for runoff, flooding and coastal inundation, any future changes in the frequency, intensity, location and paths of TCs are of importance. GCMs typically do not have a horizontal resolution fine enough to directly simulate cyclones and cyclone behaviour under enhanced greenhouse conditions. This problem has been addressed using various indirect or theoretical approaches with no clear consensus in their results.[29]

CSIRO is currently pursuing a highly promising approach to assess changes in tropical cyclone behaviour. Simulations for the Australian region with DARLAM, a finer resolution limited-area model, nested in the CSIRO slab GCM, are able to generate tropical low pressure systems which have many of the characteristics of tropical cyclones.[30]

Results under enhanced greenhouse conditions are currently being analysed. In other recent work, the theoretical approach of Holland[31] for estimating the maximum potential intensity of a tropical cyclone (MPI) in a given area of the globe, has been applied to enhanced greenhouse GCM output by Tonkin et al.[32] and Suppiah et al.[33] These studies indicate increases of MPI, in terms of maximum wind-

---

28. H.L. Davies & B.G Hunt, *The Problem Of Detecting Climatic Change In The Presence Climatic Variability*, 72 J. METEOROLOGICAL SOC'Y JAPAN 765-771 (1994).

29. *See* for example, B.F. Ryan, I.G. Watterson & J.L. Evans, *Tropical Cyclone Frequency Inferred From Gray's Yearly Genesis Parameter: Validation Of GCM Tropical Climates*, 19 GEOPHYSICAL RES. LETTERS (1992), at 1831-1834; J. Lighthill et al., *Global Climate Change And Tropical Cyclones*, 75 BULL. AM. METEOROLOGICAL SOC'Y 2147-2157 (1994); I.G. Watterson et al., *Seasonal And Interannual Variability Of Tropical Cyclogenesis: Diagnostics From Large-Scale Fields*, 8 J. CLIMATE 3052-3066 (1995).

30. K.J. Walsh & I.G. Watterson, *Tropical Cyclone-Like Vortices In A Limited Area Model: Comparison With Observed Climatology*, J. CLIMATE (in press).

31. G.J. Holland, T*he Maximum Potential Intensity Of Tropical Cyclones*, J. ATMOSPHERIC SCI. (in press).

32. H.Tonkin et al., *Tropical Cyclones And Climate Change: A Preliminary Assessment*, in., *Assessing Climate Change: Results From The Model Evaluation Consortium For Climate Assessment*, (W. Howe & A. Henderson-Sellers eds., 1997), at 327-360.

33. R. Suppiah et al., *Climate Change Under Enhanced Greenhouse Conditions In Northern Australia: Final Report*, CSIRO Division Of Atmospheric Research (in press).

speed, of up to 20% for a doubling of CO_2.[34]

4.1 SEA-LEVEL RISE

Global warming is expected to lead to mean sea-level rise due primarily to thermal expansion of the ocean and melting of mountain glaciers. The contributions of the Greenland and Antarctic ice-sheets are expected to be minimal or negative due to increases in accumulation rates.[35] The IPCC[36] estimates of global mean sea-level rise range from 5–25 cm in 2030 to 10–60 cm in 2070. These take into account uncertainties in future greenhouse gas emissions and the sensitivity of the global climate system as described above. Sea-level rise will not be uniform due to regionally variable uptake of heat by the ocean, changes in atmospheric and oceanic circulation, and vertical land movements,[37] although in Australia the latter is generally small.

Increases in sea level may increase saline intrusion into coastal groundwater and estuaries, increase flood heights in low-lying coastal areas by reducing the discharge rates of rivers in flood, and contaminate the fresh water resources of small coral islands.[38] With regard to such coastal impacts, changes in the frequency and intensity of meteorological conditions which lead to storm surge conditions (e.g., tropical cyclones, east coast pressure systems) may be as important as mean sea-level rise.[39]

5. Impact Research

5.1 HYDROLOGY

Hydrologic systems are sensitive to climate change because rainfall, evaporation and transpiration are the major determinants of factors such as soil moisture, groundwater recharge and runoff. The sensitivity of catchment hydrology to climate change was demonstrated by Chiew et al. who carried out a sensitivity analysis of runoff and soil moisture to temperature and precipitation change in 28 Australian catchments.[40] They found that changes in rainfall are always amplified in percentage terms for runoff, with larger amplifications in drier catchments. In

---

34. A. Henderson-Sellers & H. Zhang, T*ropical Cyclones And Global Climate Change: Report From The WMO/CAS/TMRP Committee On Climate Change Assessment (Project TC-2)*, World Meteorological Organization. Geneva (1997).
35. IPCC, *supra* note 2.
36. *Id.*
37. Pittock, *supra* note 21.
38. C.R. Wilkinson & R.W. Buddemeier, *Global Climate Change And Coral Reefs. Implications For People And Reefs.* Report Of The UNE-IOC-ASPIE-IUCN Global Task Team On Coral Reefs. IUCN Gland, Switzerland (1994).
39. A.B. Pittock et al., *Climatic Change, Climatic Hazards And Policy Responses In Australia*, in CLIMATE CHANGE AND RISK (T.E. Downing, R.J.S. Tol & A.A. Olsthorn eds., 1998).

wetter catchments, rainfall changes had little impact on soil moisture, but in drier catchments changes in soil moisture were sometimes larger than changes in rainfall. Under the CSIRO 1992 scenarios, possible increases and decreases in runoff were noted.

Schreider et al. applied "most wet" and "most dry" scenarios to a number of catchments in the upper Murray Basin to determine the impact of climate change on water availability.[41] For the "most dry" scenarios in 2030, they found reductions in annual streamflow of 28–38% in the four catchments surveyed. For the "most wet" scenarios, changes in annual streamflow ranged between -3 to +4% in the two snow-free catchments, with a 7-11% increase in the snow-affected catchments.[42] However, there was a large increase in flood events due to an increase in heavy rainfall events.

Two studies have used stochastic weather generator data in order to estimate the impacts of changes in the relative frequency of daily rainfall amounts.[43] Using a scenario taken from a single GCM, data was stochastically generated for temperature, evaporation and rainfall. High flow events increased in most cases, consistent with simulated rainfall increases in the GCM.

More recently, a study on water resources in the Macquarie Basin of NSW has pursued vertical integration by applying scenarios for rainfall and evaporation from the DARLAM regional climate model to a rainfall-runoff model.[44] Streamflow from this model was applied to the IQQM river management model developed by the NSW Department of Land and Water Conservation to estimate allocations for competing uses such as irrigation, industrial and domestic use and environmental flows. The output from these models was used to estimate impacts on irrigated and dryland agriculture and wetland breeding habitat in the Macquarie Marshes.

The scenarios resulted in a net reduction in rainfall of 2 to 12% on a seasonal basis. This led to a reduction in flows into the Burrendong Dam, the major site for water

40. F.H.S. Chiew, P.H. Whetton, T.A. McMahon & A.B. Pittock, *Simulation Of The Impacts Of Climate Change On Runoff And Soil Moisture In Australian Catchments*, 167 J. HYDROLOGY (1995), at 121-147.

41. S. Schreider et al., *Estimation Of Possible Climate Change Impacts On Water Availability, Extreme Flow Events And Soil Moisture In The Goulburn And Ovens Basins, Victoria*, 34 CLIMATIC CHANGE (1996), at 513-546; S. Schreider, A.J. Jakeman, P.H Whetton & A.B. Pittock, *Estimation Of Climate Change Impact On Water Availability And Extreme Events For Snow-Free And Snow-Affected Catchments Of The Murray-Darling Basin*, 2 AUSTRALIAN J. WATER RES. (1997), at 35-46.

42. Schreider *supra* note 41.

43. B.C. Bates et al., *Climate Change And Its Hydrological Implications For South Australia*, 118 TRANSACTIONS ROYAL SOC'Y S. AUSTRALIA 35-43 (1994); B.C. Bates et al., *Impact Of Climate Change On Australia's Surface Water Resources*, in *Greenhouse: Coping With Climate Change*. CSIRO Publishing (W.J. Bouma, G.I. Pearman & M.R Manning eds., 1996), at 248-262.

44. Hassall & Associates, NSW Department Of Land And Water Conservation, NSW National Parks And Wildlife Service & CSIRO Division Of Atmospheric Research, *Climate Change Scenarios And Managing The Scarce Water Resources Of The Macquarie River - Final Report*, Report To Environment Australia (1998).

storage, of 11% in the low case and 30% in the high case. The loss to the agricultural economy ranged between 6% and 22%, incorporating an allowance for the fertilization effect of CO_2. The fertilization effect of CO_2 improves the water-use efficiency of plants, increasing their net productivity as a function of soil moisture. This is expected to increase yield for many species and may also lead to decreases in transpiration. The greatest losses were in the livestock sector. The NSW National Parks and Wildlife Service estimated that reduced flows into the Macquarie Marshes had the potential to contribute to local, and possible global, extinction of water birds.

Developed catchments are vulnerable to increases in heavy rainfall events that are expected to occur if average rainfall increases. Increases in heavy rainfall events may also occur when there is a slight decrease, or no change, in average rainfall.[45] Minnery and Smith showed that the cost of a 1 in 100 year flood in the Hawkesbury-Nepean corridor in NSW increased tenfold under a 2′ CO_2 scenario.[46] This large sensitivity is due to non-linear increases in land under flooding as levees fail, and building collapse associated with increases in flood depth.

Sufficient studies have been carried out to show that Australian catchments are sensitive to climate change. However, due to a continuing uncertainty about the direction of rainfall change, prescription for policy remains uncertain. A new study funded by the Rural Industry Research and Development Corporation (RIRDC), aiming to overcome this limitation through the application of risk assessment, has recently commenced. A climate scenario generator, OzClim (see below), will be coupled to the rainfall-runoff model used in the Macquarie Basin study, forming the core of a water resources climate impact toolkit. OzClim will be used to simulate the impacts of a wide range of possible climate changes, including changes in climatic variability. Thresholds of impacts, such as levels of productivity in agriculture and the health of wetlands, will be linked to levels of climate change in order to assess the levels of risk faced by these activities.[47]

Water resource use in Australia is adapted to high climate variability, but these adaptations may not be sufficient to deal with climate change. Recent moves to reform water use through capping allocations and the introduction of the user-pays principle in the Murray-Darling Basin may have a significant impact on factors such as irrigation and environmental flows.[48] Coupled GCMs have simulated reduced rainfall over much of Australia, indicating an added pressure on already

45. P.H. Whetton et al., *Fine Resolution Climate Change Scenarios For New South Wales, Part 2: Climatic Variability*, Consultancy Report By CSIRO For The New South Wales Environment Protection Authority (1997).

46. J.R. Minnery & D.I. Smith, *Climate Change, Flooding And Urban Infrastructure*, in GREENHOUSE: COPING WITH CLIMATE CHANGE 235-247 (W.J. Bouma, G.I. Pearman & M.R. Manning eds., 1996).

47. R.N. Jones, A.B Pittock & C.M. Page, *A Toolkit For Water Resources Impact Assessment*, AUSTRALIAN J. WATER RESOURCES (in press).

48. W.Cox & P. Baxter, *Setting The Cap: Report Of The Independent Audit Group*. Murray-Darling Ministerial Council (1996)

limited water resources. Further research is needed to determine adaptation mechanisms that can account for changes in both climate and water policy.

5.2 TERRESTRIAL ECOSYSTEMS

Australia is one of the most biologically diverse regions of the world.[49] Extinctions during the period of European occupation have by far outweighed previous extinctions due to climate change.[50] This demonstrates that Australian ecosystems have been able to adapt to climate change in the past.

Predictive modelling of the distribution of individual plant and animal species shows that they can move beyond their current climatic envelope in plausible scenarios of climate change.[51] For instance, four species of Dryandra in south-western Australia are shown to move almost or totally outside their suitable range, as defined by climate and soil-type, with a warming of 2°C.[52]

Using three scenarios derived from the CSIRO climate change scenarios, Dexter et al. modelled the potential distribution of 57 nationally threatened vertebrates.[53] With regional warmings varying between 0.6–1.0°C and regional rainfall changes of -5 to +10%, 47 species were reduced in range while 10 species increased. With regional warmings of 1.0-1.4°C and current rainfall, and with the same warming and rainfall changes as above, 54 and 55 of the 57 species were reduced in range, respectively. In the last two cases, the suitable climate envelope for 10 species moved outside their current distributional ranges. This demonstrated that species that are already endangered might be placed at an even higher risk because of modest rates of climate change.

Mitchell and Williams have demonstrated that many plant species can exist outside their modelled climatic envelope once they are established.[54] This is because the climatic envelope based on a natural distribution is subject to a particular species' ecology. Accordingly, the weakest link(s) in its physiological relationship to cli-

---

49. World Conservation Monitoring Centre, *Global Biodiversity: Status Of The World's Living Resources* (1992).
50. Flannery, supra note 6.
51. D.B. Lindenmayer et al., *The Conservation Of Leadbeater's Possum, Gymnobelideus Leadbeateri (Mccoy): A Case Study Of The Use Of Bioclimatic Modelling,* 18 J. BIOGEOGRAPHY 371-83 (1991); R. Brereton, S. Bennett & I. Mansergh, *Enhanced Greenhouse Climate Change And Its Potential Effect On Selected Fauna Of South-Eastern Australia: A Trend Analysis*, 72 BIO. CONSERVATION 339-354 (1995); E.M.Dexter, A.D. Chapman & J.R. Busby, *The Impact Of Global Warming On The Distribution Of Threatened Vertebrates*, Environmental And Resource Information Network (ERIN), http://www.erin.gov.au/life/end-vuln/animals/climate/climate-change/cc-exe.html (1995).
52. Pouliquen-Young & Connell, unpublished data. Institute for Science and Technology Policy. Murdoch University.
53. Dexter et al. *supra* note 51.
54. N.D. Mitchell & J.E. Williams, *The Consequences For Native Biota Of Anthropogenic-Induced Climate Change*, in BOUMA, *supra* note 46, at 308-324.

mate may place limits upon its distribution. Established ecosystems may remain fairly stable, whereas ecosystems subject to outside stresses and extreme events may be more vulnerable to climatic change.

Therefore, although land-use change is a far greater determinant of biodiversity than climate change in the short term,[55] the processes associated with land-use change serve to increase the vulnerability of remnant ecosystems. For instance, the spread of environmental weeds and fragmentation of ecosystems are two significant agents of vulnerability under climate change.[56] Fragmentation hinders the movement of species in response to changing conditions, while environmental weeds are likely to exploit niches that have opened up due to fire, physical disturbance or stresses arising from changing climate.

Natural ecosystems are subject to climate variability on a scale from less than a day to decades. Most ecosystem turnovers happen when an extreme event, such as fire, changes due to vegetative reproduction, replacement by the species in the soil seed bank or taxa that migrate into the area. These are all likely to change under climate change. CO_2 fertilisation will also change the competitive structure between plants, altering growth rates, nutrient and soil moisture fluxes. Climate change could potentially lead to recombinant ecosystems, structurally similar to those we see today, but with very different species communities.

Wetlands, in particular, are at risk from climate change; partly due to changing inflows, as described above for the Macquarie Marshes, but also due to increased wet-area evaporation. Any increase in the moisture efficiency of wetland vegetation is likely to be countered by an increase in evaporation from open water. Wetlands will also be increasingly vulnerable to aquatic weeds of semi-tropical and tropical origin.

Adaptation is divided into two types: autonomous, where biota respond automatically to environmental factors such as climate, and planned adaptations undertaken by humans. Planned adaptation is likely to be incorporated into other programs such as those aimed towards mitigation (forestry), remnant habitat protection (Bushcare) and land degradation (Landcare). Whether climate change can be kept at a level where ecosystems can adapt naturally remains uncertain. Further research into both autonomous and planned adaptation responses is urgently needed, if only to quantify processes contributing to increased vulnerability of ecosystems under climate change.

---

55. *See* for example, D.A. Saunders & R.J. Hobbs, *Impact Of Biodiversity Of Changes In Land-Use And Climate*, in BIODIVERSITY OF MEDITERRANEAN ECOSYSTEMS IN AUSTRALIA (R.J. Hobbs ed., 1992), at 61-75.

56. R.N. Jones & A.B.Pittock, *Assessing The Impacts Of Climate Change: The Challenge For Ecology*, in FRONTIERS IN ECOLOGY: BUILDING THE LINKS (N.I. Klomp, D.C. Lunt & J.E. Williams eds., 1997), at 311-322.

5.3 AGRICULTURE

As a commodity-based economy, Australia is dependent on sectors such as agriculture, which produces about one-quarter of Australia's export earnings.[57] Australian agriculture is reasonably well adapted to climate variability and low productivity compared to most other regions. However, possible interactions between climate, land-use, plant and animal physiology, and social and economic behaviour make it difficult to estimate the impacts of climate change on agriculture.

The main limitations on Australian agriculture are water, and to a lesser degree, temperature and nutrients. As described in the section on hydrology, changes in rural water supply are likely under a number of scenarios. Changes in irrigation water supply could have severe and lasting effects on large sections of the rural economy.

With warming, it is expected that cold weather impacts such as frost will decrease, as has already been observed,[58] while heat stress associated with rises in daily maximum temperature may increase. A rise in minimum temperatures may reduce stock losses and increase pasture and animal production in southern and mountain areas. Crops that require winter chilling, such as stone- and pome-fruits could suffer a decrease in suitable range with warming.[59] An increase in temperatures during spring may restrict the period for grain-filling in winter cereals, promoting the need for short-season cultivars.[60] Heat stress may affect stock in summer, requiring a shift to heat-tolerant breeds. In summary, relief from cold stress will occur in winter in the south, while heat stress will increase in summer and in the north.

About 90% of primary production in grasslands (rangelands and pastures) is influenced by soil moisture (and hence, precipitation), however nominate changing CO_2 concentrations on grassland structure and productivity through increased plant yield, root mass and leaf area, changed nutrient cycling and increased water-use efficiency as possible positive adaptations.[61] Vertically integrated modelling of CO_2 related impacts on an Australian rangeland system under beef production showed an increase in annual accumulated plant growth, cash surplus and a reduction in the variability of annual pasture growth.

---

57. W.McLennan, *Australians And The Environment*, Australian Bureau Of Statistics (1996).
58. N. Nicholls, *Increased Australian Wheat Yield Due To Recent Climate Trends*, 387 NATURE (1997), at 484-5.
59. K.J. Hennessy & K. Clayton-Greene, *Greenhouse Warming And Vernalisation Of High-Chill Fruit In Southern Australia*, 30 CLIMATIC CHANGE (1995), at 327-348.
60. Y. P. Handoko Wang & G.M. Rimmington, *Growth Sensitivity To Air Temperature, Rainfall And Ambient Carbon Dioxide Concentration Of A Wheat Crop In Victoria, Australia: A Simulation Study*, 2(1) CLIMATE RESEARCH 31-49 (1992).
61. B.D. Campbell, D.M. Stafford Smith & G.M. McKeon, *Elevated CO_2 And Water Supply Interactions In Grasslands: A Pastures And Rangelands Management Perspective*, 3 GLOBAL CHANGE BIOLOGY177-187 (1997).

How such systems may react under climate change, particularly precipitation change, is still poorly understood. Increased water-use efficiency may be counterbalanced where bare-soil evaporation increases. Farmer adaptation is also considered to be a key determinant[62] and further research into thresholds of response for sustainable management is recommended.[63] CO_2 fertilisation effects will also benefit cereal crops in the short-term, but increasing temperatures could counter this effect over the long-term.[64]

Forestry impacts share many of the uncertainties associated with agriculture. Plantation and on-farm forestry are both projected to increase in area to allow a greater protection of natural forests, alleviate land degradation and to sequester carbon. Higher forest productivity will accompany CO2 increases, but this increase in productivity may not be sustained over longer periods.[65] Long production cycles may lead to trees on the margins of their ranges becoming vulnerable to climate change over a single life-cycle. Destructive losses, such as disease or fire, may also vary relative to climate change. These pose much greater economic threats than possible losses to annual and most perennial crops.

Pests, weeds and diseases associated with higher temperatures are likely to spread south with increased temperatures. This is expected to result in higher management costs due to the increased activity of tropical and sub-tropical pests, as they are more numerous than those in temperate environments.[66] The range of some species that prefer cool and/or damp conditions (e.g., the southern army worm) will probably contract southwards.

Agriculture is expected to be more adaptable in countries with developed technologies, and Australia's current adaptations to climate variability will serve as the basis for future adaptations.[67] A significant research effort into improving sustainable management under climate variability is now beginning to incorporate strategies for global change, including climate change.[68]

---

62. R.M. Gifford, B.D. Campbell & S.M. Howden, *Options For Adapting Agriculture To Climate Change: Australian And New Zealand Examples*, in *Greenhouse: Coping With Climate Change*, in Bouma, *supra* note 46, at 399-416.
63. Gifford, *supra* note 62.
64. Y.P Wang & D.J. Connor, *Optimal Development for Spring Wheat at Two Locations in Southern Australia Under Present and Changed Climate Conditions*, 79 AGRIC. & FOREST METEOROLOGY 9-28 (1996).
65. J.J Landsberg, *Impact Of Climate Change And Atmospheric Carbon Dioxide Concentration On The Growth Of Planted Forests*, in Bouma, *supra* note 46, at 205-219.
66. R.W. Sutherst, *Impacts Of Climate Change On Pests, Diseases And Weeds In Australia*, Report Of An International Workshop. Brisbane 9-12 October. 1995, CSIRO Division Of Entomology (1996).
67. R.F Darwin, T. Marinos, J. Landrewski & A. Raneses, *World Agriculture And Climate Change: Economic Adaptations*, Agricultural Economic Report No. 703, U.S. Dept. of Agriculture (1996), at 86.
68. B.D. Campbell et al., *Impacts Of Atmospheric Composition And Climate Change On Temperate And Tropical Pastoral Agriculture*, in Bouma, *supra* note 46, at 171-189.

A recent trend in government policy is to make the private sector more responsible for risk with respect to factors such as drought and water management. For example, the current corporatisation of water resources and capping of allocations in the Murray-Darling Basin will have financial impacts on irrigated agriculture. Changes to drought policy currently in train may also need to incorporate climate change into definitions of exceptional circumstances.[69] These trends will also need to be factored into integrated assessments of agriculture under climate change.

5.4 COASTAL AND MARINE

The coastal zone is particularly complex because of sea-level rise, potential changes to storm activity or ocean currents, invasions of exotic biota and interactions with adjacent land-uses. Two-thirds of Australia's population lives in cities near the coast and there is a great deal of infrastructure in coastal regions, particularly in Australia's south-east. Rapid development is occurring along much of the east coast, particularly in Queensland.[70]

The southern Australian coastline is highly vulnerable to changes in wave energy and current directions, as it has mostly sandy beaches and rocky shores. Estuaries and coastal swamps vulnerable to storm surges include much of Port Phillip Bay, near Melbourne,[71] the Spencer Gulf and a number of ports around the country. Northern Australian wetlands are much more vulnerable to changes in sea level interaction with terrestrial hydrology. For instance, changes in sedimentation, hydrology and sea-level have the potential to change low-lying wetlands in northern Australia from terrestrial to tidal (e.g., mangroves).[72] Where coastal wetlands are backed by built environments, sea-level rise may lead to a net loss in their coverage.

Coral reefs should be able to cope with the projected rises in sea-level of up to 5 cm per decade. However, reef islands may not be able to cope with the same growth rate,[73] with fresh groundwater supplies being particularly vulnerable.[74] Coral bleaching is a phenomenon that may be caused by a number of factors. However, it can occur when sea temperatures rise more than 2°C above normal. Thus, bleaching may become common with global warming, especially if corals are under stress from another source.[75]

69. *Indicators Of Drought Exceptional Circumstances: Proceedings Of A Workshop Held In Canberra On 1 October, 1996*, Bureau Of Resource Sciences (D.H. White & V.M. Bordas eds., 1997).
70. McLennan, *supra* note 56.
71. K.L. McInnes & G.D. Hubbert, *Extreme Events And The Impact Of Climate Change On Victoria's Coastline*, Environment Protection Authority Publication 488 (1996).
72. J. Chappell et al., *Coastal Impacts Of Enhanced Greenhouse Climate Change In Australia: Implications For Coal Use*, in Bouma. supra note 46, at 220-234.
73. C.R. Wilkinson, *Global Change And Coral Reefs: Impacts On Reefs, Economies And Human Cultures*, 2 GLOBAL CHANGE BIO. 547-558 (1996).
74. Basher & Pittock, *supra* note 5, at 105-148.
75. *Id.*.

Relatively little is known about how climate affects marine fishes, although fish populations are sensitive to climate variability. Coastal fisheries are vulnerable to a number of shorter term problems, such as over-fishing, loss of suitable habitat and coastal pollution. These fisheries may become more vulnerable under climate change, whereas those in the deep-water fisheries sector may adapt by moving away from the targeting of collapsed stocks for new targets. Scientific management of fisheries is the recommended adaptation option.

5.5 HEALTH

Most people in Australia are sheltered from temperature stress, which is likely to decrease in winter and increase in summer, especially where maximum temperature exceeds 40°C.[76] Urban air pollution, especially due to photochemical oxidants, may increase in major cities.

Australia has a number of endemic arboviruses that have already been linked with climatic variability (e.g., Murray Valley Encephalitis and Ross River Virus). Ross River Virus is currently moving southwards in Victoria, presumably due to increased temperature and human vectors. A number of tropical diseases could potentially spread in Australia due to global warming, but as the vectors currently exist without the diseases, present control methods may continue to keep these diseases at bay.[77]

5.6 BUILT ENVIRONMENT

Approximately 85% of Australians live in urban centres and utilise a relatively high degree of fossil fuel-generated energy compared to people in other countries. Temperature rises reduce energy expended on heating in winter but greatly increase summer energy demands for air-conditioning. For example, energy use in Melbourne now peaks in summer, particularly when the maximum daily temperature exceeds 35°C. The increase of heat islands around cities can further exacerbate this process.

Built-up areas are under threat of increased flash-flooding if hydrological cycles become more intense with warming. Some building materials are climate sensitive and may deteriorate if conditions change. For instance, cement emits CO_2 and deteriorates when the atmospheric concentration of CO_2 increases, creating a positive feedback. Other building materials are sensitive to changed heat and humidity and warrant further investigation under climate change.[78]

76. Hennessy & Pittock, *supra* note 19.
77. Recent assessments for climate change and health in the Asia-Pacific Region can be found in *Climate Change And Health In The Asia-Pacific Region*, AMA and Greenpeace International (P. Curson, C.S. Guest & E. Jackson eds., 1997).
78. Cole, *CSIRO Building Construction and Engineering*, personal communication to author.

The major physical threats to the built environment come from combined sea-level rise and storm surges in coastal areas, changed flood conditions and tropical cyclones. Given the relatively long life of many developments, urban and industrial infrastructure in sensitive areas may require planning that takes account of the likely impacts of climate change.

6. Impact Research Methods

As described in the Introduction, there are two purposes for assessing climate change impacts within the FCCC: mitigation and adaptation. For impact assessment to contribute to mitigation policy, levels of climate change need to be linked with critical thresholds. The need for sustainability, as articulated in the FCCC, dictates that these critical thresholds must be assessed using both monetary and non-monetary criteria. Therefore, critical thresholds need to be assessed for different sectors and regions as decided by physical, social or economic criteria. The regional climate change scenarios for Australia, summarised earlier, show that a broad range of climate change is possible; thus, the outcomes are most useful if they are described in terms of risk. This requires sampling from the fullest possible range of quantifiable uncertainty, finding the position of the critical threshold within that range, and assessing the resultant risk by calculating the probability of reaching that threshold.

Adaptation assessment can be approached through a similar process. In this case, both critical and desirable thresholds are chosen, then adaptation options are explored to reduce the risk of a critical threshold being reached. In CSIRO, both methods of assessment are being incorporated into a common assessment framework utilising the climate scenario generator, OzClim, various methods of assessing thresholds, and risk assessment by CSIRO and outside collaborators.

6.1 OZCLIM

OzClim is a software package that is being constructed to generate climate change scenarios for both average climate and climate variability for Australia.[79] A pilot version of OzClim has been developed by the CSIRO Division of Atmospheric Research in collaboration with the Centre for Environmental and Resource Studies (CEARS) at the University of Waikato, New Zealand. OzClim is grid-based, with an array of about 200 x 200 points, producing a grid-spacing of about 25 km over Australia. A region covering southeast Queensland and northwest New South Wales at a grid-spacing of about 7 km has been incorporated to demonstrate OzClim's high-resolution capability. A region covering the Murray-Darling Basin at a similar resolution is planned.

---

79. CSIRO, *Ozclim: A Climate Scenario Generator And Impacts Package For Australia,* CSIRO DAR (1996).

OzClim has three parts:

1. The MAGICC program developed by the Climate Research Unit at the University of East Anglia, UK, calculates projected global warmings from a range of greenhouse gas emissions scenarios, climate sensitivities and sulfate aerosol scenarios.[80]

2. A regional scenario generator for Australia uses these in combination with GCM or regional climate model patterns to provide regional climate change scenarios for 5-yearly intervals from 1995 to 2100. Variables currently supported are maximum, minimum and average temperature, and rainfall.

3. The capacity to "plug in" a range of simple impact models to explore the sensitivity of impacts to different emissions scenarios and climate sensitivities.

OzClim has a graphical user-interface that produces regional climate change scenarios for Australia based on various global climate and regional climate model patterns. This interface presents impact results as maps and is to be upgraded to show time-series results as graphs. The modular nature of the program allows simple impact models to be easily incorporated into the OzClim system to demonstrate the sensitivity of various factors to climate change. Conversely, the climate scenario generator can be removed and coupled to more complex models run by other groups, or can be adapted to provide output suitable for groups using non-compatible operating systems. This demonstrates the benefit of the "toolkit" approach as compared to less flexible, linear programming systems.

80. T.M.L. Wigley & S.C.B Raper, *Future Changes In Global Mean Temperature And Thermal-Expansion-Related Sea Level Rise,* in CLIMATE AND SEA LEVEL CHANGE: OBSERVATIONS, PREDICTIONS AND IMPLICATIONS 111-133 (R.A. Warrick, E.M. Barrow & T.M.L. Wigley eds., 1993).

6.2 THRESHOLDS

Threshold events may signal a distinct change in conditions or a step on a scale that has been nominated as significant (i.e., a "benchmark"). Examples of *climatic thresholds* are frost, snow and monsoon onset. Biophysical or *environmental thresholds*, represent a distinct change in conditions, such as the drying of a wetland, floods, breeding events etc. *Operational thresholds* are set by benchmarking a level of performance, e.g., yield per area of a crop in weight, volume or gross income.[81] An example of an operational threshold is where the identification of sustainable thresholds for grazed grassland systems under global change is recommended.[82]

Critical thresholds are a special category, assessed to determine the point at which the risk of an impact becomes "dangerous." This assessment involves placing values on processes and/or outcomes. The assumptions behind such valuations should ideally be transparent and should be understood by all those affected. As impacts differ within sectors and regions, and may vary over time, critical thresholds for different activities and localities will not be reached at the same time or with the same rate of climate change. Climate scenarios can be used to determine when and where "dangerous" thresholds are reached in various sectors.[83] This is then related back to rates of greenhouse gas emissions.

An example of two hypothetical "critical" thresholds defined in terms of rates of global warming based on IPCC projections[84] is shown in Figure 4. These thresholds are related to both the rate and magnitude of climate change. Threshold A is unlikely to be exceeded until after 2030, but Threshold B is extremely sensitive to climate change and is exceeded under most scenarios. Adaptation options can then be assessed for critical thresholds that are sensitive, and therefore vulnerable, to climate change. Management scenarios involving adaptation can be trialled under climate change by suitably modifying impact models to included adaptation/management feedbacks to determine which provide the greatest benefit or least cost.

The use of thresholds endorsed by stakeholders is crucial. By relating thresholds to past experience, stakeholders can identify when an activity becomes vulnerable, when adaptation options might be needed and what adaptation options should be explored. The setting of thresholds also allows risk assessment to be conducted.

---

81. Jones & Pittock, supra note 56.
82. Described in Campbell, supra note 61.
83. M.L. Parry, T.R. Carter & M. Hulme, *What Is A Dangerous Climate Change?*, 6 GLOBAL ENVTL. CHANGE (1996).
84. IPCC, *supra* note 2.

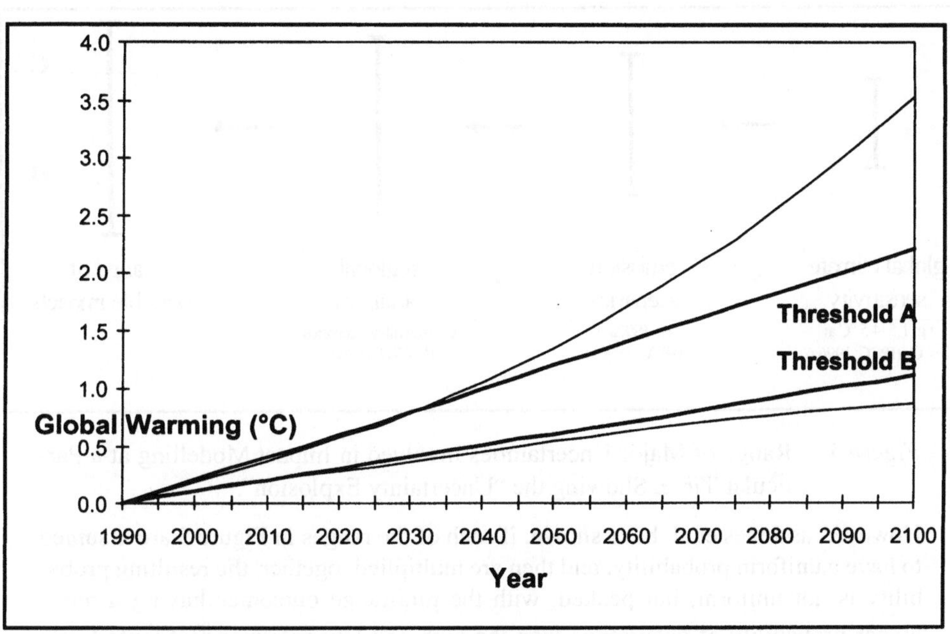

Figure 4. Comparison Of Critical Thresholds With IPCC Global Warming Projections Calculated From The IS92a–F Emissions Scenarios With Variable Sulfate Aerosol Showing Thresholds Based On A Constant Rate Of Increasing Temperature.

6.3 RISK ASSESSMENT

Risk assessment relates thresholds to a range of possible climate change and expresses the likelihood of that threshold being reached or exceeded. The simplest way of calculating risk is to assume that the range of regional climate change for any part of Australia has a uniform probability of occurrence. For example, Figure 5 shows the most commonly used technique for calculating impacts in Australia, where scenarios as described above are used to calculate the fullest possible range of impacts for a particular date, producing what Henderson-Sellers has termed the "uncertainty explosion."[85] If thresholds C and D are placed within that uncertainty and assessed according to a uniform probability of occurrence, then Threshold D has a 75% chance of being exceeded, while Threshold C has a 25% chance.

85. A. Henderson-Sellers, *An Antipodean Climate Of Uncertainty*, 25 CLIMATIC CHANGE (1993), at 203-224.

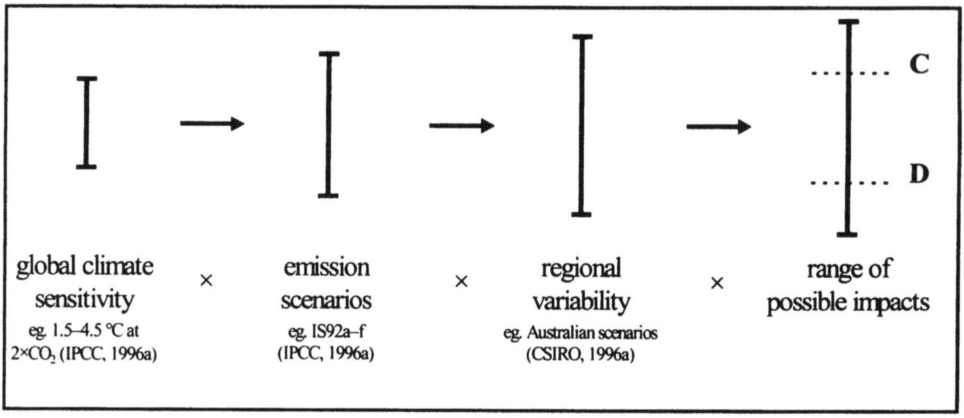

Figure 5. Range of Major Uncertainties Involved in Impact Modelling at a Particular Time, Showing the "Uncertainty Explosion".

However, as Jones et al. have shown, if each of the ranges in Figure 5 are assumed to have a uniform probability, and then are multiplied together, the resulting probability is not uniform, but peaked, with the mid-range outcomes having a much higher probability of occurrence than the high and low extremes.[86] Monte Carlo sampling of the ranges of global warming sensitivity, ranges of greenhouse gas emissions and regional climate changes, can be multiplied together to create a probability distribution which is not uniform.

For example, Figure 6 shows the probability distribution of regional warming for the northern coast of Australia in 2030. Two ranges of global warming (0.4–0.8°C) and regional warming (0.9–1.3°C per degree of global warming) taken from CSIRO[87] are multiplied, producing a regional range of 0.36–1.04°C. The resultant distribution is peaked, with the extremes of the range being far less likely than the values surrounding the median.

When a threshold is associated with a particular level of warming, it is possible to estimate its probability of occurrence by using this methodology, i.e., quantify its uncertainty. For example, the two thresholds initially assessed for risk based on uniform probability as in Figure 5 are re-assessed using the probability distribution in Figure 6. In Figure 5, they have a 25% and 75% probability of being exceeded, but in Figure 6 the probability becomes 9% and 79%, a decrease of 16% and an increase of 4% respectively.

86. Jones, *supra* note 47.

87. CSIRO, *supra* note 15.

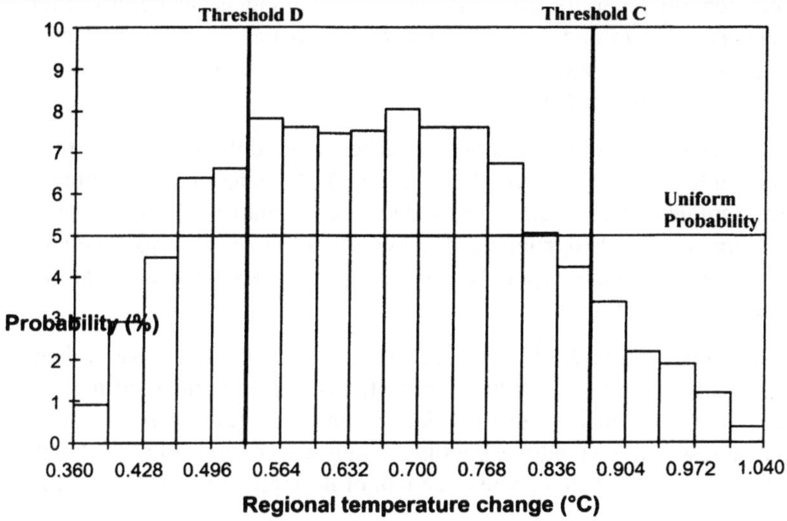

Figure 6. Regional Temperature Change

Figure 6 displays probability distribution of a regional scenario for temperature change in 2030 for the northern Australia coast,[88] showing the probability of occurrence for 5% increments within a total range of 0.36–1.04 °C, based on Monte Carlo sampling. The component ranges are 0.4–0.8°C (global warming) and 0.9–1.3°C (local warming per degree global warming), sampled randomly and multiplied 5,000 times. Two hypothetical thresholds are shown to demonstrate changing risk. The risk of Threshold C being exceeded decreases from 25% to 9%, with a change from a uniform to a non-uniform probability distribution, whereas the risk of Threshold D being exceeded increases from 75% to 79%.

The results of risk assessment can be used to define the level of "dangerous" climate change for regional impacts, thus contributing to the broader assessments required at the national level. Risk assessment can also rank the benefits of various adaptation methods, with the most successful being those that reduce risk. This technique of ranking allows stakeholders to weigh up the possibilities of investing in adaptation to counter the harmful impacts of climate change and to take advantage of any potential benefits.

7. Australian Research in Impacts Assessment

This chapter describes some of the latest climate change scenarios and impact assessments completed in Australia. It also outlines a common assessment framework that is evolving amongst climate impact researchers in CSIRO and their col-

88. CSIRO, *supra* note 79.

laborators. This development is described as a climate impacts toolkit, where a climate scenario generator, data-bases and impact models can be linked for a more efficient use of data and resources and for the better integration of models and methods. The framework and toolkit will also facilitate more uniform comparison of the results of different studies.

This is a response to the limitations of earlier methodologies, where studies often estimated a high and low range of possible impacts using the latest climate scenarios. Such studies often become superseded when climate scenarios are updated. By making impact models dynamic and linked to a climate scenario generator, where scenarios can be updated quickly and easily, impact assessments can be readily upgraded with new information.

The large ranges produced by these earlier impact assessments are limited in their usefulness to policymakers. When confronted with a wide range of future possible impacts, it is very difficult to formulate meaningful policy. By using thresholds and risk assessment it should be possible to deliver scenarios of impacts rather than scenarios of climate change, where the risk of defined outcomes is quantified.

A further reason for the climate impacts toolkit approach is that only limited funds are available for climate impact research. By adopting a modular approach, the strategies outlined above are being pursued as a means to overcome these limitations by preparing packages, such as the climate scenario generator, that are widely applicable to a variety of impact models. The integration of data-bases and impact models is also being explored.

The current challenge for impact research is to conduct assessments that can address issues associated with the mitigation and adaptation of climate change. The identification of sensitivity and vulnerability by studies[89] only partially fulfils this need. The next step is to identify the levels of climate change that are "dangerous" and to assess adaptations for natural ecosystems, socio-economic systems and human health as outlined by the IPCC[90] in time for the Third Assessment Report due in 2000/1.

89. *See* Basher & Pittock, *supra* note 5.

90. IPCC, supra note 3.

2. THE FORMATION OF AUSTRALIAN CLIMATE CHANGE POLICY: 1985-1995

HARRIET BULKELEY
Research Fellow, St. Catharine's College
Department of Geography, University of Cambridge
Downing Place
Cambridge, United Kingdom CB2 3EN

1. Introduction

Australia has proved to be something of an unknown quantity in the international climate change policy process. Having adopted a pro-active stand at Rio in 1992 during the Conference of the Parties process, it subsequently displayed an unwillingness to accept legally binding targets and timetables and advocated the merits of a 'differentiated' approach. The concept of differentiation is grounded in the belief that each nation-state should bear commensurate losses of economic welfare in the pursuit of greenhouse gas emission reduction goals. Although Australia has been criticised internationally for this approach,[1] it formed the basis of the agreement reached at the Kyoto Conference of the Parties in December 1997. This chapter undertakes an historical overview of the policy responses of the Federal government in Australia from 1985 – 1995 in order to shed some light upon this position. It will be argued that while grounded in the policy discourse of ecological modernisation, Australia's policy approach has failed to overcome some of the significant contradictions that this discourse tries to address. This analysis calls into question not only the potential for Australia to act on any outcomes achieved at Kyoto, but also the ability of the 'ecological modernisation' approach to resolve such contradictions.

Before addressing the ways in which the Greenhouse[2] issue came to be seen as a policy issue in Australia and the development and implementation of a National Greenhouse Response Strategy (NGRS), the concept of 'ecological modernisation' merits further discussion.

1. While other countries, such as Norway and Japan, have also argued for the principle of differentiated targets, their positions were based on different concepts of what would constitute a 'just' process. Australia has attracted most of the international criticism for advocating differentiation, for example in John Gummer's speech at Berlin in 1995, from President Clinton in a speech made during a visit to Australia in November 1996 at the Rio plus 5 conference in New York in 1997 and during the Conference of the Parties process.
2. Throughout this chapter "Greenhouse" is used to refer to anthropogenically induced climate change.

A. Gillespie and W. C. G. Burns (eds.), Climate Change in the South Pacific: Impacts and Responses in Australia, New Zealand, and Small Island States, 33–50.

2. Ecological Modernisation

The analysis of environmental policy development has traditionally taken as a given the environmental problem under consideration. Policy and politics enter the stage once a perceived environmental threat has been identified. Disputes arise, are mediated and settled as interest groups manoeuvre, usually through the state, to gain desired outcomes. Hajer[3] takes issue with this type of 'interest-based' analysis. He argues that in order to understand the evolution and significance of any environmental issue it is necessary to examine not only which solutions were advocated and implemented, but also how the issue came to be viewed as an 'environmental problem' in the first place. As he explains, environmental controversy "should not be conceptualised as a conflict over a predefined unequivocal problem with competing actors pro and con, but is to be seen as a complex and continuous struggle over the definition and the meaning of the environmental problem itself."[4] The policy-making process provides an important arena for this struggle, as it represents the institutionalisation, and therefore domination, of certain meanings and understanding of what environmental problems are. Environmental issues are marked by their complexity, involving a bewildering variety of claims, concerns and actors. Yet from these often contradictory and fragmented understanding of the issues, "somehow we manage to distil seemingly coherent problems"[5] which command public and political attention. Hajer describes this process of creating a narrative through which to conceptualise an environmental issue as the construction of "story-lines."[6] He is at pains to point out that an emphasis on 'discourse' does not restrict analysis to only what is said, but also involves consideration of the institutional contexts which enable and constrain what are seen as legitimate ways of understanding an issue. The creation of 'story-lines' involve actors in an 'argumentative' struggle in which each tries to attribute meanings to the 'problem' and position other actors in specific ways.[7] The resulting ensembles of 'story-lines,' actors and practices constitute 'discourse-coalitions' through which the meanings and appropriate resolution of environmental conflicts are contested. Due to their eclectic nature, such discourse coalitions do not represent shared interests, goals or values, but instead share concepts and terms through which an environmental issue is framed.[8]

---

3. *See* M. HAJER, THE POLITICS OF ENVIRONMENTAL DISCOURSE: ECOLOGICAL MODERNISATION AND THE POLICY PROCESS (1995); M. Hajer, *Ecological Modernisation as Cultural Politics*, in RISK, ENVIRONMENT AND MODERNITY: TOWARDS A NEW ECOLOGY (S. Szerszynski Lash & B. Wynne eds., 1996).
4. HAJER, *supra* note 3, at 15.
5. *Id.* at 1-2.
6. *Id.* at 56-65.
7. *Id.* at 53, 62-5.
8. HAJER, *supra* note 3, at 247.

It has been argued that during the 1980s a new discourse gained credence in environmental politics.[9] This discourse of "ecological modernisation" can be defined as one which "recognises the structural nature of the environmental problematique but none the less assumes that existing political, economic, and social institutions can internalise this care for the environment."[10] As cited in the Brundtland Report, the central tenet is that there need be no fundamental contradiction between environmental protection and economic growth.[11] This approach stems from the changing character of environmental problems and solutions over the last three decades. The cross-media, international and inter-generational character of some contemporary environmental concerns exceeds the capacities of state authorities to 'react and cure' on a single-issue basis. The move to a more integrated, 'anticipate and prevent' approach involves science in determining acceptable solutions, as well as identifying problems, shifts the burden of proof from 'damage' to 'possibility' and argues that 'pollution prevention pays' in an attempt to reconcile 'growth' and 'environment' within existing institutions.[12] Hajer's analysis of the policy making process is useful in that it shifts emphasis from the chronological inevitability of policy development towards a recognition of the contested nature of policy outcomes. This approach, which sees ecological modernisation as essentially a 'policy discourse,' is one of many which have been offered. Ecological modernisation has been variously described as anything from the more technical endeavour of achieving greater levels of economic efficiency, to a challenge to the tenets of industrial modernisation, thereby encompassing a wide range of potential values and policy outcomes.[13] The extent to which policy processes and outcomes represent 'stronger' or 'weaker' interpretations of ecological modernisation, and therefore its potential political implications, remains a question for empirical investigation and debate.

The emergence of the Greenhouse issue as a policy problem in Australia coincided with an attempt by the Federal Government to forge a national environmental strategy, guided by the principle of Ecologically Sustainable Development (ESD).[14] Control over the environment has traditionally fallen under the jurisdiction of state governments. However, during the 1980s the Federal government invoked its foreign affairs powers to intervene in environmental conflicts that had an international dimension. This resulted in a number of bitter controversies, such as the Franklin dam case in Tasmania in 1983 and the World Heritage listings in Queens-

9. *See* for example HAJER, *supra* note 3, at 3. P. Christoff, *Ecological Modernisation, Ecological Modernities*, 5 ENVIRONMENTAL POLITICS, 476-500 (1996); A. WEALE, THE NEW POLITICS OF POLLUTION (1992).

10. HAJER, *supra* note 3, at 25.

11. WORLD COMMISSION ON ENVIRONMENT AND DEVELOPMENT, OUR COMMON FUTURE (1987).

12. HAJER, THE POLITICS OF ENVIRONMENTAL DISCOURSE, *supra* note 3, at 26-29.

13. *See* Christoff, *supra* note 9, at 9, for a summary of the different uses of 'ecological modernisation'.

14. P. Toyne, *Ecologically Sustainable Development*, in AUSTRALIA: PRESENTATION TO THE PRESIDENT'S COUNCIL ON SUSTAINABLE DEVELOPMENT (1995) (then Executive director of Environment Strategies Directorate), Department of the Environment, Sports and Territories.

land in 1988.[15] These conflicts, amongst others, placed economic and environmental imperatives in opposition and led to a policy-making approach akin to fire-fighting. The ESD process represented Australia's most concerted attempt to form a broad environmental strategy and to escape these endemic environment-economy, inter-governmental conflicts.[16] As such, it represented the clearest indicator of the influence of the Brundtland report, with its associated discourse of ecological modernisation, upon environmental policy making in Australia.

The remaining portions of this chapter examine the success of the ecological modernisation approach to Greenhouse policy making in Australia. The first section examines the ways in which 'Greenhouse' became defined as a political and policy issue, internationally and within Australia. The second section takes as its focus the process through which a 'National Greenhouse Response Strategy' emerged, and in the final section issues surrounding the implementation of this strategy are considered.

3. Science, Sustainability and the Global Stage

Although there has been over a hundred years of scientific concern and investigation into the effect of greenhouse gases on the climate system, a concerted research effort was not conducted by national and international institutions until the 1970s.[17] The Villach Conference of 1985 represented an attempt to review this research under the auspices of the World Meteorological Organisation, the International Council of Scientific Unions and the United Nations Environment Programme. It is this conference that is most often credited with moving the issue of climate change from the scientific backburner into the political limelight through its articulation of the potential severity of the issue and the need for political action to address it.[18]

In Australia, (see Table 1) there had been similar scientific interest in the effects of 'greenhouse gases' in the atmosphere, with the government research organisation, CSIRO (Commonwealth Scientific and Industrial Research Organisation) having undertaken studies of carbon dioxide concentrations since the early 1970s. Members of CSIRO were active in climate change science, involved with international

15. *Id.* at 6.

16. *See* D. Downs, *Neo-Corporatism and Environmental Policy*, 31 AUSTRALIAN J. POL. SCI. 175-190 (1996); P. Kinrade, *Towards Ecological Sustainable Development: the Role and Shortcomings of Markets*, in MARKETS, THE STATE AND THE ENVIRONMENT: TOWARDS AN INTEGRATION. (R. Eckersley ed., 1995); I. Lowe, *The Greenhouse Effect and the Politics of Long-Term Issues*, in STATE, ECONOMY AND PUBLIC POLICY IN AUSTRALIA (S. Bell & B. Head eds., 1995); POLITICS AND THE ENVIRONMENT: THE AUSTRALIAN EXPERIENCE (E. Papadakis ed., 1993).

17. J. Jäger & T. O'Riordan, *The History of Climate Change Science and Politics*, in POLITICS OF CLIMATE CHANGE: A EUROPEAN PERSPECTIVE 12 (T. O'Riordan & J. Jäger eds., 1996).

18. R. Taplin, *Climate Science and Politics: The Road to Rio and Beyond*, in CLIMATE CHANGE: DEVELOPING SOUTHERN HEMISPHERE PERSPECTIVES (A. Henderson-Sellers, A. & T. Giambelluca eds., 1996); Jäger & O'Riordan, *supra* note 17.

research efforts, and instrumental in the lead-up to Villach. CSIRO also played a crucial role in translating the outcomes of Villach into policy development in Australia.[19] Together with the Commission for the Future (CFF), an organisation established by the Federal government to raise the level of public debate over key scientific and technical develop-ments,[20] CSIRO organised two Greenhouse 'information' events in the late 1980s. The first, Greenhouse '87, was an attempt to engage the interest of the scientific community, assess what the likely impacts of climate change could be, and to attract political and public attention. In a scenario-building exercise, CSIRO asked 'experts' from various fields to consider, over a twelve-month period, the effects of climate change scenarios on their respective areas of special interest. The resulting conference produced substantial media attention and acted as a source of information on the likely impacts of climate change that could be used by the policy-making community. The second event, Greenhouse '88, created a public forum for the discussion of Greenhouse issues. State capitals were linked by satellite for the opening discussion, and organisers estimated that about eight thousand people participated. These events won CSIRO and CFF a 'Global 500' award from the UN and created a lasting impression on both policymakers and the public.[21]

The concerns voiced at Villach, and echoed in Australia, had also resonated elsewhere. In 1988, the Canadian government hosted a conference on 'The Changing Atmosphere' in Toronto, with participants from a wide variety of backgrounds, though with no official governmental representation. Although it had no 'official' remit, the conference became notorious for its 'Call to Action' for developed countries to reduce emissions of carbon dioxide to 1988 levels by 2000, and by a further 20% by 2005. This target was chosen to symbolise the extent of effort that the problem merited, rather than to correspond to any proven 'acceptable' level of emissions. The conference attracted a high level of media attention, coinciding as it did with a serious drought in the United States. It placed the issue of 'global warming' firmly alongside rainforest destruction and stratospheric ozone depletion at the centre of new public concerns for the global environment. Aside from raising awareness of the issues, the conference had two major impacts. The first was that it drew attention to the need to form an international advisory body through which governments could develop policy responses. The Intergovernmental Panel on Climate Change (IPCC) was established in 1988 with the express task of reporting on the state of climate change science, impacts and responses for the second world climate conference in 1990. The second was the influence of the 'Toronto Target.' This was adopted by various nation-states, in various guises, as an interim policy measure in the absence of a co-ordinated international approach.

The impetus to develop a policy approach to the Greenhouse issue in Australia, sprung, therefore, both from gathering international momentum surrounding the

---

19. Research Interview with CSIRO representative, December 1996. *See also* Lowe, *supra* note 16.
20. *Id.* at 317.
21. Research Interview with ACA representative, December 1996.

issue and from concerns expressed domestically by the scientific community. Such efforts led to significant "Greenhouse science" commitments by the Federal government. In April 1989 the Federal Government set up a national climate change program, which included the establishment of the National Greenhouse Advisory Council (NGAC). This was primarily composed of climate experts, and provided substantial funding for climatic modelling and impacts work.[22] A Prime Ministerial Working Group was also established to assess achievable domestic targets[23] and a discussion paper released late in 1989 argued that any solutions to the 'Greenhouse problem' would need to involve the co-operation of governments, industries and the community.[24]

During 1990, the Australia and New Zealand Environment Council emphasised the need for a national Greenhouse strategy and the Federal Government took the first steps towards that end. Many state governments adopted one version or other of a 'Toronto Target' for 'planning purposes,[25] and interest groups made their voices heard. The minerals and energy industries were quick to realise the potential economic implications of the Greenhouse issue and its impacts on their sectors. The Australian Minerals and Energy Council released a report in 1990 detailing the energy dimensions of the Greenhouse issue.[26] Such concerns, combined with growing international pressure following the release of the first IPCC report and the imminence of the Second World Climate Conference, prompted the Federal government to take a policy stance. On 11 October 1990, just before the international conference, the Federal Government adopted the 'Interim Planning Target' (IPT). This target was based on that called for at Toronto and committed the government to aim to reduce emissions of greenhouse gases, not covered by the Montreal Protocol, to 1988 levels by 1990 and to cut emissions by 20% by the year 2005. Significantly, this aim was accompanied by the caveat that in attempting to reach such targets there should be no adverse effect on the Australian economy, and upon trade competitiveness in particular, in the absence of similar action by other countries.[27]

In initiating a precautionary policy approach to a global issue, through an inclusive strategy to overcome potential environment-economy and intergovernmental conflicts, this policy discourse clearly represented some elements of ecological modernisation. However, despite the proactive role taken by the domestic, and international, scientific community in urging this policy approach, the role ascribed to science was narrower, focusing on the need to reduce uncertainties in

22. G. WILKENFELD, C. HAMILTON & H. SADDLER, AUSTRALIA'S GREENHOUSE STRATEGY: CAN THE FUTURE BE RESCUED? Discussion Paper No.3. The Australia Institute (1995), at 8.

23. *Id.* at 8.

24. R. Taplin, *Greenhouse: An Overview of Policy and Practice*, 1(3) AUSTRALIAN J. ENVTL. MGT. 145 (1994).

25. *Id.* at 145; Wilkenfeld, *supra* note 22, at 8.

26. Taplin, *supra* note 24, at 145

27. *Id.*, Lowe, *supra* note 16, at 315-316

greenhouse science. This more traditional policy discourse was also reflected in the framing of the relation between the environment and economy, which indicated that pollution prevention might incur costs. In the early construction of the 'greenhouse story-line,' tensions between uncertainty and precaution, and between economic objectives and environmental values were written into, rather than out of, policy discourses. How these tensions played out in the formation of a national response strategy is examined in the next section.

TABLE 1. A chronology of key Greenhouse policy events in Australia; 1985-1995.[28]

Year	International	National	Intranational
1985	Villach Conference		Scientific community involved in international and domestic research
1986			CSIRO and the Commission for the Future initiate the 'Greenhouse project'
1987			Greenhouse '87: scientific assessments of the implications of Greenhouse
1988	Toronto Conference Formation of the IPCC		Greenhouse '88: Public information event
1989		Greenhouse Funding designated National Greenhouse Advisory Committee established Prime Ministerial Working Group established ESD Process initiated	State Governments adopt versions of the Toronto Target
1990	IPCC First report Second World Climate Conference, Geneva. Formation of the INC	ANZEC report Interim Planning Target adopted ESD Greenhouse working group established	AMEC Report Industry Commission asked to consider Greenhouse issue
1991	INC Negotiations towards Rio	ESD Greenhouse Report	Industry Commission Report
1992	Framework Convention on Climate Change signed at Rio	Draft NGRS released NGRS adopted by COAG FCCC ratified by COAG	
1993			
1994			Carbon Tax debate is reopened
1995	IPCC Second report Conference of the Parties, Berlin	Greenhouse 21C adopted Greenhouse Challenge Programme established	

28. Although assigned to different 'scales' these policy 'events' were significant at more than one level of policy-making, this is discussed in the text.

4. Designing Responses

During 1991-2, the Federal government set about exploring possible policy options for meeting the Interim Planning Target, and their associated costs and benefits,. This was pursued through two major channels, the ESD working groups and the Industry Commission.

4.1 ECOLOGICALLY SUSTAINABLE DEVELOPMENT?

In late 1989, the Federal government prepared a working paper that set out the terms under which Ecologically Sustainable Development was to be pursued. The remit was to involve governments, industry, environmental and community groups in a consideration of the possibilities for sustainable development in nine sectors of the economy, primarily those concerned with either resource development or energy use.[29] While such an approach represented a considerable break with past attempts at fire-fighting environmental issues, the emphasis in the discussion paper on certain sectors and issues "may have pre-empted the possibility that . . . [it] could seek guidance from the community on a different kind of approach and different issues."[30] Within this framework, working groups were established in an attempt to provide a forum through which consensus could be achieved by broad consultation with stakeholders. The working groups were dominated numerically by government officials, but each also had representatives from industry, environment and community groups.[31] The Federal government played a key role in selecting these participants, choosing them from what it deemed "key interest organisations," and thereby encouraging the "monopolistic representation" of interests.[32] Such a strategy has been labelled 'neo-corporatist' intermediation by Downes, who argued that it provided a means through which the Federal government could manage potential conflict and legitimize outcomes.[33] This means of consultation was not acceptable to some environmental organisations; for example, the Wilderness Society and Greenpeace were notable for their absence from this process. In the formation of the ESD process the legitimate terms of, and participants in, debate were delimited, thereby establishing the contours of "the problem" in advance. Although this process may have been initiated as an attempt to reconcile environmental and economic issues, the terms in which it was grounded appear to be antithetical to such aims.

During 1990, the Federal Government asked the ESD Working Groups to form a Greenhouse Working Group and prepare a report, with the remit to assess options for meeting the IPT and the cost-effective combinations of doing so. As the only cross-sectoral issue considered in any depth, aside from the overall strategy of

29. *See* Lowe, *supra* note 16; Kinrade, *supra* note 16; Toyne, *supra* note 14.

30. Toyne, *supra* note 14, at 11

31. Downes, *supra* note 16, at 176; Kinrade, *supra* note 16.

32. Downes, *supra* note 16, at 184

33. *Id.* at 184

ESD, the report was closely tied to the fortunes of the whole process. Along with the other working groups, the Greenhouse group met through the remainder of 1990 and 1991 and released its report in late 1991. The outcomes of the ESD process were seen to represent a 'remarkable' degree of consensus about the issues that needed to be tackled, and the principles to guide this process.[34] So much so that the whole process has in fact attained somewhat 'mythical' proportions amongst those who participated. While not all aspects of the process can be regarded as successful, Toyne argued that some of this mythologising is deserved. To many of the participants it was a valuable and progressive approach, involving the efforts of hundreds of people in a process they believed to be "akin to nation-building."[35] The outcomes of the ESD reports were deemed to contain:

> enough substance to provide the steps for taking concrete steps towards ESD in Australia, providing the working group recommendations were implemented vigorously and as an integrated package.[36]

Specifically, the Greenhouse working group found:

> that there (were) a large range of actions which would be cost-effective on energy grounds alone, so that additional benefits in greenhouse gas reduction would be free.[37]

The working group made a number of recommendations and called for the government to draw up timetables and clearly defined responsibilities[38] for acting to reduce emissions of greenhouse gases. In articulating the presence of 'win-win' Greenhouse policy options, the ESD working group recommendations can be seen as representing an approach to ecological modernisation that Hajer has termed the rationalisation of ecology.[39] Far from challenging the basis of industrial modernity,[40] the working group's recommendations emphasised the prospects for action through existing institutional and social structures. Ecological modernisation as expounded here was neither 'weak' nor 'strong,' but instead represented a compromise between these two goals.[41]

The degree of consensus apparently took government representatives involved in the process by surprise.[42] The governments faced the imposing task of considering more than 500 recommendations.[43] From this morass, two documents were

34. Lowe, *supra* note 16, at 319; Toyne, *supra* note 14, at 12.
35. *Id.* at 17.
36. Kinrade, *supra* note 16, at 88.
37. Wilkenfeld, *supra* note 22, at 9.
38. Taplin, *supra* note 24, at 146.
39. Hajer, *supra* note 3, at 261.
40. Downes, *supra* note 16, at 186.
41. Christoff, *supra* note 9.
42. Lowe, *supra* note 16, at 6; Toyne, *supra* note 14, at 17.
43. *Id.*

drafted: a National Strategy for Ecologically Sustainable Development (NSESD) and a National Greenhouse Response Strategy (NGRS). Two hundred of the recommendations were chosen by a total of thirty-seven interdepartmental and intergovernmental committees for incorporation into the NGRS and NSESD.[44]

In an apparent attempt to find politically viable policy options, these negotiations effectively excluded the original working group participants.[45] The resulting draft NGRS bore few similarities with the conclusions of the working group, representing instead a 'lowest common denominator' approach as to what governments and bureaucracies were prepared to accept. A two-day forum held to discuss the draft NGRS ended abruptly on the first day, as the environmental NGOs refused to attend and those stakeholders who did attend refused to move into the discussion sessions until their concerns had been addressed. By the end of the first day, the level of discontent was running high and the planned second day had to be abandoned. The Institute of Australian Engineers, an organisation not known for its radicalism, issued a press release condemning the report and voicing concern that changing the conclusions of the working groups gave the impression that a small group of officials knew better than the wider community what action should be taken.[46]

4.2. CALCULATING THE COSTS

The outcomes of the National Greenhouse Response Strategy policy development process were significantly influenced by other 'story-lines' of the Greenhouse issue. During 1991, the Industry Commission had also been asked to comment on the costs and benefits of stabilising Greenhouse emissions for Australian industry. This process took the form of an inquiry with public hearings and submissions from interested parties.[47] It concluded that compliance with the interim planning target of stabilising emissions by 2000 would lead to a reduction of 1.5% in national output, and that while most sectors of the economy would be affected, the costs would be disproportionately borne by some sectors and regions more than others. However, the Commission contended that it had a number of difficulties with its task. It argued that it was difficult to assess the implications for the economy while future Greenhouse impacts were still surrounded by uncertainty. Calculating the costs of inaction was, therefore, problematic, and uncertainty was used as justification for delaying action in this context.[48] It also conceded that it had been unable to quantify the benefits of taking Greenhouse action.[49] This stood in stark contrast to the findings of the ESD Greenhouse Working Group that there were a number of 'no regrets' options that

44. Lowe, *supra* note 16.
45. Kinrade, *supra* note 16; Toyne, *supra* note 14.
46. Lowe, *supra* note 16, at 322.
47. Taplin, *supra* note 24, at 145.
48. *Id.* at 145.
49. Lowe, *supra* note 16, at 320.

could be taken that would have, at the least, engendered no net economic losses.

The calculations of the Industry Commission relied upon the application of 'top-down,' neo-classical, economic modelling. This approach assumes that the market for energy is essentially perfect, and that behaviour is profit maximising. Further, the appropriate instrument for effecting reductions of greenhouse gas emissions is assumed to take the form of a market instrument, such as a carbon tax.[50] The possibility that the market can be made to take the environment into account aroused considerable interest during, and subsequent to, the ESD process, and numerous other studies have been made of the potential impacts on the Australian economy of such a policy. The most influential, such as the report from The Industry Commission itself, a report from London Economics, which analysed the impacts on key industrial sectors, and the Tasman Institute, have all shown that such a carbon tax would have a detrimental effect.[51] Also, importantly, they demonstrate that this effect would be disproportionately borne by some sectors, such as energy and minerals processing. While the assumptions behind such models have been criticised extensively elsewhere,[52] it is important to draw attention to the fact that such top-down analysis fails to capture the full extent, or nature, of market failures, or of the non-rational components of energy decisions.[53] Some have also focused attention solely on the energy intensive sector of the Australian economy which, significant as it is, does not capture the range of responses to market interventions, such as a carbon tax. Small businesses in the renewable energy sector, for example, might profit from such a policy measure. Therefore, such approaches have usually failed to consider other means than a carbon tax for achieving greenhouse gas reductions, or those groups whom might stand to gain from policy actions. However, they have had significant influence on the Federal government policy position. Representatives of the large and well organised industries have attracted considerable support for their concerns from governments, in particular within finance and development departments where there is considerable sympathy for approaches grounded in 'economic rationalism.[54]

The Industry Commission also found that the study could not take full account of environmental or sustainability objectives within policy guidelines that emphasise economic efficiency, and that equity issues would have to be left for political judgement.[55] The Industry Commission conclusion that emission reduction measures could be taken without imposing significant burdens on the Australian economy in general, and specific sectors and regions in particular, has to be seen in the light of its methodological approach and narrow remit. It marked a clear departure

---

50. Kinrade, *supra* note 16, at 101-103.
51. *Id.* at 103; Lowe, *supra* note 16, at 320-1.
52. Kinrade, *supra* note 16; Wilkenfeld, *supra* note 22.
53. Lowe, *supra* note 16; Wilkenfeld, *supra* note 22.
54. Kinrade, *supra* note 16, at 90.
55. Taplin, *supra* note 24, at 145-6.

from the discourse of ecological modernisation, articulated by the ESD working groups, through its emphasis on the potential conflict between economic and environmental goals. Nevertheless the report, and others like it, had a far-reaching influence on Australia's Greenhouse response strategy.

4.3. REACHING RIO

The rationale for undertaking such an exhaustive policy formation process stemmed not only from a desire to take a more strategic domestic approach to environmental issues, but also in response to international developments. Since the IPCC report and second climate conference in 1990 the process of designing an international response to Climate Change had intensified. The Intergovernmental Negotiating Committee was formed by the United Nations General Assembly in late 1990 with the aim of establishing a framework convention for signing at Rio in June 1992.[56] Australia has been keen to be viewed as an active participant on the world environment stage and, despite concerns on the part of key industrial sectors in Australia, the Federal Government joined other nation-states at Rio to sign the Framework Convention on Climate Change (FCCC). While the FCCC did not contain any legally binding targets or timetables, it called for nations to aim to reduce their emissions of greenhouse gases to 1990 levels by 2000 and adopt both the precautionary principle and the goal of sustainable development. Significantly, it also committed signatories to the ongoing process of designing policy responses. In December 1992, Australia became the eighth nation to ratify the convention. This early ratification was meant to highlight the seriousness with which Australia was taking its international environmental responsibilities. The fact that it was able to do so with such confidence reflects both the apparently innocuous nature of the Convention's commitments, and the adoption by the Council of Australian Governments (COAG) of the National Greenhouse Response Strategy (NGRS).

5. Implementation: the Real Greenhouse Challenge?

Despite the fiasco associated with the publication of the draft NGRS, a final version was assembled by the Federal government and ratified by the COAG in December 1992. Its implementation has proved equally contentious and its perceived failure has led to additional policy measures.

5.1 THE NATIONAL GREENHOUSE RESPONSE STRATEGY

Labelled as the first phase of a strategy in an evolutionary process, the NGRS concentrated on the need for 'insurance' measures, including research and adaptation to reduce uncertainties. It also called for adoption of 'no regrets'[57] measures in the

---

56. M. PATTERSON, GLOBAL WARMING AND GLOBAL POLITICS (1996); O'Riordan & Jäger, *supra* note 17; R. Taplin, *International Co-operation on Climate Change and Australia's Role*, 26(1) AUSTRALIAN GEOGRAPHER 16-21 (1995).

interim. 'No regrets' refers to measures that entail net benefits, or at least no net economic losses, in addressing the Greenhouse issue. The influence of the concerns expressed by the Industry Commission was evident in that the pursuit of 'no regrets' was accompanied by some significant caveats. No regrets measures should have low total social and economic costs and meet equity requirements by not imposing an undue burden on any sector of the economy or any region of Australia.[58] This represented a significant deviation from the perspective of the Greenhouse ESD WG that noted:

> some industrial decline and closure, and some restructuring, would be necessary to achieve cuts in greenhouse gas emissions but that this would be countered by growth in industry orientated towards energy efficiency and renewable energy.[59]

It is clear, therefore, that the principles through which the environment and economy are to be reconciled differed in the NGRS from those articulated in the ESD process. Eckersley argues that there was in fact a clear failure to link the principles of ESD to policy measures and strategy formation in the NGRS. It is evident that while the National Greenhouse Steering Committee "performed the standard invocation of ESD principles ... (they) thereafter applied a different set of principles that were seen to be specific to a greenhouse strategy."[60]

While the pursuit of the IPT represented both the foundation and goal of the NGRS, governments did not commit themselves to any definite targets and timetables for achieving reductions of emissions. The NGRS concentrated on reviewing measures already underway at the various tiers of government and called for few new commitments. Actions were not clearly defined or prioritised and it is difficult to see where responsibilities lay.[61] Indeed, "it was highly questionable whether the IPT could be achieved by implementing the response actions listed in the strategy."[62] Other commentators went further with their critique. Wilkenfeld et al. argued that by early 1995 governments had failed to implement actions from the NGRS, establish any new strategies or programmes, or assign clear responsibilities to a single authority. Instead, responses were left to ad hoc government processes and commercial decisions. They suggested that actions taken in the energy and resource sectors actually ran counter to the NGRS in both principle and practice.

---

57. Commonwealth of Australia, *National Greenhouse Response Strategy*, Department of the Arts, Sport, Environment, Tourism and Territories. ACT (1992).
58. *Id.*
59. Taplin, *supra* note 18, at 391.
60. R. Eckersley, *Re-interpreting No-regrets: A Green Economic Critique of the Greenhouse Response Strategy*, in GREENHOUSE AND THE ENERGY REGIONS: PROCEEDINGS (S. Pfuller, S. M. Hooper & D.H. Harvey eds., 1993), at 59.
61. National Greenhouse Advisory Panel, *Report on the National Greenhouse Response Strategy* Environment Australia, Commonwealth of Australia. ACT (1996), at 26; Taplin, *supra* note 18; Wilkenfeld. *supra* note 22.
62. Taplin, *supra* note 24.

The commissioning of coal-fired power stations in both the Hunter Valley and Western Australia stand as testament to this. Efforts by Greenpeace and the Total Environment Centre to challenge the building of more coal-power capacity in the Hunter Valley through the Land and Environment Court were unsuccessful. While the court recognised that such projects might produce substantial greenhouse gas emissions, it argued that the NGRS contained no obligations or directives to limit the actions of individual operators.[63] Far from reconciling environment and development objectives, the end result of the NGRS has been to maintain the status quo.[64] As such, it was asserted that:

> There is no evidence that the NGRS has saved one single tonne of greenhouse gas emissions which would not have been saved in any case for other reasons. In other words, there has been no departure from 'business as usual.[65]

The extent to which any version of ecological modernisation could be said to have influenced policy outcomes is therefore highly questionable.

The difficulties in implementing the NGRS have also been recognised in official assessments. In early 1994, a progress report on the first year of the NGRS was released. It argued that budget and staff restrictions had impeded the implementation of the NGRS, and would continue to do so.[66] This recognition, together with renewed international developments surrounding the design of a climate 'protocol' and impending release of the IPCC's second assessment report, placed Greenhouse concerns back on the domestic agenda during 1994.

5.2 REVISITING NO REGRETS

The debate over the best means to achieve greenhouse gas reductions was reopened in 1994 by Senator Faulkner, then Environment Minister.[67] Initially the possibility of implementing a Carbon Tax was considered, but concerted industry opposition soon scuppered this. Instead, industry proposed that it could adopt a voluntary approach that would focus on 'no regrets' measures. The possibility of a carbon tax seemed to have refocused minds on the availability of cost-effective measures, something the neo-classical models of the early 1990s had ignored. A significant difference in this approach to 'no regrets' was, however, that they were defined within 'firm' boundaries rather than at a national scale. The resulting 'Greenhouse Challenge' Programme encouraged companies and industry associations to sign up for voluntary reductions in emissions, which would be externally audited. These reductions could take the form of reducing the emissions of green-

---

63. Wilkenfeld, *supra* note 22.
64. Taplin, *supra* note 18.
65. Wilkenfeld, *supra* note 22, at 4. (Emphasis in the original).
66. Taplin, *supra* note 24, at 153.
67. Taplin, *supra* note 18, at 392.

house gases, through the adoption of energy efficient technologies, or enhancement of sink capacities, through for example, tree planting programmes.[68] The Programme is run jointly by the Departments of the Environment, Industry and Technology and Primary Industries and Energy. This represents a significant difference from the NGRS, which was administered through the Environment portfolio, though in consultation with other departments. Physically located in the Primary Industries building, the core principles reflect the Department's ideology. Publicity material made it clear that in pursuing such no regrets' options, the growth paradigm should not be questioned:

> Co-operative agreements are voluntary. The Greenhouse Challenge *is not intended to compromise the business objectives of development and growth* and participants may withdraw from the program at any stage without penalty.[69] (emphasis added)

This belief that there were 'win-win' options that could be adopted in pursuit of Greenhouse goals represented a significant shift in position from the more 'traditional' emphasis of the Industry Commission report. It seems that the principles of ecological modernisation, a precautionary approach and the possibility of a reconciliation between economic and environment objectives, have had some impact. However, the scale at which these principles are applied still only creates a 'weak' version of the rational approach to ecological modernisation. Not only was the 'Greenhouse Challenge' programme an important means for the energy and resource community to deflect the debate over a carbon tax, it also provided a key part of an additional Greenhouse strategy administered by the Environment portfolio, Greenhouse 21C, which was implemented in 1995.[70]

Greenhouse 21C also encouraged enhanced levels of reporting on reduction measures, and states that it will commit funding to sustainable energy technologies and initiate a White Paper on a national sustainable energy policy. However, the main thrust of the strategy can be "viewed as an attempt to implement reductions more effectively,"[71]and emphasises increasing co-operation between governments in pursuit of that goal. Even if fully implemented, it was "estimated that Australia will still overshoot the FCCC aim of reducing greenhouse gas emissions to 1990 levels by 2000 by 3% CO_2 equivalent."[72]

68. Commonwealth of Australia, *The Greenhouse Challenge*, Greenhouse Challenge Office, ACT (1995).
69. Commonwealth of Australia. (1996), *The Greenhouse Challenge: Implementation Plan*, Greenhouse Challenge Office, ACT, at 8
70. Commonwealth of Australia, *Greenhouse 21C*, Department of the Environment, Sport and Territories. ACT (1995).
71. Taplin, *supra* note 18, at 392.
72. Commonwealth of Australia, *supra* note 70; Taplin, *supra* note 18, at 392.

6. Taking a Backseat

By 1994, it was becoming clear that, in the light of substantial difficulties in implementing the NGRS, Australia would not be able to meet its international commitments. This persuaded the Federal government to adopt a more cautious attitude in the international negotiation process.[73] Despite the involvement of many Australian delegates in the Intergovernmental Negotiating Committee and the IPCC process, by mid-1995 the Federal government had aligned itself with the so-called 'JUSCANZ'[74] group, taking the position that any further commitment under the FCCC should be minimal. Although the second report from the IPCC did much to mute the voices that held up scientific uncertainty as the rationale for inaction, in the light of the purported threats to the Australian economy the Federal government was keen to pursue a go-slow policy.

During the decade that had passed since the Villach Conference 'Greenhouse' had gained a firm hold on the Australian political agenda. Although initiated through a policy discourse that represented broad principles of ecological modernisation, the implementation and development of policy responses has taken place in the context of alternative, competing interpretations of the Greenhouse issue.

7. Conclusion

Many explanations have been offered to account for Australia's specific difficulties in creating a domestic Greenhouse policy response and meeting its international obligations during the period 1985-1995. Australia is economically reliant on a number of greenhouse emission-intensive industries that have been quick to calculate the potential costs of action. Moreover, they have framed the issue in terms of energy policy and trade, which resonated strongly in some key sectors of the Australian bureaucracy. A division between those responsible for designing domestic responses, mainly the Federal Environment portfolio, and those responsible for many of the sectors and regions through which such a strategy has to be implemented, the Department of Primary Industries and Energy and state governments, also proved problematic. The manifold uncertainties surrounding the Greenhouse issue, and the fact that its ramifications may not be manifested for decades or more, also contributed to a lack of political will to address the Greenhouse problem.[75] Despite its origins in a discourse that advocated the need for precautionary policy approaches and the possibility of overcoming divisions between economic and environmental objectives, Greenhouse appears to represent a "throwback to the previous era of confrontation"[76] within environmental policy-making. The very problems that appear at the heart of Australia's Greenhouse policy represent those that the ESD process were instigated to overcome.

---

73. Taplin, *supra* note 24, at 153-54.

74. During the first Conference of the Parties in Berlin in 1995 the broadly similar positions adopted by Japan, the United States, Canada, Australia and New Zealand earned them this collective title.

75. Lowe, *supra* note 16; Taplin, *supra* note 24; Wilkenfeld, *supra* note 22.

76. Lowe, *supra* note 16, at 330.

The failure of Greenhouse policy to adhere to ESD principles agreed to through community consultation, the integration of economic, environmental and social objectives, a precautionary approach and recognition of the global dimensions of environmental problems, has been outlined above. This failure stemmed not only from the inability to implement ESD policies once agreed upon, but also, more fundamentally, from a struggle over what ESD might mean. From the outset, the ESD process was orientated towards an economic division of socio-environment concerns. Working groups were established to address economic sectors rather than environmental issues, and much was made of the ways in which the economy could take account of environmental concerns. Greenhouse represented an exception, addressing as it did a cross-sectoral, *environmental* issue. However, it is hardly surprising that, given the circumstances, it too became seen as a predominantly economic concern.

Far from reconciling economy-environment tensions, the ESD process seems to have perpetuated them. Moreover, the terms of legitimate debate failed to allow some fundamental issues to be addressed, as "from the outset the government considered that the strategy for ESD should be achieved without threat to continued economic growth."[77] The Australian experience of Greenhouse policy-making aptly illustrates the difficulties in achieving 'strong' ecological modernisation in practice. Instead, the approach is in keeping with an ecological modernist policy discourse that favours the rationalisation of ecology through the existing institutions of modernity rather than seeking to challenge industrial modernity as part of the ecological problem.[78] Hajer has argued that such a project of rationalisation may lead to more radical outcomes.[79] However, it can be seen in this case that the adoption of "rational ecological modernisation" by the business and government sectors through the principle of "no regrets" led to the scaling down of potential Greenhouse responses during this period.

It is tempting to conclude that efforts at ecological modernisation in Australian Greenhouse policy were an abject failure, and that hopes for any reconciliation between economic and environmental values are thus remote. In terms of reducing greenhouse emissions this seems a safe conclusion. However, by creating arenas and 'discursive' spaces where issues of equity, responsibility, efficiency and effectiveness surrounding Greenhouse responses remain contested, the discourse of ecological modernisation retains some potential. This can be seen, for example, in the creation of the New South Wales Sustainable Energy Development Authority, attempts to create and sell 'Green Power' in NSW, and various activities concerning energy efficiency and renewable energy being promoted by Newcastle City Council, amongst other local authorities. While these efforts are still small-scale, it

---

77. Downes, *supra* note 16, at 186.
78. For further discussion of this point, *see* U. BECK, RISK SOCIETY: TOWARDS A NEW MODERNITY (1992); H. Bulkeley, *Global Risk, Local Values: Risk Society and the Greenhouse Issue*, in 2(3) LOCAL ENV'T (1997).
79. Hajer, *supra* note 3.

remains to be seen what influence such practices and discourses will have in the post-Kyoto era.

Acknowledgements

This chapter was prepared during research for my Ph.D. and I am grateful for financial support for this research project from the University of Cambridge (W.A. Meeks Scholarship, The Smuts Fund and The Philip Lake Fund) and from the Sir Robert Menzies Centre for Australian Studies (Australian Bicentennial Scholarship). I would also like to thank those people I interviewed for sparing their valuable time to talk with me, and those at the Climatic Impacts Centre, Macquarie University, and the Department of Geography, University of Newcastle, for their hospitality. Thanks also to my Australian family for making research trips possible and enjoyable. Last but by no means least, thanks to Dr. Susan Owens, Paula Meth and Tim Rayner for their valuable comments on earlier drafts of this chapter. Errors, of course, remain my own.

3. CLIMATE CHANGE POLICIES IN AUSTRALIA

CLIVE HAMILTON
The Australian Institute
PO Box 72
Lyneham, Australia ACT 2602

1. Introduction

In the period leading up to the Kyoto Conference in December 1997, the Australian Government took an increasingly intransigent stance in negotiations to tackle the problem of global climate change. While claiming that he recognised that climate change is a reality and that nations must take measures to reduce emissions, the Prime Minister, Mr. Howard, stated publicly that Australia should never have signed the Framework Convention on Climate Change (FCCC).[1]

The Foreign Minister Alexander Downer stated "the only target that Australia could agree to at Kyoto would be one that allowed reasonable growth in our greenhouse emissions."[2] The Minister for Resources, Senator Parer, who has de facto carriage of the Government's greenhouse policy, actually stated that he does not believe that climate change is occurring.[3]

Members of the Government and industry representatives were muttering darkly about withdrawing from the Convention if Australia's position on differentiation was not accepted. Despite increasingly forceful and cogent arguments from experts in Australia, the Government's position hardened over the year leading up to the Kyoto Conference. There was no distinction between the Government's position and that of the fossil fuel lobby and the Government became so entrenched in its beliefs that it seems incapable of stepping back and assessing what was in the national interest.

The basic position of the Australian Government was as follows: Australia is heavily dependent on fossil fuels for export revenue and relies on fossil fuels as the chief source of domestic energy. Thus, uniform emissions reduction targets would be very costly and would impose a disproportionate economic burden on Australia compared to other Annex I countries. Australia advocated 'differentiation,' that is, allocation of different targets for Annex I countries based on the economic costs caused by emission reductions. It argued that targets should be set to impose equal economic cost per capita for each Annex I country. Under this proposal, Australia would have more lenient targets than most other countries. Australia also called for developing countries to be

1. J. Howard, Interview, ABC Radio, 28 April 1997.
2. A. Downer, Speech to an Australian Business Seminar, 7 July 1997.
3. "We are now going through all this greenhouse stuff. I don't have any figures to back this up, but I think people will say in 10 years that it [greenhouse] was the Club of Rome." Quoted in SYDNEY MORNING HERALD, 14 March 1997. Senator Parer is a former executive with the Queensland Coal Association.

A. Gillespie and W. C. G. Burns (eds.), Climate Change in the South Pacific: Impacts and Responses in Australia, New Zealand, and Small Island States, 51–77.

required to adopt mandatory targets sooner rather than later.

There were many flaws in the Australian Government's position and many Australians – including scientists, economists, energy experts and ordinary citizens – protested at what they saw as its short-sighted and self-defeating stance. In June 1997, a statement signed by 131 professional economists, including 16 full professors of economics, called on the Government to reverse its position. Summarising, the statement said that:

> 1. Global climate change is a serious environmental problem.
>
> 2. Policy options are available that would slow climate change without harming employment or living standards in Australia, and these may in fact improve Australian productivity in the long term.
>
> 3. The economic modelling studies on which the Government is relying to assess the impacts of reducing Australia's greenhouse gas emissions overestimate the costs and underestimate the benefits of reducing emissions.
>
> 4. Economic instruments will be an important part of a comprehensive climate change policy.
>
> 5. Since OECD countries are responsible for over 80% of increased greenhouse gases in the atmosphere, and are in a stronger economic position to reduce their emissions, they should take the lead in cutting emissions.
>
> 6. Affluent countries that are responsible for high per capita emissions should do more to reduce their emissions.
>
> 7. Withdrawal from the FCCC could seriously harm Australia's long-term interests, especially if Australia became locked into a fossil fuel-based economic system while the rest of the world shifted to low-emission energy sources.

The signatories to this statement represented a large proportion of Australia's professional economists, and the issuing of it reflected a deep concern among economists about the advice on which the Government based its policies.

This paper is arranged as follows. The next section considers arguments and evidence about the economic costs of emissions reductions in Australia. It concludes that the Australian Government exaggerated the costs of meeting emissions reduction targets, and that the economic modelling results and estimates of costs to Australia have little credibility. Despite this, the economic modelling results that were used to justify Australia's position actually showed that the costs of meeting targets would be very low, although they were used to argue the opposite.

Section 3 considers the issue of carbon leakage and concludes that the amount of potential leakage was overestimated by the Australian Government. Section 4 details some of the major flaws in the Australian position, including its abrogation of the polluter pays principle, the perverse consequences if the Australian position were adopted and the implications of differentiation for Australia. It argues that if differentiation had been accepted by the Kyoto conference then it would have resulted in more stringent targets for Australia than uniform targets and concludes that Australia's differentiation position was untenable and contrary to Australia's interests.

Section 5 contrasts the claim by the Australian Government that it is taking measures to reduce emissions with its actual performance. It also considers the opportunities for Australia to reduce its emissions and concludes that Australia is in a strong position to reduce its emissions sharply through energy efficiency at no or low economic cost. Section 6 analyses the 'Australian deal' agreed to at the Kyoto Conference and asks whether it can be regarded as a 'victory.' The final section draws some conclusions.

2. The Economic Costs Of Reducing Emissions

2.1 ECONOMIC IMPACTS ON AUSTRALIA

Australia will potentially experience two types of economic cost as a result of mandatory reductions in greenhouse gas emissions. The failure to distinguish between the two has led to a confused debate.

The first is the cost of reducing Australia's own emissions to the agreed level over the agreed period. Many people – including the 131 economists who signed the recent statement – believe that Australia could reduce its greenhouse gas emissions to the original European target of 15% below 1990 levels by 2010 without any cost to the Australian economy.

The second is the cost to Australia of activities by other countries in fulfilling their obligations under the FCCC. As other countries reduce their emissions, they will shift away from coal and towards natural gas, energy efficiency and renewables. As worldwide demand for coal slows and then declines, Australia's terms of trade with the rest of the world will deteriorate. Although it is not obvious from the reported results, this second set of costs is the dominant input in the Government's economic modelling reviewed below.[4]

The question arises as to what Australia can do to reduce these costs. In relation to the costs of reducing its own emissions, there is a great deal that can be done through energy efficiency, fuel switching and controlling emission sources, such as land clearing. These are discussed in Section 5 below.

In relation to the costs imposed by the actions of other countries, there is nothing Australia can do other than attempt to persuade other countries to reduce their emissions more slowly. However, Australia could have little influence on the overall outcome of the Kyoto negotiations, and the simple fact is that Australia will need to adjust to the changing pattern of world trade in response to climate change. The real issue for Australia is whether it will either continue to attempt to expand markets for coal in a world that will increasingly turn away from that form of energy or whether it will focus attention and resources on developing the forms of energy that the world will increasingly seek out, in particular energy efficiency technologies and renewable energy technologies.

To this point, the Government has pursued policies to protect the coal industry. There is a widespread suspicion that the Australian Government had decided that the best strategy at Kyoto Conference of the Parties was to attempt to sabotage the negotiations and prevent any agreement on mandatory emissions reductions.

2.2 THE GOVERNMENT'S ECONOMIC MODELLING

The Government used the results of the MEGABARE economic model to support its belief that uniform targets would seriously damage the Australian economy. The model was developed over 3-4 years by the Australian Bureau of Agricultural and Resource Economics (ABARE), a public research agency. On the basis of these modelling results the Australian Government made some very strong claims about the high costs to Australia of any international agreement to reduce emissions. For example, it claimed that:

- Australian wages would be cut by 20% below business-as-usual levels by the year 2020;
- GDP would be cut by 2% by 2020;
- Each Australian would lose $9,000 from their savings accounts;
- Tens of thousands of jobs would be lost; and
- The economic cost for each Australian would be 22 times higher than for each European.

Despite the evident absurdity of some of these claims, they were accepted uncritically by the key ministers involved, including the Prime Minister. Even if one accepts the results of the MEGABARE model, a close look demonstrates that that

4. This is suggested by an experiment reported in Appendix B of ABARE. (1997). *The Economic Impact of International Climate Change Policy*, Research Report 97 (1997), at 4. In this experiment, a rate of price-induced fossil fuel energy efficiency improvement in Australia is assigned a value of 0.5% per annum higher than in the business-as-usual scenario, representing a 70% increase in the annual rates of efficiency gain. The growth rates of labour and capital productivity are adjusted downwards so that the overall rate of increase in efficiency of input use remains unchanged. The impact of this large improvement in technical progress in energy use is to reduce the estimated impact on gross national expenditure from 3.4% to 3.1% in 2020.

the economic impact of stabilising Australia's emissions on the economy would in fact be extremely small.[5] This is despite the fact that the MEGABARE model was constructed in a way that tends to exaggerate the costs of greenhouse gas reduction measures. First, it assumes there are no 'no regrets' energy saving measures, i.e., the absence of costless energy savings; and that emissions abatement measures would induce no additional energy-saving technological change. Most non-economists (and indeed most economists) are puzzled, to say the least, when it is suggested that policies designed to sharply reduce emissions will not induce any technological change.

The model also ignored the benefits of cutting emissions and took no account of non-energy sources of emissions, especially land-use change, the emission reduction costs of which would be very small.[6]

The absence of no-regrets measures arises from the assumption that the economy is operating on the 'production frontier' (i.e., at maximum efficiency), and this reflects the ABARE belief that markets work perfectly. Clearly, in the absence of no-regrets measures to reduce emissions in the Australian economy, any required emissions reductions will come at a higher cost.

The economic costs of reducing emissions depend on the level of emissions reductions required and the marginal costs of emissions abatement. The data and assumptions used in the MEGABARE model see Australia emerging with marginal costs of abatement at about the average for Annex 1 countries.[7] Clearly, if there are substantial reductions in emissions that can be attained at little or no cost, then the marginal abatement cost will be lower. The information presented in Section 5 indicates that ABARE has seriously overestimated marginal abatement costs for Australia, and thus the economic costs to Australia of emissions reduction policies.

Welfare changes in the MEGABARE model are measured by changes in annual per person real gross national expenditure (GNE). The model results indicate that real GNE falls below the 'business-as-usual' path by amounts ranging from -0.27% in the year 2000 to -0.49% in 2020.[8] It is most important to recognise that this does not mean that the growth rate of GNE is lower by these amounts, but that the absolute levels of real GNE are lower by these amounts.[9] It is also important to recognise that these estimates have been made using the modelling assumption that

---

5. DFAT & ABARE, *Global Climate Change: Economic Dimensions of a Cooperative International Policy Response Beyond 2000* (1995); ABARE, *supra* note 4.
6. For a full discussion of the MEGABARE model and its pitfalls *see* C. Hamilton & J. Quiggin, *Economic Analysis of Greenhouse Policy: A layperson's guide to the perils of economic modelling.* Discussion Paper Number 15, The Australia Institute (1997).
7. ABARE, *supra* note 4, at 41. The reason why ABARE estimates relatively high total costs on the basis of average marginal abatement costs is twofold: since Australia is projected to have higher emissions growth under business as usual, its abatement task is bigger, and, because a large share of the costs are due to declining terns of trade rather than domestic abatement policies.
8. ABARE & DFAT, *supra* note 5.

these policies do not increase the unemployment rate. This has not prevented Australia's political leaders from simultaneously claiming welfare losses as estimated by ABARE and massive job losses.

One way of understanding the size of the costs predicted by MEGABARE is to compare them to income levels in the future. If the Australian economy grows on average by 3.5%, then *per capita incomes* will reach double their current levels around 1 January 2025. By contrast, if Australia adheres to its international commitments and reduces its emissions, according to the MEGABARE estimates the doubling of per capita incomes will have to wait until around 1st March 2025, a delay of two months.

These MEGABARE results were very embarrassing for the Government. The Government thus resorted to distorting the figures to support its case. The Minister for Resources, Senator Parer, went on record as saying that stabilisation of greenhouse gas emissions at 1990 levels by the year 2000 would:

> have a cumulative effect upon our economy equivalent to a loss of some six per cent of our 1996 gross national expenditure . . . [and that this] would be equivalent to a loss in the savings of every Australian of $1,900 in 1996 dollars or a reduction in the savings of a family of four of about $7,600.[10]

This extraordinarily misleading statement was based on advice from ABARE, which manufactured these figures out of the results of its MEGABARE model.

The only way to make the numbers 'look big' is to take a series of very small numbers over a very long period (25 years from 1996-2020) and aggregate them. Thus, Senator Parer aggregated projected short-falls in GNE for each year over the period 1996-2020 to arrive at a total figure (after discounting at 5%) of $1,900 per person or $7,600 'for a family of four'. Senator Parer and ABARE claimed that this is 'equivalent to' taking $1,900 from every person's savings account or $7,600 from the savings of a family of four.

ABARE used present value lump sums rather than annual flows, thus multiplying the estimated impact by a factor of between 15 and 20. Then it estimated the impact on a family of four rather than per person. Both of these deviate from normal practice in reporting modelling results. In effect, ABARE takes the number and multiplies it by 60 before putting it into a media release. The $7,600 per average family should in truth be compared to the accumulated income over the same period which, in present value terms, would be around $1.75 million.

---

9. The MEGABARE model results are obscured in the ABARE/DFAT report and are presented in a way that makes it difficult for the casual reader to understand what is being measured. The report focuses on the sectors that experience the largest declines in output (coal, oil and gas) and relegates to Appendix C output changes in sectors that would expand. Thus, under the stabilisation scenario, output would expand in agriculture (3.73% by the year 2020), processed agricultural goods (1.85%), manufacturing (2.58%) and services (0.38%).
10. S. Parer, *Senate Hansard*. 26 Nov. 1996.

These results were from an earlier version of the MEGABARE model, which was revised in 1997. The new results multiplied the costs to Australia fourfold, so that Senator Parer could claim that the expected targets would cost each Australian $9,000.[11] The principal reason for this is a change in the model, which limited the options to switch to hydroelectricity. If making a small change to the assumed response by the economy to abatement measures can quadruple the estimated economic costs, the whole structure of the model must be questioned.

It is worth noting that while the Government based its international arguments on cost estimates that assume the *absence* of no regrets reductions, the Government's domestic greenhouse strategy is built almost entirely on the *existence* of no-regrets measures. The Greenhouse Challenge Program of the Department of Primary Industries and Energy (DPIE) – discussed in more detail in Section 5 below – comprises voluntary agreements with major emitting companies. An examination of the agreements already signed, and the Government's literature on the Greenhouse Challenge, indicate that they are based on no-regrets measures. Thus, the Government was telling domestic industry that substantial emissions reductions can be attained virtually for nothing, but was telling the international community that emissions reductions will impose large costs on Australia.

2.3 THE FUNDING OF CLIMATE CHANGE POLICY MODELLING

The fossil fuel industry provided most of the funding for ABARE's modelling work. A financial contribution to the MEGABARE modelling work entitled the contributor to a seat on the Steering Committee and immediate access to all standard simulation results of the model. In providing this information about funding, extracted from the Government by persistent questioning in the Senate, Resources Minister Senator Parer assured Parliament that members of the Steering Committee "have no influence over results and the manner in which the results are reported publicly."[12] The fossil fuel industry had rapidly increased its funding of ABARE's climate change modelling over the previous four years, to the point where in 1996-97 it provided 80% of the funding.

While it is not suggested that the economic modellers in ABARE changed their model construction or results under pressure from industry, they certainly shared a worldview with the fossil fuel industry. One must ask the question: if the fossil fuel industry had not believed that the results of ABARE's modelling would work in its commercial interests, would it have continued to fund the work at ever-increasing levels? In private conversations, coal industry executives made it very clear that they believed that their contributions to ABARE climate change modelling were an excellent 'investment,' although once the model came under attack at home and abroad some of them began to doubt the wisdom of the strategy.[13]

11. For the modelled 'less stringent' target of stabilisation at 1990 levels by 2010 and a 10% reduction by 2020, ABARE, *supra* note 4, at 6.
12. S. Parer, *Senate Hansard*, 2 May 1997.

In response to complaints, the Commonwealth Ombudsman carried out an investigation into the issue of the transparency of funding of ABARE modelling and the influence of the fossil fuel industry. The Ombudsman's report in February 1998 severely criticised ABARE, stating that the credibility of its modelling work had been "compromised" and that ABARE management had displayed "poor judgement" in accepting industry funding because it could create a reasonable public perception that the research was "weighted in favour of the interests of Australian industry."[14]

3. Carbon Leakage

"Carbon leakage" refers to the phenomenon whereby carbon-intensive industrial activities such as aluminium production shift to non-OECD countries because OECD countries lose a competitive advantage as a result of emissions reductions measures. The problem arises if the countries to which industry moves have higher levels of emissions per unit of output.

How much of Australia's greenhouse gas emissions occur in sectors that might have their international competitiveness affected by emissions reduction policies? Table 1 shows the shares of emissions from all sources except land clearing and forestry. Emissions from electricity generation (but not gas processing and oil refining) have been allocated to the sectors of final use. The percentages in Table 1 do not give any indication of the energy-intensity of output. For example, non-ferrous metals will be more affected than pulp and paper because energy use accounts for a higher proportion of production costs.

---

13. After sustained criticism at home and abroad, including withering comments from Tim Wirth, US Undersecretary of State for Global Affairs, industry began to have doubts about the political value of the modelling work.
14. Ombudsman, *ABARE and ACF: Report of the investigation into ABARE's external funding of climate change economic modelling*, Commonwealth of Australia (1998).

TABLE 1. Summary of Australian Greenhouse Gas Emissions by End-use Sector, 1993/94.[15] (excluding forestry and land-use changes)

Sector	**Percent of GWP total***
Energy	
Coal mining	5.2
Gas processing and supply	5.0
Oil refining	1.7
Non-energy mining	2.1
Iron and steel	3.8
Non-ferrous metals	8.9
Chemicals	1.8
Pulp and paper	1.2
Non-metallic minerals	1.8
Food, beverages and tobacco	1.7
All other manufacturing	2.1
Construction	0.7
Civil aviation	0.9
Road transport	13.3
Rail and pipeline transport	0.8
Navigation	0.4
Commercial/institutional	7.4
Agriculture etc.	1.4
Residential	11.1
Military and other	0.4
Total energy	71.5
Industrial processes	1.8
Solvents etc.	0

15. *See* H. Saddler, *Analysis of Australia's Greenhouse Gas Emissions*, Unpublished, Energy Strategies (1997).

Sector (continued)	Percent of GWP total* (continued)
Enteric fermentation	15.4
Manure	2.3
Rice	0.1
Agricultural soils	2.1
Burning of savannas and field residues	2.5
Agriculture	22.4
Waste	4.3
Total	100.0

* GWP = Global Warming Potential

The sectors whose competitiveness may be affected by emissions reduction policies are non-energy mining, iron and steel, non-ferrous metals, chemicals and pulp and paper. Excluding forestry and land-use change, these account for 17.8% of total emissions. The coal mining industry is also a big emitter, but capturing methane from coal mines is likely to improve the competitiveness of the coal industry. In the case of gas processing and supply, a traded good, while a measure such as a carbon tax would increase the relative price, this effect would be swamped by increasing demand for natural gas as its price fell relative to coal. Other sectors are either not traded (residential and road transport), or changes in energy prices would have no significant impact on costs (such as elaborately transformed manufactures) or are unlikely to be affected significantly by greenhouse policies (agriculture).

Since emissions from land-use changes are being revised, it is not possible to say *precisely* how much of Australia's total emissions are accounted for by activities that produce traded goods. The figure will be in the region of 14-15%. Thus, around 15% of Australia's total greenhouse gas emissions are generated in sectors that may be subject to carbon leakage. The issue of carbon leakage arises only for those industries that both produce traded goods *and* are intensive in the use of fossil fuels.

Within these sectors, to what extent are emissions reduction measures likely to lead to carbon leakage to non-OECD countries? Decisions by corporate investors in Australia that result in carbon leakage depend on several factors:

- The emissions reductions levels and timetables agreed to at the climate change negotiations;
- The types of policies adopted by the Australian Government to meet greenhouse gas reduction obligations;

- Whether domestic policies provide relief to export industries that may be significantly disadvantaged;
- The extent to which policies result in an increase in the prices of these goods in Australia compared with major competing nations;
- The extent that domestic industries suffer a loss in price competitiveness compared to non-Annex 1 countries, whether investors expect those countries to agree to mandatory emissions reductions at some stage in the future; and
- The extent to which emissions per unit of output in non-Annex 1 countries exceed those of Annex 1 countries.

Different modelling exercises show differing levels of likely leakage, including some that show it to be very low, perhaps only a few percentage points.[16] The MEGABARE model estimates a rate of carbon leakage of 13.9% for its more stringent scenario, i.e., for every million tonnes of CO2 reduction in Annex 1 countries, an additional 139,000 tonnes will be emitted by non-Annex 1 countries.[17]

Thus, ABARE estimates a rate of carbon leakage of around 12-13% and this varies little with the severity of the cuts modelled. ABARE's estimate is based on a number of assumptions that all result in an inflation of the estimated level of carbon leakage. The analysis assumes:

- That a carbon tax is the sole measure adopted by governments to meet targets. Such a tax would need to be very high to achieve the reductions required by the modelled scenarios. Many studies have shown that a combination of policy measures, of which a carbon tax would be one component, can reduce emissions at much lower cost than a stand-alone carbon tax;
- That this single tax rate is applied uniformly across the economy without any adjustment for industries that might be subject to pressures to move off-shore;
- That producers are operating at full efficiency so that none of the price impact of the carbon tax could be offset by increased energy efficiency, i.e., it is assumed that there are no 'no regrets' measures available that will allow aluminium producers in particular, or more precisely electricity generators and distributors that supply the aluminium industry, to reduce energy consumption at no or very low net cost;

16. *See* Bureau of Industry Economics, *Economic Impact of Reducing Greenhouse Gas Emissions in Australia: A Survey of Recent Studies*, Occasional Paper 24, AGPS (1995).

17. ABARE, *supra* note 4, at 30-31.

- That there would be no technological response to the policy change that would result in a fall in the relative price of low-emissions energy sources;
- That the major competing countries lie outside Annex 1 *and* that investors assume that these countries will not be subject to emissions reductions obligations over the next several decades; and
- That competing producers use energy less efficiently than Annex 1 countries.

Quite clearly, these are strong assumptions. In practice, the following are likely to prevail:

- Governments will adopt a mix of policies and these policies may, for a time at least, give some relief to energy-intensive exporters;
- In the face of emissions-reduction policies the metal-processing companies will take measures to reduce their energy costs, including development of cheaper low-emission sources and energy-efficiency;
- The Kyoto agreement will clearly foreshadow the application of obligations to reduce emissions to developing countries by 2005 or 2010, albeit at lower levels than for Annex 1 countries. This will reduce any emergent price differential for energy-intensive exports; and
- Any newly installed metals processing capacity in developing countries will use globally available state-of the-art technology.

In the face of these more likely outcomes, the level of carbon leakage would be much lower than estimated. In short, with sensitive policies the issue of carbon leakage becomes a minor one. However, those who are attempting to prevent any measures to reduce emissions frequently exaggerate the problem in order to undermine confidence in the effectiveness of mandatory obligations by Annex 1 countries.

The Government has argued that Australia should be permitted to become a global specialist in the use of fossil fuels because Australia uses fossil fuels much more efficiently than Asian countries in producing, for example, metals. Therefore, global emissions will be lower if Australia, rather than developing countries, carry on such activities. It is true that current electricity production from coal in Australia is 15-25% more efficient in the use of coal, and thus carbon dioxide emissions, than typical installed capacity in Asian countries. However, the efficiency of electricity production in many of the relevant developing countries is rapidly catching up with that of Australia, and any new capacity installed anywhere to take up the slack offered by Australian withdrawal from new metal smelting is likely to use globally available state-of-the-art technology. Thus, current relative efficiencies are no guide to future relative efficiencies.

Moreover, the efficiency advantage of Australian electricity production is often over-stated. The newly privatised Hazelwood power station in the Latrobe Valley has had its life extended by 30 years. Yet the thermal efficiency of Hazelwood is only 24%, which is less efficient than many existing coal-fired stations in developing countries.

4. Flaws In Australia's Differentiation Position

Australia's position in the lead up to Kyoto – which was little understood even within policy circles – was to oppose uniform emissions reductions targets in favour of differential targets. It was argued that since Australia is more heavily dependent on fossil fuels for domestic production and exports, imposing uniform targets would impose a higher economic cost on this country.

Australia put forward a number of indicators that it said should be used to determine each country's target for emissions reductions. These indicators would allow more lenient targets for countries that have high GDP growth, high population growth, high emissions per unit of GDP, export fossil fuels heavily, and depend heavily on fossil fuels to produce exports.

It is not surprising that the international community reacted so negatively to Australia's proposal. It would have meant that countries that have the highest per capita emissions, such as Australia,[18] would be required to do the least to reduce them, while countries that have done most to reduce their emissions already would be required to do more. What were the major problems with the Australian position from the perspective of equity and public policy?[19]

4.1 THE POLLUTER PAYS PRINCIPLE

The Australian position was wholly contrary to the polluter pays principle, which has been the basis of environmental policy in Australia and abroad for years. Not only is this principle seen as equitable – those who cause the problem should be responsible for cleaning it up – but it is also maximizes economic efficiency because it is based on the internalisation of external costs and therefore brings about an efficient allocation of resources. Thus, the argument that targets should be differentiated so as to impose equal economic costs per capita on each country is fundamentally inequitable, especially if developing countries were – as the Australian Government argues – brought into the process sooner rather than later.

18. Australia probably has the highest per capita levels of greenhouse gas emissions in the world. C. Hamilton, *A Comparison of Emission Sources and Emission Trends Among OECD Countries: Background Paper No. 1*, The Australia Institute (1994).

19. The following paragraphs draw heavily on the Australia Institute, *A Policy Without a Future: Australia's International Position on Climate Change*, Background Paper No. 8 (1997).

4.2 EFFECTS OF FOSSIL FUEL TRADE

The Australian proposals for the treatment of fossil fuel trade were especially perverse. As a result of emissions targets, two types of economic costs would be imposed on Australia – the costs of meeting domestic emissions targets and the costs of losses in revenue from fossil fuel exports as other Annex I countries reduce their emissions. The effect of the Australian 'equal economic cost' proposal would have been to make domestic targets in Australia even more lenient, as economic costs are increased if other countries import less coal in order to reduce their own emissions. Under the Australian position, to offset lower coal exports, higher emissions in Australia would be permitted. Thus the more ambitious are other countries' greenhouse measures, the less demanding Australia's would be. Depending on the formula used, this could lead to demands by Australia that the other Annex I countries actually pay compensation to Australia, a position that has been explicitly advocated by OPEC countries.

If it had been adopted, the effect of the Australian position would have been that as the rest of the industrialised world progressively shifted away from fossil fuels, Australia would become ever more firmly entrenched as a carbon dioxide emitter. We would actually see carbon leakage from the rest of the world into Australia. In time, instead of being among the highest polluters, Australia would be the outstanding global emitter of greenhouse gases.

4.3 LOCKING IN THE PAST

When the estimated costs of reducing emissions are as small as those estimated by MEGABARE, the relative costs become largely irrelevant. Nevertheless, part of the reason that Australia appears to have relatively high costs of abatement is that the modelling assumes away the existence of low-cost and no-cost energy savings, as well as technological progress. As soon as we allow for these – including energy efficiency, energy conservation and demand management, and the growth of renewables – then Australia is, in fact, in a strong position to reduce its emissions.

The Government's thinking and ABARE's modelling assumed that Australia's 'comparative advantage' in fossil-fuel based industries is fixed and given. This reflects the rigid and short-term thinking about Australia's economic future from which we have spent 20 years trying to escape. A major theme of Australia's official position is that Australia's international competitive advantage lies in the use of fossil fuels, especially coal, to provide low-cost electricity to smelt metals from Australia's mineral reserves. It was argued that this is the optimal course both for Australia and the world. This advantage is part of the reason emissions reduction costs are said to be higher in Australia than in other industrialised countries which specialise in less energy-intensive activities, such as elaborately transformed manufactures. However, a longer-term view is required. In a decade or so, in the second commitment period after 2012, much more demanding greenhouse targets are likely and all industrialised and industrialising countries will be faced with the need to turn decisively to alternative energy sources. At this point a very different picture of comparative advantage emerges, one in which any country that has specialised in fossil fuels will be at a severe disadvantage.

4.4 ROLE OF POPULATION GROWTH

The Australian position also sought higher emissions for Australia because of higher projected population growth. In effect the Government argued that, as a result of our past and expected immigration policies, we expect higher population growth in the future than other developed countries and this will make achieving a fixed reduction target more difficult. It therefore sought to change the target to a per capita goal so as to escape the responsibility for global climate impacts of Australia's higher population growth. It is as though one part of Australia's emissions should not be counted because of a completely unrelated policy. The Australian policy accepts the economic benefits of immigration but wants to absolve Australia of responsibility for the costs. How could other Annex I countries have found this a credible position?

The Australian position was mired in contradictions. The Government argued that developing countries should be brought into any agreements to reduce emissions because they will be responsible for the most of the emissions in the future due to their high rates of population and economic growth. Yet, these same factors were used to argue that Australia should be given more lenient targets.

4.5 OTHER DIFFERENTIATION POSITIONS

Pursuit of differential targets by some other countries was hailed by the Australian Government as a vindication of its position. However, proposals by other countries for differentiation would have resulted in more stringent targets for Australia compared to other countries. Other countries that opposed uniform targets wanted *larger* reductions for countries like Australia with high emission levels.

Norway, for instance, which Australia seemed to regard as a major ally in seeking differentiation, took the following position:

> Parties should take their share of the burden in proportion to their relative contribution to the climate change problem. Those who currently emit more than their fair share should thus contribute more. Also, Parties that have greater capacity, economic or otherwise, to deal with the problem, should in principle do more than other Parties to reduce emissions.[20]

---

20. H. Dovland, *Climate Change - Some Views From a Norwegian Perspective*, in COMMODITY MARKETS AND RESOURCE MANAGEMENT, Papers from the Outlook 97 Conference, ABARE (1997).

This is wholly contrary to the Australian position. Under the Norwegian proposal, a country would be required to reduce its emissions by *more* if it has higher emissions per capita, higher GDP per capita, and higher emissions per unit of GDP.

The first and third of these three indicators were the precise opposite of Australia's proposed indicators, and the second (GDP per capita) is not helpful to the Australian position. These differences are shown in Table 2. Japan, which Australia saw as a potential ally, has low emissions per capita and per unit of GDP and is entering an era of lower economic growth. Under the Norwegian differentiation position it would have more lenient targets, while under the Australian position it would have more stringent targets.

TABLE 2. Indicators Leading To Higher or Lower Targets, Australia And Norway.

Indicator	**Australia**	**Norway**
High GDP growth per capita	Lower	Higher
High population growth	Lower	Irrelevant
High emissions per unit of GDP	Lower	Higher
Export fossil fuels heavily	Lower	Irrelevant
High emissions intensity of exports	Lower	Higher
High emissions per capita	Irrelevant	Higher

The European Union (EU) was heavily criticised by Australian representatives for adopting differentiation among its member countries while seeking uniform reductions for other countries, as though this justified the Australian stance. However, this was a shallow criticism for two reasons. Firstly, the EU negotiated as one party for the purposes of negotiations under the Convention and other international agreements. At Kyoto, each EU country signed up to a uniform target, but there is provision in the Protocol to allow 'burden sharing' within designated groups of states. Thus, the richer countries of the EU have agreed to subsidise the poorer countries of Europe. Australia is now permitted under the Protocol to enter a similar arrangement if it can find some Annex 1 countries that are effectively willing to subsidise it.

Secondly, the EU countries that have more lenient targets within the EU are not chosen for the reasons that Australia sought more lenient targets. Generally they are low GDP, low emitting, non-fossil fuel exporting countries, such as Portugal, Greece, Spain and Ireland. If Australia were a member of the EU, it would probably be treated more like Britain and Germany – with high emissions, high GDP per capita and relatively low marginal abatement costs – than Portugal and Greece, and would therefore be required to reduce its emissions by *more* than the average.

5. Domestic Greenhouse Response

The Australian Government mounted a major public relations offensive to convince the world community that Australia should be given special dispensation to pollute more than other Annex 1 countries. Part of its campaign was to attempt to gain credibility by arguing that Australia is already making considerable efforts to reduce its emissions at home. Yet from the time of taking office in March 1996 the Coalition Government has wound back or abolished several of the already meagre federal programs aimed at reducing emissions.

5.1 BACKSLIDING BY THE FEDERAL GOVERNMENT

The National Energy Efficiency Program, although relatively small, had been the Commonwealth's major contribution to stimulating more efficient use of energy throughout the Australian economy. In the 1997-98 budget its appropriation was cut by 60%, from $4.7 million in 1996-97 to $1.9 million. The largest single consequence of the cut was the abolition of the Enterprise Energy Audit Programme, which subsidised the cost of energy audits for businesses.

The Government abolished the Energy Research and Development Corporation (ERDC). The rationale – that there are other sources of support for energy R&D – is unconvincing, and fails to recognise the unique role played by the ERDC. The ERDC's work has been significant in concentrating on research projects that offer strong prospects of successful commercialisation and investing in projects in collaboration with the private sector.

The Government argues that deregulation and restructuring of the electricity and gas industries will have a significant impact on emissions. However, while deregulation has stimulated gas cogeneration it is likely that any gains have been offset by falling prices and aggressive marketing that increased demand for electricity and generate higher emissions. The Government is now pursuing policies to cut fuel prices.

5.2 THE GREENHOUSE CHALLENGE PROGRAM.

The Greenhouse Challenge Program (GCP) is a federal government scheme involving voluntary agreements with major emitting companies. The agreements specify measures that companies agree to undertake to reduce their greenhouse gas emissions. The GCP is complex, its objectives are unclear and no evaluation of its effectiveness has been carried out to date. The GCP is based explicitly on 'no regrets' measures, i.e., measures that are worthwhile undertaking for commercial reasons irrespective of their impact on greenhouse gas reductions. It should be noted that the Federal Government spends only $2.4 million per year on the GCP.

Major emitting firms in Australia have been enthusiastic about signing up to voluntary agreements, and there is no doubt that some of the firms that have signed agreements have made serious efforts to develop emissions reductions plans —

EnergyAustralia, a NSW electricity distributor, is a case in point. In addition, the GCP serves the important function of raising awareness among the business community about the importance of climate change and the need for the corporate sector to play a major role.

However, the overall effectiveness of the GCP is very limited for three reasons. First, many of the measures specified in the agreements to reduce emissions are measures that the companies planned to undertake irrespective of the agreement, i.e., part of the 'business-as-usual' scenario. Therefore, they cannot be attributed to the GCP. The second concern is that in calculating the emissions reductions due to the GCP, a 'frozen efficiency' assumption is made, i.e., it is assumed that the companies in question would make no improvements in energy efficiency over the next several years. The effect of this is to exaggerate the claimed reductions in emissions attributable to the GCP. The third and final consideration is that the details of the agreements are treated as commercial-in-confidence so it is not possible for the public to make a proper assessment of how real the claimed emissions reductions are.

A strong indication of the manipulation of figures is provided by the GCP's own 1996 Progress Report,[21] where estimates are made of emissions by signatory companies to the year 2000 under 'frozen efficiency' and under the impact of voluntary agreements. The document claims that emissions will be reduced by 15 Mt of CO2 equivalent by the year 2000, down from a projected 93 Mt to 78 Mt – see Figure 1. This represents a cut of 16%. However, a simple extrapolation of the emissions of these companies based on emissions in the years 1990 (71 Mt) and 1995 (73 Mt) would see these companies emitting only 75 Mt by the year 2000, less than projected under the voluntary agreements.[22] In sum, although the details of the agreements remain shrouded in secrecy, there are good grounds for believing that the emissions reductions claimed by the Program are gross exaggerations, and that the GCP as it is currently constituted will not significantly reduce Australia's emissions below the levels that would be anticipated in the absence of the Program.

---

21. Greenhouse Challenge Program, *Progress Report* (1997).

22. The document attempts to explain this anomaly as follows: "The explanation of this is a fairly flat economic period with little to no growth which inhibited significant emissions growth. Some sectors also made progress in reducing their greenhouse gas emissions." Greenhouse Challenge Office. *id.*, at 11. In fact, the opposite has been the case, with real GDP growth in the order of 3.5 to 4.0% since the recession in 1990-91.

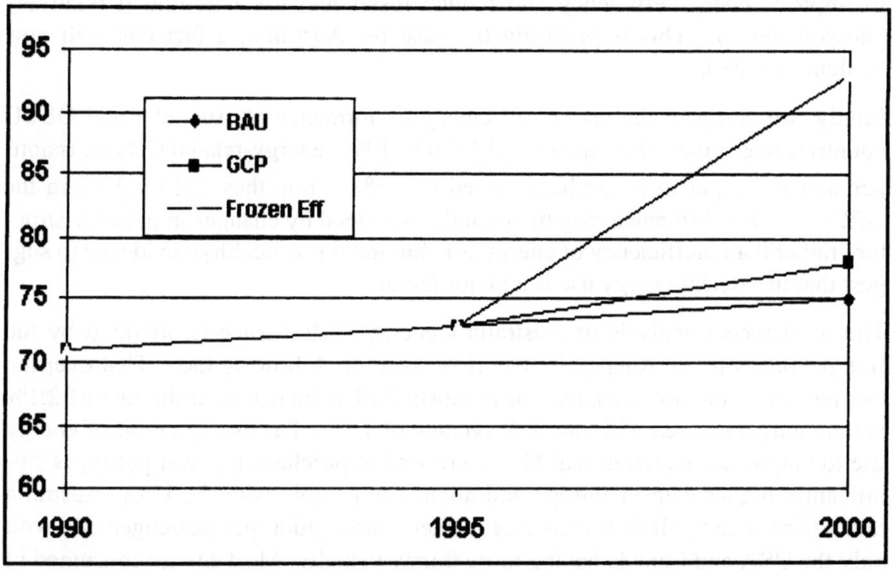

Figure 1. Greenhouse Gas 'Reductions' Due To the Greenhouse Challenge Program

5.3 OPPORTUNITIES FOR LOW-COST EMISSIONS REDUCTIONS

For a country that needs to reduce its greenhouse gas emissions, heavy dependence on fossil fuels may be a blessing rather than a burden. The key question is not so much whether the country consumes a large quantity of fossil fuels, but how efficiently it consumes them. The more inefficient a country has been in the past, the easier and cheaper it will be to reduce consumption. This is precisely why Australia has supported the development of joint implementation, whereby Annex I countries can claim emissions reduction credits for projects that reduce emissions in non-Annex 1, especially developing countries, where it is believed there is more inefficiency in the combustion of fossil fuels.

ABARE acknowledges this basic fact in explaining why the MEGABARE model shows that the costs of uniform targets would be very high for Japan:

> Japan experiences high costs because Japanese industries have already taken major steps to improve energy efficiency and reduce fossil fuel use. Further action to reduce emissions by significant amounts in Japan would imply further structural adjustment to the Japanese economy, carrying large costs.[23]

23. ABARE, *supra* note 4, at 5.

Of course, this cuts both ways. Countries that have not already taken major steps to improve energy efficiency and reduce fossil fuel use will find it relatively cheap to do so. This is precisely the case for Australia, a fact that will now be demonstrated.

Firstly, we compare the energy efficiency performance of Australia and OECD countries over time. Over the period 1970 to 1992, energy-related CO_2 emissions per unit of output in Australia declined by 13%, while they fell by 36% in the OECD.[24] This difference may be partially explained by changes in industry structure rather than inefficiency of energy use, but there is a wealth of evidence to suggest that inefficient energy use is a major factor.

The most recent analysis of Australia's energy performance is provided by the International Energy Agency.[25] The IEA noted the following facts. That over the last ten years, energy consumption in Australia has increased at the rate of 2.1% per annum, compared with the IEA average of 1.1%. The average ratio of energy use to output (measured in real $US converted at purchasing power parity) is significantly higher than in Europe and about comparable with the USA. Australia ranks third among all IEA countries in fuel consumption per passenger car, with only the USA and Canada having more thirsty vehicles. Most energy consumed in Australia is fossil fuel, and therefore energy savings can reduce the level of CO_2 emissions. Finally, energy is cheap in Australia because of the abundant supply of low-cost energy resources and low energy taxes. "This low level of energy prices makes investments in energy efficiency less profitable than in the average IEA country."[26] With such concerns in mind, the IEA concluded:

> the results of the industrial energy audits, the absence of general mandatory standards for buildings and domestic appliances and the high level of fuel consumption by passenger cars all indicate that there is a *great potential for improvements in energy efficiency.*[27] (emphasis added).

The IEA then went on to recommend, *inter alia*, mandatory energy efficiency codes for buildings and electrical appliances, including air conditioning, stronger fuel efficiency targets for passenger and commercial vehicles, higher fuel taxes, and improvements in interstate public transport.[28]

In Australia, there have been several studies estimating the potential for reducing greenhouse gas emissions in cost-effective ways. In other words, these studies estimate the reductions in emissions that would repay the investments required in capital equipment. While the studies vary widely in the assumptions about levels

---

24. Hamilton, *supra* note 18, at Chart 8.
25. International Energy Agency, *Energy Policies of IEA Countries: Australia 1997 Review* (1997).
26. *Id.*
27. *Id.*
28. *Id.*

of energy savings, end-use fuel substitution, cogeneration and supply-side fuel substitution, the conclusions are remarkably similar. Wilkenfeld divided the studies into two groupings and concluded that one group estimates that the potential for cost-effective CO_2 reductions is in the range of 20-30% and the other group estimates reductions in the range 40-48%.[29] The results are summarised in Table 3. The largest savings are in the residential sector, but the industrial, commercial and transport sectors all have substantial potential for reductions. The conclusion from this evidence is that there are many cost-effective opportunities for Australia to cut emissions and that Australia will find it easier to reduce emissions than most other OECD countries.

TABLE 3. Estimates of Cost-Effective Energy Measures to Reduce CO_2 Emissions from the Australian Energy System (All Sectors) (%)[30]

Reference	**Potential reduction**
ABARE 1991	30
AMEC 1990	25
ESD 1992	22
ESD 1992	29
GWA 1991	20
GWA 1991	30
Range, this group	20-30
Average, this group	26
Greene 1990	47
Greene/NIEIR 1990	48
NIEIR 1990	40
Range, this group	40-48
Average, this group	45

29. G. Wilkenfeld, *Energy Efficiency Programs in the Residential Sector*, in *Greenhouse: Coping With Climate Change* (W.J. Bouma, G.I. Pearman & M. Manning eds., 1996).

30. Wilkenfeld, *id.* at Table 1

6. Australia and the Kyoto Protocol

6.1 THE AUSTRALIAN DEAL

The Australian Government celebrated a victory after winning large concessions at the Kyoto Conference. The Prime Minister claimed that it was a vindication of the position his Government had taken in international fora and that the outcome was fair because it recognised Australia's special difficulties in reducing emissions of greenhouse gases.

The Kyoto Protocol requires an average reduction in emissions from the Annex B Parties (i.e., the developed countries and essentially the same as Annex 1) of 5.2% below 1990 levels to be achieved over the period 2008-12. This includes an 8% cut for European Union countries, 7% for the United States, and 6% for Japan and Canada.

The main outcomes for Australia at Kyoto were as follows:

- It was agreed that net emissions from land use change could be added to emissions from other sources to determine 1990 base year emissions measured in tonnes of carbon dioxide equivalents. The second sentence of Clause 7 of Article 3 of the Protocol immediately became known as the 'Australian clause'; and
- While most of Annex 1 countries agreed to cut their emissions below 1990 levels by the budget period 2008-2012, with an average reduction of 5.2%, Australia secured a target of 8% above 1990 levels.

Using the best current estimates of emissions, the effect of the inclusion of emissions from land clearing is to increase Australia's 1990 emissions from an 380 million tonnes (Mt) of carbon dioxide equivalent to 496 Mt with the addition of 116 Mt from land clearing emissions (NGGI 1997a). The Protocol sets Australia a target of 8% more than this, that is, 536 Mt a year averaged over the period 2008-2012.

6.2 HOW TOUGH IS AUSTRALIA'S TASK?

In the lead up to Kyoto, the Government announced a package of energy measures that it predicted would limit emissions (excluding those from land clearing) to 18% above 1990 by the year 2010. Thus, energy and industrial emissions were expected to rise to 448 Mt by 2010, a level even the Government conceded could be improved upon.

This leaves room for at least 88 Mt to come from land clearing in 2010 in order to come in at the target of 536 Mt. However, according to the official greenhouse gas inventory, by 1995 emissions from land clearing had already *fallen* to 78 million tonnes from 116 Mt in 1990. Thus Australia could *increase* emissions from land clearing and still meet the Kyoto target. The situation is summarised in Table 4.

TABLE 4. Australia's Emissions Task after Kyoto[31]

	Mt CO_2-e
Emissions in 1990	
Energy and other sources	380
Land clearing	116
TOTAL	496
Emissions in 2010	
TOTAL Kyoto emissions target	536
Expected emissions other than land clearing	448
Balance due to land clearing	88
Actual land clearing emissions in 1995	78

If emissions from land clearing continue their natural decline and stabilise at around 48 Mt then Australia will have 40 Mt of surplus emission savings. Under the emissions trading system agreed by the Kyoto Protocol, these emission credits can be sold to other signatory countries. Thus far from easing an unfair emissions reduction burden, the concessions won by Australia probably represent a substantial wealth transfer from other developed countries, a transfer to the country with the worst record of per capita greenhouse gas emissions.

In per capita terms, the inclusion of land clearing emissions for Australia means that official emissions per head in 1990 rose from about 21 tonnes per year to about 29 tonnes, making Australia officially by far the highest greenhouse emitter per capita. With an expected population increase of 23% between 1990 and 2010,[32] it might be expected that there would be a fall per capita by 2008-12, in either total emissions per capita, or non-land clearing emissions per capita. The impression might be given that we would show a moderate improvement in fossil-fuel energy efficiency.

However, land clearing emissions have already fallen by 33% from a net of 116 Mt in 1990 to 78 Mt in 1995 and are likely to continue falling without any actions by governments. Consequently, there is ample scope for per capita emissions from energy related uses to rise within the total target allowed for Australia. If, for instance, net land clearing emissions were to disappear by 2010 – note that it is stated government policy to achieve this by 2001[33] – emissions per capita from all

31. National Greenhouse Gas Inventory (1997) and Government statements
32. *See* Australia Institute, *A Poisoned Chalice: Australia and the Kyoto Protocol*, Background Paper No. 13 (1998).

other sources could rise from 21 tonnes in 1990 to 26 tonnes in 2010 while Australia remains within the Kyoto limit. We would therefore be increasing our emissions per capita from fossil fuels, the source of most non-land clearing emissions.

However, there is even greater scope to increase our energy related emissions per capita because, within the non-land clearing emissions, there are significant elements which are not energy-related. Emissions from the latter (which are mainly methane emissions from agriculture) fell slightly between 1990 and 1995. If we assume conservatively that this will stabilise at its 1990 level, and assume the same for the other minor emissions, it is possible to see energy-related emissions rising an extraordinary 29% per capita between 1990 and 2010, from 17 to 21 tonnes per capita.

In any case, using almost any basis for comparison, the concessions made at Kyoto will see Australia become the world's outstanding per capita emitter. Previously it vied with the US and Canada for this title, but with the addition of land clearing emissions, and Australia's 8% growth versus their 6 or 7% fall, Australia's preeminence as a polluter will be unchallenged.

6.3 DID AUSTRALIA WIN THE ARGUMENT?

To put the Australian 'win' into perspective, it is instructive to compare the outcome with the position put forward by the Australian Government in the period leading up to the Kyoto conference. Australia advocated the following:

- Voluntary rather than binding emission limits;
- Emission limits should be based on estimates of equal economic costs per capita between developed countries;
- Rather than uniform targets, there should be differentiated targets between developed countries based on equal economic costs. While the actual degree of differentiation was not spelled out, the criteria put forward by Australia would have involved at least a 30% range above and below 1990 levels;
- Differentiated targets should be based on five indicators – economic growth rates, emission intensities of the economy, population growth rates, exports of emission-intensive goods, and emission intensity of total exports;
- Emission limits should apply to developing countries in any agreement; and
- Emissions trading between countries should be adopted.

---

33. Speech by Senator Robert Hill, Minister for the Environment to the *Bridge to the Future* Forum, , 23 October 1997

The unreality of the overall Australian position is shown by the fact that what the Government agreed to so eagerly in Kyoto accords with only one of these conditions, the adoption in the Protocol of emissions trading. Australia's advocacy of equal economic costs, the use of indicators, and voluntary rather than legally binding targets never looked remotely a possibility, while Australia was earnestly trying to sell them around the world. Developing country participation is still to be worked out, and almost certainly will not involve targets in the shorter term.

The idea of differentiation was the cornerstone of the Australian position before Kyoto. But before Kyoto and in the Kyoto Protocol itself, differentiation was never accepted as a basic concept or major influence on targets in the way advocated by Australia. Almost every target was within the narrow range of 1% increase to 8% reduction, and 32 out of 38 countries (with widely differing characteristics) accepted cuts between 5% and 8%. Under any feasible differentiation criteria (including those put forward by Australia), Japan and the USA would have been given markedly different targets, whereas in practice they differed by only 1%. It was made quite clear by the Conference Chair, Raoul Estrada, that Australia would get nothing like the 'headline' increase it sought, with a maximum increase for any country of 10% at the very most.[34] The outcome was therefore very close to uniform reductions for almost all countries, with a few deviations of a few percentage points.

The only Australian position adopted was emissions trading, but this was the least emphasised Australian aim, with senior bureaucrats suggesting that emissions trading would take 20 years to implement. ABARE had been an early advocate of emissions trading, but was largely silent on the issue for the two years prior to Kyoto. The reason was that while the Government argued vociferously that Australia would suffer huge economic costs as a result of uniform emission targets, ABARE's own economic modelling showed that emission trading would reduce the estimated costs of emission reductions by around 75%. After Kyoto, ABARE is now attempting to take control of the emissions trading debate in Australia.

6.4 DIFFERENTIATION REJECTED

The Kyoto outcome finally puts to rest the Australian Government's contention that the EU arrangement of varying targets within the EU was equivalent to Australia's differentiation position, and that the Europeans were being hypocritical in pressing for uniform reductions for other countries. By the end of the Conference, most commentators recognised the conceptual equivalence of the 'EU bubble' and the 'emissions trading bubble' for Annex B countries generally. This acknowledged that the variations within the EU are essentially a form of emissions trading among those countries. Parties may emit more if they provide the wherewithal in a bargaining process with other parties to allow them to do so.

---

34. THE AUSTRALIAN, 4 December 1997, at 6

This is not the free ride for higher emitters that was the essence of the Australian proposal. Appendix B of the Protocol shows each EU member, and the EU as a whole, committed to an 8% reduction. Increases will be allowed under the Protocol to EU countries only if other countries make up the difference. No country negotiating an emission limit in the future can base a claim for a more lenient target on the EU position because any EU country increase is strictly within a trading 'bubble.' On the other hand, the 8% increase for Australia is a pure free ride and will undoubtedly be used by other countries in pursuit of lenient targets. Indeed, there is evidence that this is already the case.

How does Australia's 8% increase compare with increases for some other countries? Note first that the whole purpose of the Conference was to start *reducing* emissions from wealthy countries. Thus Canada, in many respects like Australia, but with a higher population growth, accepted a 6% reduction. New Zealand accepted 0%, even though it has far lower per capita emission levels than Australia (and hence *prima facie* will have greater difficulty reducing emissions proportionately, because fossil fuel use is more specialised), and similar population growth. The only other countries to get an increase were Iceland and Norway. However, Iceland is a country with a population of 250,000, and Norway obtained an increase of only 1%, for a population of 4.5 million and per capita emissions about half those of Australia. Australia's increase of 8%, not to mention the inclusion of land clearing emissions, is therefore the outstanding anomaly from Kyoto.

6.5 AUSTRALIA'S NEGOTIATING STRATEGY

There is not the slightest evidence that other countries accepted the key contention of the Government, that the cost of uniform targets would be unfairly high for Australia. How then did Australia win concessions if its arguments carried no weight internationally? The answer to this question is clear from a survey for this paper of 260 press reports published in Australian newspapers over a two week period spanning the Kyoto Conference. On the basis of observations by members of the Australian delegation, the Conference Chair, the Secretary, and many delegates and observers from other countries, it is apparent that Australia won concessions by threatening to withdraw from the Convention if its demands were not met.

While all countries negotiated with the national interest in mind, none was quite so irresponsible in both threatening to withdraw, thus destroying consensus, and in seeking an increase in emissions. The Secretary of the Convention, Michael Zammit Cutajar, for instance, referred to every country except Australia as being committed to its success.[35] The Chair of the Conference, Raoul Estrada, stated that Australia had been allowed to have its way only in the interest of obtaining unanimous agreement.[36] The Australian negotiating strategy was no surprise: the Howard Government had been making these threats for some months. If any

35. SYDNEY MORNING HERALD, 1 December 1997, at 1.

36. AUSTRALIAN FINANCIAL REVIEW, 13 December 1997, at 31.

larger power, or a small number of countries, had acted like Australia, agreement would never have been reached. Australia therefore took advantage of the more responsible approach adopted by other countries and exploited the fact that agreement on mandatory targets by all Annex I countries was essential to obtaining a protocol.

After some hesitation, Environment Minister Senator Hill signed the Kyoto Protocol in New York, but there is little evidence that the Federal Government has grasped the significance of global developments since Kyoto. It believes that it deserves praise for its greenhouse program amounting to $180 million over five years. This amounts to the cost of one bus ticket per person per year, a paltry amount for perhaps the gravest threat to Australia's natural systems.

The precedents established to keep Australia in the Kyoto negotiations will bedevil future negotiations. An 8% increase for a country that is wealthy and the world's highest per capita polluter will make it difficult to gain the agreement of developing countries to begin cutting their emissions, one of the Australian demands at Kyoto.

The land clearing clause may be even more damaging, especially as developing countries are brought into the target-setting process. Land clearing in developing countries, as in Australia, will probably be declining for other reasons. Thus, the inclusion of land clearing allows emission cuts that would occur in the energy sectors to be 'transferred' to land clearing, thereby delaying cuts in emissions from industrial processes in exchange for reductions that would happen anyway.

The Australian clause opens up a large loophole in the Protocol because, unlike energy emissions that can be reduced only gradually, land clearing emissions can change sharply from year to year. It may be feasible to stop land clearing only for the target period 2008-2012, and then to resume it after the target is met.

In the longer term, the shape of global climate change controls beyond 2012 has become clearer. The Kyoto conference foreshadowed a move towards equal per capita emission rights and the institutionalisation of the polluter pays principle. These bode ill for Australia, the country with the highest per capita emissions, and one now not obliged to begin purposeful action on emission reductions. The effect of using the Kyoto concessions will be to undermine investments in greater energy efficiency and renewable alternatives, the only feasible long-term solutions.

The Australian deal at Kyoto was 'a poisoned chalice' both for those seeking a global response to climate change, and for Australia's economic future. For the former, pursuit of consensus has come with the destructive precedents established for future negotiations. For Australia, the pursuit of a lenient target will come at the long-term cost of being unprepared for much tougher anticipated targets after 2012.

4. A CRITIQUE OF THE AUSTRALIAN GOVERNMENT'S GREENHOUSE POLICIES

MARK DIESENDORF
Institute for Sustainable Futures
University of Technology, Sydney
PO Box 123, Broadway, NSW 2007, Australia

1. Introduction

The Australian government is a signatory to the Framework Convention on Climate Change (FCCC), which has the ultimate objective of "... stabilisation of greenhouse gas concentrations in the atmosphere at a level that would prevent dangerous anthropogenic interference with the climate system ..." Since the present rate of change in global average temperature appears to be greater than any natural changes observed over the past 150,000 years, this objective may require stabilisation of greenhouse gas (GHG) concentrations at present levels in carbon dioxide (CO_2) equivalents, or lower levels. According to the calculations of the Intergovernmental Panel on Climate Change (IPCC), this would require a reduction in annual global CO_2 emissions by at least 60 per cent and substantial reductions in the other GHGs as well.

In the lead up to the Third Conference of the Parties to the FCCC at Kyoto in December 1997, a number of targets were proposed. The target proposed by the European Union, the strongest apart from that proposed by the Alliance of Small Island States (AOSIS), was a reduction of 15 per cent below the 1990 level by 2010. The EU believed that this was a technically feasible and inexpensive proposal. However, its target fell far short of what is actually required in order to stabilise atmospheric concentrations at any level. Nevertheless, if implemented it would have at least set in motion a process of continuing reductions.

In the Kyoto Protocol, Parties in Annex I of the FCCC agreed to a much weaker target: to reduce their overall annual emissions of six greenhouse gases (GHGs) to at least 5% below 1990 levels between 2008 and 2012. However, as the result of the Australian government's special pleading, Australia will be permitted to increase its total GHG emissions by 8 per cent. Since the Conference of the Parties also agreed to include emissions from land clearing in the total (at the instigation of Australia) and since this contribution appears to be declining slowly in Australia without government intervention, it appears that Australia could reach the 8 per cent target without taking any additional action to reduce the emissions from the energy sector. The Australian government's position was/is based on the assertion that Australia is a fossil fuel dependent country and that to reduce emissions would cost about $9,000 per person over the 22 years from 1998 to 2020 and would also result in the loss of thousands of jobs (one source suggesting 150,000 jobs).

This chapter offers a critique of government policies by examining the assumptions,

A. Gillespie and W. C. G. Burns (eds.), Climate Change in the South Pacific: Impacts and Responses in Australia, New Zealand, and Small Island States, 79–93.

arguments and methods upon which they rest and the effectiveness of actions taken to reduce emissions. If the assumptions are at variance with observation, if the arguments are at variance with logic, if the methods fail to carry out generally accepted procedures of verification, and if the actions are largely ineffective, then the policies are poorly based and there is justification for questioning whether the government is acting in the interest of the Australian public and the people of the world.

The focus of this chapter is upon the energy sector. This is chosen because energy consumption constitutes the largest and most accurately measured GHG emissions. Although the Australian government's current policies on emissions have not been presented in a White Paper, they can be inferred by its budget and its actions and lack of actions. This will be done below.

2. Government Actions to Reduce Emissions

Table 1 outlines some of the main events in Australia's greenhouse response strategy. The Labor government of Bob Hawke started quite strongly in 1989 by setting an Interim Planning Target to return emissions to 1988 levels by the year 2000 and to reduce them by a further 20 per cent by 2005. However, the target was subject to Australia "avoiding actions with net adverse economic effects or adverse effects on trade competitiveness, unless similar actions were taken by major greenhouse emitting countries."[1]

The above caveat led to two separate processes to assess the costs and benefits of greenhouse response. This occurred with the ecologically sustainable development (ESD) process, of which greenhouse response was an important part, and an inquiry by the Industry Commission. The latter failed to consider the economic benefits of greenhouse response,. producing only an estimate of costs.[2] However, the reports of the ESD Working Groups on Energy Production and Energy Use, and the Greenhouse report by the ESD Chairs, acknowledged that there were potential economic benefits, especially from the efficient use of energy.[3] Between them, the reports (especially ESD Energy Use) made some valuable recommendations which, if implemented promptly, could possibly have met the Interim Planning Target while saving money for Australians.[4]

1. Ecologically Sustainable Development Working Group Chairs, *Greenhouse Report* (1992), at xv.
2. Industry Commission, *Cost And Benefits Of Reducing Greenhouse Gas Emissions* (1991).
3. Ecologically Sustainable Development Working Group Chairs, *Greenhouse Report* (1992); Ecologically Sustainable Development Working Groups, *Final Report: Energy Use* (1991); Ecologically Sustainable Development Working Groups, *Final Report: Energy Production* (1991).
4. M. Diesendorf, & T Kinrade, *Integrated Greenhouse Policies for Energy and Transport*, Australian Conservation Department (1992).

However, very few of the recommendations were actually implemented. In early 1992, Paul Keating replaced Bob Hawke as the Labor Prime Minister, the ESD recommendations were handed over to 37 interdepartmental committees of federal and state officials, and the resource industries commenced a strong media and lobbying campaign to oppose action on greenhouse concerns. In the resulting National Greenhouse Response Strategy (NGRS)[5] and the National Strategy for Ecologically Sustainable Development (NSESD)[6] many of the ESD recommendations were weakened or discarded. This can be verified by referring to the Compendium of ESD Recommendations,[7] which tracks the fate of each ESD recommendation. In the National Strategy for ESD even the original ESD principles, provided to the ESD Working Groups by Prime Minister Hawke had been modified.[8]

Furthermore, NGRS severely constrained action by providing that "first phase measures will meet equity objectives by causing minimal disruption to the wider community, any single industry sector, or any geographical region."[9] This was interpreted by some commentators as special protection of the coal industry. It was, in effect, replacing the concept of 'no regrets' measures -- those with no net costs to society as a whole -- with one that excluded actions that would impose costs on any interest group. The government would oppose any change in economic structure, such as the gradual replacement of fossil fuels with renewable sources of energy, as well as the implementation of energy efficiency programs. The National Audit Office, concerned about the economic losses of failing to implement cost-effective energy efficiency measures, criticised the government's failure to reduce emissions from its own buildings and nationally.[10]

The fate of eight key ESD recommendations on energy are tracked over the period 1991-1997 in Table 2. It illustrates how the retreat from sustainable energy commenced under the Labor government and was accelerated under the Coalition government that took power in 1996.

Because the ESD Energy Use Working Group recognised the existence of market failure in efficient energy use, it recommended several regulatory measures. The most notable of these was mandatory Minimum Energy Performance Standards (MEPS) for appliances and equipment. The recommendations spanned electrical, gas and solar appliances and equipment, and the residential, commercial and industrial sectors, and were to be implemented in 1993. By 1996, MEPS had been

5. Australia, *National Greenhouse Response Strategy* (1992).
6. Australia, *National Strategy for Ecologically Sustainable Development (*1992).
7. Australia, *Compendium Of Ecologically Sustainable Development Recommendations* (1992).
8. Compare ESD, Appendix 1, *supra* note 3, with Australia *supra* note 6, at 8-9.
9. National Greenhouse Advisory Panel, *Report on the National Greenhouse Response Strategy*, Environment Australia (1996), at 12-13.
10. Auditor General, *Efficiency Audit: Energy Management of Commonwealth Buildings*, Audit Report N. 47 (1992).

reduced drastically in scope, and some of the actual standards were weaker than the corresponding US standards for 1993 and would only be implemented in 1999 (Table 2).

Integrated least-cost energy planning (ILCEP) involves the provision of energy services at least-cost. It is based on the understanding that people do not really want to buy energy per se - a kilowatt-hour of electricity or a megajoule of natural gas - but rather the services that energy provides, i.e., a warm building in winter, cold drinks and hot showers. The least cost provision of that service generally includes a substantial role for efficient energy use - e.g., a well insulated building, an energy efficient refrigerator and a water efficient shower head - and this results in a significant reduction in GHG emissions. Thus, ILCEP puts efficient energy use and the supply of energy onto an equal footing.

However, the restructuring of the electricity industry was used as an excuse for not implementing ILCEP. After this restructuring, it was claimed that the alleged competitive market for electricity would somehow encourage cost-effective efficient energy use to be implemented automatically.[11] The economic theorists who made such claims did not recognise that the restructuring failed to remove several barriers to efficient energy use and actually introduced some new ones. For energy services, market failure was and is endemic.[12] Only the New South Wales government has tried to encourage electricity retailers to become energy service companies. It has done this by placing the requirement to reduce greenhouse gas emissions in the utilities' licence conditions and by placing a cap on the revenue per customer that each utility can earn selling electricity. The revenue cap discourages ever-increasing sales of electricity and encourages ILCEP.

The ESD recommended a substantial increase in funding for research, development, demonstration and early commercialisation of energy efficient and renewable energy technologies. The government responded by cutting funding.

The government's attitude to ecologically sustainable energy was reflected in the composition of the so-called Business and Ministerial Roundtable on Sustainable Energy Policy set up by the Minister for Resources and Energy in 1996. Only one of the 12 members represented a genuine sustainable energy organisation.[13] All the other business representatives were from fossil fuel interests, which suggests a peculiar conception of sustainable energy.

The preparation of a Sustainable Energy White Paper commenced in 1995 under a Labor government, but progressed very slowly. A draft version, which was issued in 1996 by the Coalition government, reinforced the impression that the government perceived 'sustainable energy' to be synonymous with 'fossil fuels.' At this

11. Australia, *Climate Change: Australia's National Report Under The United Nations Framework Convention On Climate Change*, Commonwealth of Australia (1994), at 4, 48.

12. M. Diesendorf, *How Can A 'Competitive' Market For Electricity Be Made Compatible With The Reduction Of Greenhouse Gas Emissions?*, 17 ECOLOGICAL ECON. (1996), at 33-48.

13. Namely, the Sustainable Energy Industries Council of Australia.

time, the final version of the White Paper has still not appeared and the preparation process appears to have collapsed.

The retreat from the ESD recommendations and the Interim Planning Target was also manifest in three separate actions. The first of these was the commencement of construction of a new coal-fired power station at Collie, W.A. This ran counter to the recommendation of the official committee of inquiry, which supported natural gas. The second was the NSW government's approval for the construction of the Redbank power station that would burn coal washery waste, a fuel which produces even greater GHG emissions per kilowatt-hour of electricity generated than coal. The third was the construction of urban freeways in most major cities, thus bringing more motor vehicles onto the roads. Although these were all State government projects, they all receive Federal tax concessions and infrastructure funding.

3. Government's Policies On Greenhouse Response

Based on the actions of the present and previous Australian governments, and numerous public statements, it can inferred that these governments have pursued the following de facto greenhouse policies:

- To increase the production of coal and its consumption in Australia and overseas.
- To expand resource processing industries, which are intensive in their emission of greenhouse gases, such as aluminium smelting.
- To maintain the market barriers to efficient energy use and thus limit its widespread implementation.
- To foster the image of action through publicity and programs with little funding and limited effectiveness, such as Greenhouse Challenge, a voluntary greenhouse gas reduction scheme for business.
- To keep renewable energy technologies as a tiny niche market.
- To discourage the use of natural gas for electricity generation and cogeneration, where it could reduce GHG emissions substantially, while encouraging its use for motor vehicles, where it could reduce GHG emissions only slightly (but could reduce local air pollution significantly).

The government's international policy on greenhouse response is clearly to oppose any agreement that would reduce the production of coal, aluminium and steel in Australia.

4. Government Assumptions, Arguments And Methods

The intellectual basis for the Australian government's greenhouse policies is considered in this section. This involves a scrutiny of the assumptions, arguments and methods of analysis it promotes and publicises.

In the early 1990s, the resource industries funded a number of economic models of greenhouse response.[14] The results of these model runs, especially in those cases when they were unfavorable for reducing emissions, were widely publicized in the media and used as a tool for lobbying federal and state governments. However, in the mid- and late 1990s, the limelight was taken over by a model of the world economy developed within a federal government agency, the Australian Bureau of Agricultural and Resource Economics (ABARE). The model is called MEGABARE, and its most recent versions, GIGABARE.

MEGABARE is funded, at least in part, by members of a Steering Committee that oversees the modelling, with each member paying $50,000 to join. The minutes of its meeting held in July 1997 reveal that committee members represented fourteen organisations. These were ABARE, Australian Aluminium Council, Australian Coal Association, Business Council of Australia, BHP Co Ltd (with substantial coal and steel interests), Dept of Environment, Sport & Territories, Dept of Foreign Affairs & Trade, Dept of Primary Industries & Energy, Electricity Supply Association of Australia, Exxon Corporation, Mobil Oil Australia Ltd, Rio Tinto Ltd, Statoil (Norway) and Texaco Inc. All the non-government businesses are, or represent, very large greenhouse gas emitters. Presumably, as members of the Steering Committee, they can influence the assumptions fed into the model. Concern was expressed about this situation by the ombudsman, who concluded that ABARE has left itself open to allegations of undue influence by industry.[15]

Any computer model is only as good as the assumptions fed into it. In the case of MEGABARE, the assumptions (few of which are spelt out explicitly in the modellers' report)[16] have the effect of making the reduction of emissions unnecessarily expensive. The model does this in the following ways:

4.1 BUSINESS-AS-USUAL SCENARIOS WITH HIGH GROWTH IN EMISSIONS

A business-as-usual (BAU) scenario is one in which there are no policies designed to limit the growth of CO_2 emissions. MEGABARE assumes that its BAU scenarios involve high continuing growth in CO_2 emissions resulting from the combus-

---

14. *See* M. Diesendorf, *Australian Economic Models Of Greenhouse Abatement*, 1(1) ENVTL. SCI. & POL'Y (in press), Table 1.
15. Ombudsman, *ABARE and ACF: Report Of The Investigation Into Abare's External Funding Of Climate Change Economic Modelling*, Commonwealth Ombudsman (1998).
16. S. Brown et al., *The Economic Impact Of International Climate Change Policy*, Research Report 97.4, ABARE (1997).

tion of fossil fuels, even in countries where there is low population growth or little dependence upon fossil fuels.

For example, although New Zealand is low on both population growth and fossil fuel use, it is given the highest projected growth in emissions of the rich countries listed, 2.2 per cent p.a., leading to a 88 per cent growth between 1990 and 2020. This is based on the assumptions that future GDP growth is 2.43 per cent and that energy growth must be fossil fuel-based. However, New Zealand has enormous wind energy potential that could be coupled very effectively with the existing hydro-electric system.

In Australia, MEGABARE assumes that CO_2 emissions will grow exponentially at an average rate of 1.63 per cent p.a., leading to a 62 per cent growth between 1990 and 2020. This is driven apparently by GDP growth, assumed to be 2.31 per cent p.a., and population growth of 0.94 per cent p.a. However, the relationships assumed between these three variables are not revealed in the latest MEGABARE report.

4.2 SIMILAR EMISSION REDUCTION SCENARIOS

MEGABARE presents only two, similar scenarios for reducing CO_2 emissions - its 'less stringent' and 'more stringent' scenarios. Despite their names, the two scenarios lead to almost identical rates of CO_2 emissions in 2020, the end-point of the study, namely 10 per cent and 15 per cent, respectively, below the 1990 level. They differ mainly in that the 'more stringent' scenario reduces emissions more rapidly in the period from 1998 to 2010. However, the report does not reveal any of the energy producing and consuming technologies assumed in its 'more stringent' scenario and only reveals the electricity generation technologies in its 'less stringent' scenario. Thus, the two scenarios do not offer the range of political choice needed for informed public debate.

MEGABARE's scenarios assume that, to meet even modest greenhouse targets, the overwhelming majority of electricity generation must come from renewable energy. This assumption follows from the exclusion of the two cheapest ways of reducing emissions (as discussed in the next subsection). However, the scenarios appear to ignore the possibility that, once renewable energy is a substantial component of Australia's electricity generation, large emission reductions can be achieved by replacing part of the motor vehicle fleet with electric vehicles charged from the grid. This could be encouraged by even a modest carbon tax.

MEGABARE's scenarios require a disproportionate component of GHG reductions to come from electricity generation, ignoring the large unutilised potential for providing low-temperature heat from solar energy. There are good reasons for believing that electricity prices are currently at rock bottom levels and will increase steadily over 5-10 years, even without a carbon tax. Natural gas consumers will have to pay for new pipelines and gas prices will follow electricity prices upwards. This will provide an expanding market for solar heat.

4.3 EXCLUDING THE CHEAPEST TECHNOLOGIES

MEGABARE excludes from its emission reduction scenarios the two cheapest and fastest ways of reducing CO_2 emissions, namely the efficient use of energy and greatly expanded use of natural gas for electricity generation and heating. Hence, neither scenario is anywhere near the minimum cost for achieving the specified emission reductions.

Although MEGABARE allows for some improvement in future efficient energy use in its BAU scenario, this is a small fraction of the potential for cost-effective efficient energy use. MEGABARE fails to consider the full potential by ignoring the large body of evidence of failure of the market for energy services. Thus it excludes the substantial additional cost-effective improvements that could be achieved in efficient energy use, both now and in the future.[17] The MEGABARE modellers seem to be unable to recognise market failure (or, more correctly, failure of the model to describe a non-competitive market). Market/model failure points to the need for government intervention to remove market barriers or to compensate for market failure.

MEGABARE's conclusion that increased use of natural gas for electricity generation is also excluded occurs because even natural gas cannot produce enough emission reduction to bridge the huge gap that has been constructed between the BAU and emission reduction scenarios. This is a peculiar situation, running counter to the existing time trend in Australia's energy use and toMEGABARE's own BAU scenario. Natural gas-fired power stations have lower capital costs but higher fuel costs compared with coal. Depending upon the level of interest rates and the amount of competition for natural gas, electricity generated from natural gas can cost about the same or slightly higher than from coal. However, when burned in combined cycle gas-fired power stations, natural gas produces 50 per cent less emissions per unit of electricity generated than black coal.

Cogeneration, the production of electricity and useful heat from small, gas-fired power stations located close to the point of use, can lead to emission reductions of 70-80%. For users who require both electricity and heat (e.g., offices, hospitals, hotels and some manufacturing plants), cogeneration is competitive with coal-fired electricity and boilers, but is held back by inappropriately high network charges.

The exclusion from the MEGABARE scenarios of new natural gas-fired power stations and substantial amounts of cogeneration dictates that renewable energy must take up a much heavier load, both in percentage and absolute contributions, than is necessary or feasible at reasonable cost by 2020. This artificially boosts the costs of MEGABARE's emission reduction scenarios. However, much less expensive scenarios are proposed elsewhere.[18]

---

17. Thirteen studies supporting this statement are cited in Diesendorf, *supra* note 14.

4.4 IGNORING ECONOMIC BENEFITS OF EMISSIONS REDUCTIONS

MEGABARE ignores the economic benefits of shifting to an energy system that is more labour intensive and more suitable for local manufacture than the existing centralised fossil fuel based system, which is shedding jobs at a rapid rate.

This is a consequence of the peculiar assumption that MEGABARE's BAU scenario is the least-cost option. The assumption is wrong because the scenario enshrines energy waste, and assumes that heavily subsidised, capital-intensive, energy-intensive industries such as aluminium smelting are more beneficial to the Australian economy than industries based on efficient energy use and renewable energy. It also ignores the economic risks of Australia remaining dependent upon fossil fuels and aluminium smelting.

Experience with the Danish and US windpower industry has shown that there are substantially more jobs created per unit of energy in renewable energy technologies than in the traditional large centralised energy technologies. Energy auditing is a rapidly expanding field in Australia, especially for engineering graduates. The products and services involved in creating a sustainable energy future tend to require small amounts (e.g., $20k-50k) of capital investment per direct job and tend to create local jobs rather than overseas jobs. Sustainable energy jobs are long-term, while jobs in the resource industries tend to be temporary construction jobs, leading to boom and bust conditions in local economies.[19]

Among Australia's greatest CO_2 emitters are the coal/electricity, aluminium, oil, and iron and steel industries. Aluminium smelting is highly capital intensive, directly creating only one job per $2M invested. The other sectors are all shedding jobs rapidly as the result of increasing automation, not environmental constraints.

4.5 ASSUMING THAT ACTION BY INDUSTRIALISED COUNTRIES ALONE IS USELESS

MEGABARE assumes that concerted action by the Annex I (i.e., industrialised) countries to reduce emissions would not influence the poorer countries to join in. This political assumption could possibly become a self-fulfilling prophecy now that the Australian government has been successful in avoiding commitment to an emissions reduction target at Kyoto. In this case, non-Annex I countries could use Australia's example as an excuse not to participate. US opponents of the agreement are already using this as an argument not to ratify the protocol. However, if Annex I countries agreed to strengthen the target in the future, there would be great pressure on developing countries, notably China and India, to take commensurate measures.

---

18. *See* M. Diesendorf, *Scenarios For Reducing Greenhouse Gas Emissions From Australia's Energy Sector*, Australian and New Zealand Society for Ecological Economics, National Conference (1997).

19. ACF/ACTU, *Green Jobs In Industry: Research Report*, Australian Conservation Foundation & Australian Council of Trade Unions (1994).

4.6 ASSUMING CARBON LEAKAGE

MEGABARE assumes that binding targets in Annex I countries would increase emissions in non-Annex I countries by 12 to 14 per cent. This notion, termed 'carbon leakage,' is that fossil fuel intensive industries, such as aluminium smelting, would relocate to countries where there are no greenhouse constraints and allegedly higher emissions to manufacture the same products. There are several flaws in this argument:

(i) Relocation of a large industry is a very expensive operation. Relocation to a developing country may also entail increased risks for the industry, such as those resulting from political instability.

(ii) In a number of developing countries, electricity is generated from large hydroelectric facilities that release much lower levels of greenhouse gas emissions than coal. Australia is actually the only *major* aluminium producer to use coal-fired electricity for this purpose.

(iii) Aluminium production in Australia is *de facto* subsidised through the provision of cheap electricity and free infrastructure by state governments and tax concessions by the Federal government. This undermines the claim that Australia has a comparative advantage based on economic efficiency.

(iv) Future international agreements under the FCCC could include emission constraints on particularly energy intensive industries, such as aluminium, and even emission constraints upon the particular multinational corporations involved in these industries. This may be feasible because there are not many of these large corporations, their head offices are located generally in Annex I countries, and their turnovers are greater than those of some countries.

(v) An international agreement could be made to ban the use of coal-fired electricity for the production of aluminium and other energy intensive products.

4.7 ASSUMING ONLY A CARBON TAX

MEGABARE assumes that a carbon tax, applied indiscriminately across the economy, is the only response measure to be implemented, instead of a package of mutually reinforcing measures. If consumers are not offered alternatives to emissions-generating activities, a very large carbon tax would be required to change behaviour and this could be expensive for many interests. However, because markets fail and BAU scenarios are not economically optimal, a small carbon tax together with regulatory and other measures, would offer a low-cost response strategy.[20]

20. M. Diesendorf, *Greenhouse Response In The Energy Sector*, in HUMAN ECOLOGY, HUMAN ECONOMY: IDEAS FOR AN ECOLOGICALLY SUSTAINABLE FUTURE (M. Diesendorf & C. Hamilton eds., 1997), at 197-242.

4.8 ASSUMING ONLY ONE USE FOR CARBON TAX REVENUE

MEGABARE assumes that the revenue from a carbon tax is returned to the economy in a lump sum. However, modelling by others suggests that there can be substantial economic benefits from using a portion of the revenue to reduce payroll taxes.[21] Another part of the revenue could also be used to facilitate the transition to a sustainable energy system, thus increasing the rate of emission reductions. It is inappropriate for the MEGABARE modellers to attempt to dismiss this by saying that "estimates of such benefits are highly sensitive to the type of models used and the underlying assumptions,"[22] since that statement is an excellent summary of criticisms of MEGABARE. When such sensitivity exists, an important response should be to perform sensitivity analyses, but the MEGABARE modellers fail to do this.

4.9 OTHER QUESTIONABLE ASSUMPTIONS.

Other questionable assumptions are discussed in detail elsewhere.[23] A particularly important one is that, like almost all other 'top-down' models, MEGABARE substitutes dubiously derived parameters (e.g., 'elasticities' and 'autonomous end-use energy-intensity improvement') for specifications of technologies.

4.10 SUMMARY OF FLAWS IN MODELLING

MEGABARE's assumptions have the effect of creating an imposing task in order to achieve quite small reductions in emissions. It excludes the two most cost-effective technologies, thus leaving only an expensive means of achieving these small reductions (namely, a rapid, massive shift to renewable energy at a time when only a smaller contribution from renewable energy is cost-effective or close to it). It also omits recognition of the economic benefits of making the transition.

Even with all these and other assumptions, which make greenhouse gas reduction appear expensive, MEGABARE projects a total cost to rich countries of only 1.0 to 1.5 per cent of projected Gross National Expenditure (GNE) in 2020, somewhat higher (3.3 per cent of GNE) in Australia. Because this is smaller than, or comparable with, the uncertainty in the modelling results, it would be actually impossible to measure.

---

21. C. Hamilton, T. Hundloe & J. Quiggin, *Ecological tax reform in Australia*, Discussion Paper No. 10, Australia Institute (1997).

22. Brown, *supra* note 16, at 29.

23. M. Diesendorf, *The Ecologically Sustainable Development Process In Australia*, in Diesendorf, *supra* note 20, at 285-301. *See also* Diesendorf, *supra* note 14.

5. Government's Re-Interpretation of the Modelling Results

The Australian government has represented the results of the modelling and other studies in ways which are misleading and potentially frightening to many people due to alleged financial and employment concerns. For example, on the basis of the MEGABARE report, the government creates the false impression that each Australian would have to withdraw $7,000 to $9,000 out of their bank accounts to pay for reductions in GHG emissions. In reality, the amount does not come from savings in people's bank accounts, but rather is a tiny reduction in the projected accumulated increase in gross national expenditure (GNE) per Australian. Over the 22 year period, GNE per person would have increased by hundreds of thousands of dollars per person and so the alleged amount lost, *if true*, would be a tiny fraction of this. Furthermore, the economic benefits to each Australian of improved energy efficiency in a Sustainable Business Scenario could be comparable in magnitude with these alleged losses.

In a similar fashion, separate from the MEGABARE report, the government claims that Australia would lose 90,000 jobs by reducing emissions. However, these jobs do not exist at present. They are based on highly optimistic estimates by industry of mostly temporary construction jobs that might be created under a scenario of untrammelled resource development. The claim ignores the many thousands of *permanent* jobs that would be created in a Sustainable Business Scenario.

These and other misleading government claims raise the question as to whether the government is capable of leading the Australian economy as a whole into the 21st century, or whether it is tied to the few declining[24] smokestack industries of the past with a vested interest in the production of greenhouse gas emissions. Moreover, as a foreign policy device, the government appears to be attempting to create divisions between rich and poor countries; between individual rich countries; and between supporters of uniform targets, differentiated targets and tradeable emission permits. All of these divisions may help to undermine the creation of an international agreement. Unwittingly the MEGABARE report provides the (rather shaky) intellectual basis for this attempt.

6. Conclusion

In 1989, the Australian government set an Interim Planning Target for reducing greenhouse gas emissions, and shortly afterwards set up processes to move the nation onto a pathway towards ecologically sustainable development in general and a reduction in greenhouse gas emissions in particular. However, since early 1992 federal and all state governments states except New South Wales, whether Labor or Coalition, have retreated from action in international, national and state arenas. This retreat has included making a large percentage reduction in the ini-

24. 'Declining' in terms of employment. Also, the profitability of the coal industry is low and aluminium smelting receives large de facto subsidies.

tially low levels of funding for sustainable energy options, i.e., efficient energy use and renewable sources of energy. It has been bolstered by economic modelling, initially carried out and publicised by the industries that are Australia's greatest greenhouse gas emitters, and now carried out and publicised by the Australian government with funding from those industries. The assumptions underlying this modelling are severely flawed and lead inevitably to the incorrect conclusion that greenhouse response would be necessarily expensive for Australians.

There are four specific concerns in this regard. First, the assumptions underlying the Australian government's policy on reducing GHG emissions, such as the existence of a perfect market for energy services, are contradicted by empirical evidence. Second, arguments such as on carbon leakage appear to be at variance with logic. The failure to carry out generally accepted procedures of verification, such as sensitivity analysis, is the third concern. Fourth, its actions are largely ineffective in reducing emissions, but are highly effective at undermining the prospects of sustainable energy.

Furthermore, there are reasons for questioning whether the government really believes in its own case. This assertion can be made because its Greenhouse Challenge program contradicts its basic assumption that the market for energy services is perfect, because, if businesses are already obtaining optimal cost-effective energy savings, there will be no further gains through a voluntary agreement that only requires the pursuit of cost-effective measures. The government also misrepresents the results computer modelling and other studies. However, modelling with alternative, more realistic assumptions suggests that there could be economic as well as environmental benefits in making the transition to an ecologically sustainable energy system. Finally, despite its rhetoric and actions to remove market barriers to the greater use of (coal-fired) electricity, it has failed to remove barriers to efficient energy use, renewable energy and cogeneration. Indeed, in some cases, it has raised new barriers.[25]

The greenhouse response policies of Australian governments, both Labor and Coalition, appear to be based on the desire to support the resource industries -- especially coal, oil, aluminium, and iron and steel — at the expense of the industries which could lead to an ecologically sustainable future, notably the sustainable energy and cleaner production industries. Yet the resource industries are shedding jobs and, in the case of oil, seriously depleting the resource at the same time.[26]

The fundamental issue is whether the Australian government will allow the economy to make a transition (which is already occurring slowly) towards one based primarily on service industries and light manufacturing, or whether it will continue to attempt to hold back the economy in its resource development and processing stage of the 1960s.

---

25. Diesendorf, *supra* note 12.

26. B. FLEAY, THE DECLINE OF THE AGE OF OIL (1995).

TABLE 1.Summary of Australia's greenhouse response strategy, 1989-1997

Year	Event
1989	Interim Planning Target set
1990	ESD process commences; Industry Commission inquiry into costs and benefits of reducing GHG emissions commences
1991	ESD Working Groups report; Industry Commission reports
1992	ESD Chairs report on greenhouse; NGRS & NSESD reports; Implementation of NGRS entrusted to ANZMEC and COAG; Australia signs and ratifies FCCC
1994	Australia's first national report under FCCC
1995	ICESD report; 1st MEGABARE report on cost of GHG reductions
1996	NGAP report; Labor budget cuts sustainable energy
1997	Coalition budget cuts sustainable energy further; second MEGABARE report; Aust. govt opposes uniform targets; December: 3rd Conference of Parties to FCCC at Kyoto where Australia in effect receives international endorsement of business-as-usual GHG emissions

TABLE 2. Status of several key greenhouse response measures by Australian Government

Measures recom-mended by ESD reports in 1991	**Measures recom-mended by NGRS or NSESD**	**Situation after Labor Gov't budget of 1995**	**Situation under Coalition Gov't budgets**
Mandatory MEPS from 1993 for elec., gas & solar; and for residential, commercial & ind. sectors	Mandatory MEPS for major residential appliances only, 'after considering costs & benefits'	Mandatory MEPS agreed for fridges, freezers & some water heaters only ; 1993 US standards for 1999	As 1995
Utilities to adopt ILCEP & spend 1% revenue on least-cost demand-side	'High priority' to ILCEP, no action, no 1% spending	Restructuring of elec. industry used to justify inaction in all states except NSW	As 1995
'Substantial increase' in R, D & D and early commercialisation of energy efficiency and renewable energy	No action. ESD recommendation changed to 'strengthen energy R, D & D, especially for renewables & energy efficiency; Ethanol fuel bounty (introduced 1994).	NSW funding PV and SEDA;	ERDC budget cut by 50% in 1996 and closed down in 1997; R & D tax concession cut from 150% to 125% in 1996 and from 125% to 100% in 1997; AusAID's DIFF scheme terminated; Ethanol fuel bounty discontinued in 1996.
Voluntary targets for fuel efficiency of motor vehicles to be backed up with legislation	Voluntary target, no mention of regulation	Weak voluntary target agreed to for new cars only	As 1995
Mandatory nationwide House Energy Rating Scheme (HERS)	Voluntary scheme proposed		HERS almost ready for national implementation; voluntary in all states & territories except Victoria and Australian Capital Territory
Commercial Building Energy Code (BECA)	Process commenced	No agreement	No agreement
Inquiry into incorporation of environmental costs into energy prices		Inquiry stopped: committee failed to agree on objectives; Labor Gov't opposes Carbon Tax	No change from 1995

5. CLIMATE CHANGE POLICY FORMATION IN AUSTRALIA: 1995-1998

R. TAPLIN AND X. YU
Dept. of Human Geography
Macquarie University
New South Wales, Sydney, Australia

1. Introduction

Australia was one of the earliest ratifiers of the Framework Convention on Climate Change in 1992. However, in the years since then, there has been slow progress in Australia reducing its greenhouse gas emissions. In 1997, Australia's National Greenhouse Gas Inventory Committee published the *National Greenhouse Gas Inventory: 1995*. The results show that in 1995, excluding the land use sector, Australia's net greenhouse gas emissions had increased by 6 per cent above 1990 levels and that in the energy sector, emissions had risen by 8.2 per cent above 1990 levels (see Table 1). According to another Australian government publication, Australia's CO_2 emissions from the energy sector were estimated to be 11 per cent above 1990 levels in 1995-1996 and without further measures, they could rise to 40 per cent above 1990 levels by 2010.[1]

TABLE 1. Australia's Greenhouse Gas Emissions in 1990 and 1995[2] (million tonnes CO_2 equivalent)

Sector	1990	1995	Change	%
Energy	293.3	317.3	24.0	8.2
Agriculture	88.8	87.4	-1.4	-1.6
Waste	14.8	16.4	1.6	10.8
Industry processes	12.1	9.0	-3.1	-25.6
Forestry & others	-29.4	27.6	1.8	-6.1
Net emissions	379.6	402.4	22.8	6.0

This chapter examines climate policy formulation in Australia between 1995 and mid-1998. International influences on Australia's climate policy formation, commitments

1. Department of Foreign Affairs & Trade, *Australia and Climate Change Negotiations: An Issues Paper*, Department of Foreign Affairs and Trade (1997).
2. National Greenhouse Gas Inventory Committee, *National Greenhouse Gas Inventory 1995: Australia*, Environment Australia (1997). Note: land use changes are excluded from this table.

A. Gillespie and W. C. G. Burns (eds.), Climate Change in the South Pacific: Impacts and Responses in Australia, New Zealand, and Small Island States, 95–112.

from all levels of government in Australia and industry actions are reviewed. Finally, some of the major barriers to effective Australian climate change policy formulation and implementation are examined.

2. Reflections on the Impact of the 1992 National Greenhouse Response Strategy

After the Rio Earth Summit Conference, the Council of Australian Governments reached agreement on Australia's *National Greenhouse Response Strategy* (NGRS) in December 1992.[3] As the primary Australian policy mechanism, this strategy provided the basis for the federal government, the states and territories and local government authorities to work together on climate change issues under the obligations of the United Nations Framework Convention on Climate Change.[4]

However, the 1992 NGRS was never effectively implemented and many key stakeholders were not sufficiently involved in putting greenhouse response actions into place. As Ian Carruthers, Assistant Secretary, Air Pollution and Climate Change Branch of Environment Australia has commented: ". . . with the benefit of experience and feedback, the message has been conveyed from a range of quarters that the 1992 Strategy was seen as a product of Government for implementation by Government."[5]

The 1992 NGRS proved to have many shortcomings.[6] For example, the National Greenhouse Advisory Panel (NGAP) in its 1996 report criticised the NGRS on several fronts. It suggested that policy coordination between jurisdictions (federal, state/territory and local governments) was very poor. There was a general lack of coordination across policy and program within jurisdictions, so that greenhouse issues were usually not considered.[7] In particular, the panel concluded that the NGRS was slow in terms of integrated least cost planning, demand side manage-

---

3. Commonwealth of Australia, *National Greenhouse Response Strategy* (1992), at 4. For background on the formulation and implementation of the 1992 NGRS *see* R. Taplin, *Greenhouse: An Overview of Australian Policy and Practice*, 1(3) AUSTRALIAN J. ENVTL. MGMT. (1995), at 142-155. *See also* the chapter by Bulkeley in this volume.
4. In the NGRS, an interim planning target was adopted for the reduction of greenhouse gas emissions. The goal was "to stabilise greenhouse gas emissions (not controlled by the Montreal Protocol on substances that deplete the ozone layer) based on 1988 levels, by the year 2000 and to reduce these emissions by 20 per cent by the year 2005. . . subject to Australia not implementing response measures that would have net adverse economic impacts nationally or on Australia's trade competitiveness, in the absence of similar action by major greenhouse producing countries". . . . This ambitious target was ignored in implementation of the NGRS.
5. I. Carruthers, *Australia's Greenhouse Response Strategy: The Way Forward*, in THE CHALLENGE FOR AUSTRALIA ON GLOBAL CLIMATE CHANGE: SUMMARY OF PROCEEDINGS, National Academies Forum, 29-30 April 1997, at 633-67.
6. P.F. Greenfield, *Global Warming: Australia's National Response*, in THE CHALLENGE FOR AUSTRALIA, *id.* at 63-67.
7. National Greenhouse Advisory Panel, *Report on the National Greenhouse Response Strategy*. Environment Australia (1996).

ment and development of renewable energy and cogeneration. The micro-economic reform emphasis was on marketing, competition and reducing prices and not on energy efficiency and greenhouse outcomes.[8] The panel pointed out that the NGRS did not devote sufficient attention to the transport sector, which accounts for 24 per cent of national energy end use. Also, it contended that government urban and transport planning was carried out without sufficient greenhouse considerations.[9] The panel suggested that land clearing is a major contributor to greenhouse gas emissions, but the controls over clearing have not received prominent consideration.[10] Australia's greenhouse responses have been largely based on measures of a "no regrets" nature. In many cases, "no regrets" has been misinterpreted as "no losers."[11] As a result, greenhouse reduction measures have been deferred or diluted because some industries or interest groups perceived themselves to be losers. Finally, the panel noted the absence of targets for controlling greenhouse gas emissions. Thus, progress in the implementation of greenhouse responses could not be measured and tracked.[12]

Professor Paul Greenfield, Chairperson of NGAP, also observed that Australian national goals for greenhouse gas emission reductions are often below international best practices, for instance, with respect to fuel economy targets for motor vehicles. He has highlighted the lack of commitment and involvement from governments, as well as very low levels of community education and awareness about greenhouse response.[13] NGAP also suggested priority should be accorded to development of projections under various emission scenarios and the impacts by different response measures; development of strategies for areas such as agriculture and infrastructure sector to adapt to climate change; greenhouse research effort focussing on energy efficiency, renewable energy and sustainable agriculture; bringing local government, small to medium sized enterprises, community and regional groups into greenhouse response partnerships; further research on potential health impacts of climate change; ensuring future greenhouse response addresses the issues of environmental quality and social equality; providing appropriate and timely information for the community, and training and advice within industry; ensuring informed and responsive access by governments to stakeholder, community and scientific opinions; altering the emphasis of response strategies from process to action and supporting this action with well considered and balanced regulatory and or economic instruments.[14]

These are all important policy actions, but unfortunately, implementation of many

---

8. *Id.* at 13.
9. *Id.* at 14.
10. *Id.* at 14.
11. *Id.* at 15.
12. *Id.* at 13.
13. Greenfield, *supra* note 6.
14. NGAP, *supra* note 7, at 16.

of them still falls far short.

3. Greenhouse 21C

In early 1995, a supplement was formulated to the 1992 NGRS, Greenhouse 21C: A Plan of Action for a Sustainable Future.[15] This embodied Australia's domestic climate policy response for the Berlin Conference of the Parties and was distributed at that meeting. Greenhouse 21C's main approach was to build a broad partnership between government and all stakeholders, including industry, academic institutions, professional associations, the commercial sector and the general public, to mitigate Australian greenhouse emissions. Greenhouse 21C consisted of five key areas. These were: Energy 21C, Urban 21C, Biosphere 21C, Global 21C and Greenhouse Partnership 21C.[16] The objectives in focussing on these policy areas are described in the following sections.

3.1 ENERGY 21C

As energy use accounts for more than 50 per cent of all greenhouse gas emissions in Australia, the efficient use of energy was recognised to be central to Australia's greenhouse response. It was argued in Greenhouse 21C that increased energy efficiency and greater use of new energy sources would reduce the overall costs to the national economy and create export opportunities.

3.2 URBAN 21C

In Australia, about half of greenhouse gas emissions are directly or indirectly produced from urban activities including urban energy use. Many aspect^ of urban development can affect greenhouse gas emissions, such as urban planning, building materials, energy use, transport systems and waste management.

3.3 BIOSPHERE 21C

Approximately 40 per cent of Australia's net greenhouse gas emissions derive from land management and agriculture. Sustainable land and forestry management was considered an opportune area for Australia in terms of reducing greenhouse gas emissions. It was envisaged that ecologically sustainable vegetation programs would enhance greenhouse gas sinks.

---

15. Department of Environment, Sport and Territories, *Greenhouse 21C: A Plan of Action for a Sustainable Future*(1995).

16. *Id.* at 2.

3.4 GLOBAL 21C

It was explicitly stated that Australia would continue to play its part in the global efforts to curb greenhouse gas emissions. In the Asia-Pacific region, in particular, Australia recognised its role and responsibilities to other nations. The document stated that Australia would develop its technological base to support environmentally sound development in the region.

3.5 GREENHOUSE PARTNERSHIP 21C

Partnership 21C highlighted the role of the federal government in coordinating efforts to reduce greenhouse gas emissions. That is, government was to work in partnership with stakeholders. In this partnership, the federal government reaffirmed that it planned to integrate greenhouse considerations in all relevant federal policies and programs. The major outcome of Greenhouse Partnership 21C has been the establishment of the Greenhouse Challenge Program. The Australian government agreed to commit A$9.7 million over four years to facilitate cooperation with industry to reduce greenhouse gas emissions.[17]

The Greenhouse Challenge Office[18] was set up as a 'shopfront' in Canberra to facilitate this interaction and, since then, active partnerships between government and industry through cooperative agreements have been negotiated. The Australian government estimated this program could reduce Australia's greenhouse gas emissions by 15 million tonnes annually up to the year 2000.[19]

Two pivotal groups in relation to Partnership 21C, were the National Greenhouse Advisory Panel (NGAP), established in mid-1994, and the Intergovernmental Committee on Ecologically Sustainable Development (ICESD), established during the development of Australia's 1992 *National Strategy for Ecologically Sustainable Development.*[20] Together, they represented the major stakeholders and government bodies responsible for monitoring and developing Australian greenhouse responses. Joint meetings of NGAP and ICESD were designed to ensure formal stakeholder involvement in government policy development.

---

17. *Id.* at 7.
18. The Greenhouse Challenge Office is a joint venture of the Department of Primary Industries and Energy, the Department of Industry, Science and Tourism, and the Department of Environment, Sport and Territories.
19. Greenhouse Challenge Office, *Implementation Plan* (1997), at 3.
20. Commonwealth of Australia, *National Strategy for Ecologically Sustainable Development*. Australian Government Publishing Service, Canberra (1992), at 110.

4. Greenhouse 21c and The National Greenhouse Response Strategy

Greenhouse 21C was based on the 1992 NGRS. However, it was anticipated that it would be a more effective policy instrument than the NGRS in reducing greenhouse gas emissions. *Greenhouse 21C* was expected to facilitate fundamental changes in long-term emission patterns. The initiative focused upon advancing and expanding the NGRS in seven areas. The first and second of these were: promotion of active partnerships between government and industry through cooperative agreements; and support for renewable energy through industry development programs and the Commonwealth commitment to a new renewables Cooperative Research Centre. The third was a greenhouse information network to support action by all sectors of society. Enhanced cooperation with the states and territories to address a number of issues related to greenhouse, including land management and energy sector reform, was the fourth area of activity. The fifth focus was on the Australian government's development and implementation of environmental best practices. The penultimate concern was a continuing emphasis on micro-economic reform in the energy sector, with a greater focus on reform in the gas sector and the delivery of energy efficiency services. Finally, the expansion of tree planting programs to further develop greenhouse sinks was to be encouraged.[21]

Greenhouse 21, as a policy package, faced similar implementation problems to the 1992 *National Greenhouse Response Strategy*. An additional complication arose one year into its implementation, in March 1996, when the Liberal and National Parties won the Australian federal election and formed a Coalition Government, under Prime Minister John Howard, taking over from the former ALP[22] Government. The Greenhouse Challenge program was certainly the most successful initiative in the Greenhouse 21C package and has been embraced by the Howard Government. Other initiatives in Greenhouse 21C did not fare as well.

5. Climate Policy Formation under the Coalition Government

Since March 1996, Australia's climate change policy has faced considerable transformation. The new Coalition Government was confronted with formulating a position on the international negotiations leading to Kyoto and also with domestic policy issues with respect to climate change. To date, multiple policy initiatives have been introduced by the Howard Government. These include introduction of the Activities Implemented Jointly program in late 1996, major revision of the National Greenhouse Strategy during 1997 and 1998, establishing a House of Representatives Standing Committee Inquiry into trading greenhouse gas emissions in late 1997, release of the Prime Minister's Greenhouse Package in November 1997 and establishment of the Australian Greenhouse Office in March 1998.

---

21. Department of Environment, Sport and Territories, *supra* note 15, at 15.

22. Australian Labor Party

5.1 ACTIVITIES IMPLEMENTED JOINTLY

Activities Implemented Jointly (AIJ) is a co-operative international approach to addressing climate change under the Framework Convention on Climate Change (FCCC). The AIJ falls within the pilot phase of the FCCC joint implementation program. Joint implementation under the FCCC involves industry in developed nations making specific investments in developing countries that result in carbo offsets.

Australia's pilot AIJ initiative was initiated in late 1996. The federal government established Australia's AIJ Office in Canberra[23] as a 'one-stop shop' for government advice and support services for industry. The role of the office is to identify and help facilitate projects that satisfy AIJ criteria and guidelines. The office's task is also to assist where appropriate with negotiating project agreements and to help with .linkages between financing organisations and domestic and international industry interested in pursuing AIJ initiatives. The aim of the Australian AIJ program is to assess the potential of international co-operation through commercial activities to reduce greenhouse gas emissions. Four goals have been formulated to assist this objective. The first of these is facilitating cost-effective greenhouse gas emissions abatement, especially in the Asia-Pacific region. The second objective is enhancing Australian trade and investment links in environmental technology and services areas. Facilitating co-operation with developing countries to address climate change is the third objective. Finally, the government hopes to encourage investment in capital, technologies and know-how in developing countries.

To date, there has been no provision of Australian government funds for financing AIJ projects and the result has been that there has been little incentive for Australian industry to become involved in the AIJ programme.[24] This differs from other nations. For example, provision of AIJ capital by the United States government has stimulated company involvement in the United States.[25] Nevertheless, the widespread feeling about AIJ is uncertainty both within and outside Australia. As Stuart has remarked: "The current market for AIJ/JI is representative of any marginal market, where the vast majority of players either don't know about the market, don't care about the market, or feel that the returns for investing in the market are not worth the effort."[26]

---

23. The AIJ Office is housed in the Department of Primary Industries and Energy.

24. *Closing the Communication Gap: Proceedings of a Workshop to Promote Communication and Consultation on Two Key Australian Climate Change Programs: National Greenhouse Response Strategy and Activities Implemented Jointly*, Climatic Impacts Centre, Macquarie University, Sydney, (O. Balashov & R. Taplin eds., 1997).

25. M.D. Stuart, *Anticipating Future Business Opportunities and Needs in Joint Implementation: The Roles of Investment and Hedging in Current Policy Initiatives*, in Balashov & Taplin, *supra* note 24.

26. *Id.*

5.2 REVISED NATIONAL GREENHOUSE STRATEGY

As a result of the criticism of the 1992 *National Greenhouse Response Strategy*, and particularly with the pressure of the international negotiations as a backdrop, the NGRS was placed under review. An updated national policy response, the *National Greenhouse Strategy* (NGS), was planned to be ready for release in Australia before the Kyoto meeting at the end of 1997.

The NGS was expected to highlight the need for greater and more effective actions from all levels of government, stakeholders and Australian society in general. More importantly, it was hoped that the NGS would enable Australia to seriously respond to the greenhouse challenge into the next century.[27] A discussion paper, Future Directions for Australia's National Greenhouse Strategy: Discussion Paper, was released for public comment by the Intergovernmental Committee on Ecologically Sustainable Development (ICESD).[28] The sections contained in the paper reflect the efforts of stakeholder working groups. The negotiation process was slowed by the necessity for the strategy to be developed in conjunction with all the state and territory governments and the Australian Local Government Association - a requirement being that all of these parties agree on the final document. Many redrafts of the strategy have been produced to date. The 1992 National Greenhouse Response Strategy was subjected to this same process and in that case, a less than desirable end product was the result. The revised National Greenhouse Strategy (NGS) is now expected to be released towards the end of 1998, 12 months later than originally envisaged.

5.3 ESTABLISHING A HOUSE OF REPRESENTATIVES STANDING COMMITTEE INQUIRY

In November 1997, Ian Causley, Chair of the House of Representatives Standing Committee on Environment, Recreation and the Arts, announced the Inquiry into the Regulatory Arrangements for Trading in Greenhouse Gas Emissions by the Committee. The terms of reference for the inquiry were multiple. These included the measurement, verification and monitoring of emissions and compliance with what is permitted; and the integration of emissions trading with the development of carbon sinks. The allocation of the right to emit greenhouse gases, and regulatory mechanisms to support a national market, and potentially an international market, in emissions trading were also terms of reference. The final three ambits of focus were possible emissions traders, administration and transaction costs; roles and responsibilities of governments and other stakeholders; and the impact of emission trading on the environment and industry and

27. Intergovernmental Committee on Ecologically Sustainable Development, *Future Directions for Australia's National Greenhouse Strategy: Discussion Paper*. Department of Environment, Sport and Territories, Canberra. (1997) at v.

28. *Id.*

the economic and social welfare of the Australian community.[29] Over 50 written submissions were received from stakeholders and public hearings were held in May 1998.[30] The inquiry is expected to report its findings in mid 1998.

5.4 PRIME MINISTER'S GREENHOUSE PACKAGE

In November 1997, just prior to Kyoto, the Australian Prime Minister, John Howard, issued a statement to the nation, "Safeguarding the Future: Australia's Response to Climate Change," and launched a five year, A$180 million package of greenhouse measures. The package's objective was to significantly reduce Australia's greenhouse gas emissions below projected 'business as usual' levels and target the energy sector as a continuing, major contributor to national greenhouse gas emissions.

The package included plans for the distribution of new funds, the creation of new laws and the encouragement and co-ordination of new incentives. New funds were allocated for the creation of a Renewable Energy Innovation Investment Fund to invest A$21 million in renewable projects. Additionally, the package included a loans and grants scheme worth A$30 million, a plan for a Renewable Energy Showcase to highlight leading edge projects and the Greenhouse Challenge Program was provided with extra funding to extend the program to smaller companies.

New legislative mechanisms mandated that electricity generating companies source an additional two percent of their power generation from renewable energy. Likewise, the automotive industry was targeted through mandatory fuel efficiency labelling, a fifteen percent fuel efficiency improvement goal by 2010 and a phase-out of leaded petrol.[31] Codes and standards, related to voluntary energy efficiency ratings for housing and commercial buildings, appliances and equipment were formulated. Other notable areas in the package were the 'Bush for Greenhouse Programme' which will help create carbon 'sinks' and the establishment of the Australian Greenhouse Office within the federal Environment Department to coordinate domestic greenhouse policy.[32]

29. Committee Secretariat, House of Representatives Standing Committee on Environment, Recreation and the Arts, *Inquiry into the Regulatory Arrangements for Trading in Greenhouse Gas Emissions: Full Terms of Reference*, http://www.aph.gov.au/house/committees/era/greenhse/grhsepre.htm, 30 April 1998.

30. *Id.*

31. The phasing out of leaded petrol was included in the package notwithstanding that lead in petrol is an air pollution rather than a greenhouse issue.

32. *Greenhouse Package Unveiled*, SCITECH., 28 November 1997, at 1.

Notwithstanding these and other measures in the package, Mr. Howard revealed that domestic emissions would increase by 18% beyond 1990 levels by 2010. Nevertheless, it was seen as an attempt "... to convince domestic and international critics it [Australia] is not seeking a 'free ride'... in Japan."[33]

Although some worthwhile initiatives were introduced with the package, overall it was seen by many commentators as a disappointment because the actions in relation to renewables and energy efficiency merely reinstated axed programs such as the controversial removal of funding in the 1997/98 federal budget for the Energy Research and Development Corporation. As Australian Greens Senator Bob Brown concluded:

> A$65 million for renewable energy over five years ... does not even retrieve the A$75 million (over five years) lost when the Energy Research and Development Corporation and Renewable Energy Industry Programs were abolished. . .The target of an extra 2% of electricity from renewables (making a total of 11% including current large-scale hydro electricity generation) compares miserably with international standards (e.g., Britain's target of 20% from renewables by 2010). . .There are no targets for energy efficiency. . .There is no move to halt clearing of native vegetation which accounts for 23% of emissions. . .[34]

Additional concerns were that the package ignored the fact that Australia is unprepared for climate change impacts and no climate change research funding was included in the package. This is regrettable, as Australia needs an R&D effort if it is to effectively prepare for climate change. As such, the package was yet another disappointment for Australian climate change researchers, as no new government funding has been made available for university researchers since March 1996. CSIRO and Bureau of Meteorology researchers have also faced severe funding cutbacks for climate change research programs. Sources of climate change funding from non-government research sponsors are negligible in Australia.

5.5 FOUNDING THE AUSTRALIAN GREENHOUSE OFFICE

As announced in the *Prime Minister's Greenhouse Package*, the Australian Greenhouse Office (AGO) was established in March 1998 with Gwen Andrews being appointed as Chief Executive Officer on 9 March 1998. The AGO, being the lead Australian government agency on greenhouse issues is housed in the Department of the Environment but has interdepartmental linkages with all areas of federal government focussing on climate change. The AGO has responsibility for coordination of domestic climate change policy and delivery of greenhouse response pro-

33. *Id.*
34. R. Brown, *PM's Greenhouse Package -- 18% Increase!*, Media Release, Australian Senate, 20 November 1997.

grams. These responsibilities were largely handled by the Atmospheric Protection Branch of the Department of the Environment before the inception of the AGO.

5.6 STATE/TERRITORY AND LOCAL GOVERNMENT INITIATIVES

In contrast to the federal government's behaviour, some state, territory and local governments have positively committed themselves to policies and measures to address climate change issues. State and territory governments have major responsibility for tackling greenhouse emissions, such as decisionmaking on energy production, transport, land use, infrastructure provision and urban planning. Local governments can influence many activities including waste management, land use, transport planning and forestry management. It has been estimated that the policies administered by local government have an impact on more than 50 per cent of greenhouse gas emissions in Australia.[35]

The New South Wales government, for example, has moved considerably ahead of the federal government in greenhouse response. NSW has made advances in the four major energy sectors, land management, transport and waste management. In December 1995, for example, the NSW State Parliament passed three Sustainable Energy Bills. These require, *inter alia,* that electricity suppliers in NSW cut their greenhouse gas emissions by 20% by the year 2005.[36] This made NSW the first State in Australia to set legislative targets for reducing greenhouse gas emissions.

In 1996, the NSW Government established the Sustainable Energy Development Authority (SEDA) with a budget of A$39 million over three years. SEDA was set up to reduce greenhouse gas emissions related to energy production and consumption in NSW. SEDA is one of the first government agencies in the world whose job is specifically to reduce greenhouse gas emissions.[37]

In electricity market reform, the NSW Government has become the only Australian government to legislate greenhouse gas reductions by imposing licence requirements on electricity retailers.[38] In the transport sector, the NSW government is running more buses and more trains. As a result, in 1996 there were 18 million more passenger journeys by public transport than in 1995. In addition, the State's conservation reserve system has increased by half a million hectares since March 1995, thereby contributing as carbon sinks.[39]

---

35. *See* J.M. Lumb, *Greenhouse Action and Local Government: The New Directions.* Prepared for the National Environmental Law Association by Enviro-Futures (1994), at 14. *See also*, Department of Environment, Sport and Territories, *supra* note 15, at 13.
36. Or to any level agreed by the Council of Australian Governments in the interim.
37. The SEDA Office: An Environmental Showcase, http://www.seda.nsw.gov.au/html/office.html/ (1997).
38. NSW Environment Protection Authority. *NSW Tackles Greenhouse* (1997), at 3.
39. *Id.* at 2.

The NSW Government is also addressing greenhouse emissions from waste facilities including landfills, waste water treatment works, incineration and non-energy use industrial processes. The NSW Waste Service plans to install, as a pilot project, a methane-based generator of 35 MW capacity, utilising methane emissions to supply power to 35,000 homes.[40]

In the lead up to Kyoto, as the Australian government strongly rejected legally binding and uniform targets for greenhouse gas reduction in the international negotiations, the NSW Government condemned the federal government's policy stance[41] and urged that the Australian government "show domestic leadership on greenhouse."[42]

At the local level, councils have instituted strategies and become involved in the Cities for Climate Protection (CCP) Australia Campaign. This campaign has international links with other similar CCP initiatives worldwide under the auspices of the International Council for Local Environmental Initiatives (ICELI).[43] It aims to reduce greenhouse gas emissions at the community level by means of development approval processes, land use strategies and energy efficiency.

The Brisbane City Council in Queensland has sought to implement this programme. It is Australia's largest metropolitan local authority and provides services for more than 900,000 people with over a A$1 billion annual budget.[44] Also, it is the only local authority to own and operate its own bus and ferry service in Australia. In Brisbane, 230 petajoules (PJ) of primary energy are used annually and over 97% of electricity is generated by coal.[45] With regard to greenhouse gas abatement, the main approach of the Brisbane City Council has been use of a policy mix to encourage energy efficiency. According to the Lord Mayor's Environment Vision and the Brisbane City Council's Corporate Plan, the Brisbane City Council plans to reduce its greenhouse emissions to 20% below 1991 levels by 1998.[46]

---

40. *Id.* at 6.

41. P. Organ, *NSW Condemns Howard's Greenhouse State*, Press Release, 26 September 1997.

42. P. Organ, *Howard Must Show Domestic Leadership on Greenhouse*, Press Release, 14 October 1997.

43. The City of Newcastle, which is actively involved with ICLEI hosted an international conference in June 1997, Pathways to Sustainability: Local Initiatives for Cities and Towns, where urban sustainability issues including greenhouse were discussed. *See*, City of Newcastle, *Pathways to Sustainability: Local Initiatives for Cities and Towns.* International Conference Proceedings, 1-5 June 1995.

44. V. McLeod, Environment Officer - Energy, Natural Environment, Urban Management Division, Brisbane City Council, personal communication, 30 October 1997.

45. *Id.*

46. *Id.*

Another local level example is the Southern Sydney Regional Organisation of Councils and its *Greenhouse Strategy for the Southern Region* (1995). The Strategy provides a greenhouse policy framework for local government within the region in line with the 1992 *National Greenhouse Response Strategy* and NSW government encouragement to councils to address climate change. The twelve SSROC member councils that jointly agreed to the strategy are Bankstown, Botany Bay, Canterbury, Hurstville, Kogarah, Marrickville, Randwick, Rockdale, South Sydney, Sutherland, Waverley and Woollahra. Together these councils have jurisdiction over a population of approximately 1.1 million.[47]

5.7 INDUSTRY COMMITMENTS

Australian industry has a major obligation to cut greenhouse gas emissions and Australian companies are increasingly realising that climate change is a serious environmental problem. In August 1997, a survey of members of the Australian Institute of Company Directors on key environmental issues was carried out by ACNeilsen-McNair. The results show that 69% of the 630 directors who responded to the survey agreed that Australia should have a global greenhouse gas reduction targets. Of those directors in favour of global targets, 70% thought the targets should be legally binding.[48] However, fossil fuel and energy intensive industry groups have had a powerful influence on policies in Canberra.

Australia's current industry commitments to greenhouse response largely rest on voluntary agreements with government and 'no regrets' measures. The 'no regrets' measures mean that these measures have other benefits or at least no net costs apart from reducing emissions. For instance, energy saving measures fit this category. Notwithstanding this, the Greenhouse Challenge Program, in particular, appears to be having some success. Several major corporations, CRA, BHP, ICI and Shell Australia, for example, have signed agreements covering 109 sites around Australia and have identified 200 actions that are expected to reduce those companies' emissions by 18% by the year 2000.[49] Also many industry associations have agreed to facilitate program implementation by encouraging member participation, developing codes of practice, providing technical assistance and documentation, supporting emission inventories and best practice and reporting outcomes to the federal government.[50]

Under the Greenhouse Challenge program, a total of 104 enterprises became 'Greenhouse Challengers' by June 1998.[51] These 104 companies and associations

---

47. SSROC, personal communication, 16 April 1998.
48. AC Nielsen-McNair, *Environmental Realism: Australian Company Directors Views on Key Environmental Issues, The Boardroom Report*, Prepared for Australian Institute of Company Directors/ KPMG (1997), at 28-44.
49. Greenfield, *supra* note 6, at 66.
50. Greenhouse Challenge Office, *supra* note 19, at 4.
51. Greenhouse Challenge Office, *Greenhouse Challenge Update*, Issue 8 (1998).

have signed cooperative agreements with the Australian government to reduce greenhouse gas emissions. About one-quarter are industry associations that signed facilitative agreements, while the remainder represent a broad cross-section of Australian industry — representing about 45% of the emissions from the resource, mining, manufacturing, transport, electricity generation and distribution, and services sectors.[52] By June 1998, an additional 152 organisations formally indicated their intention to develop a cooperative agreement.[53]

Despite these achievements, it must be emphasised that the above agreements are purely *voluntary* and so industry participation can be motivated by corporate public relations concerns rather than concerted greenhouse action. Some Greenhouse Challengers are more proactive participants than others. The report by George Wilkenfeld and Associates for the Greenhouse Challenge Office released in November 1997, *Evaluating the Greenhouse Challenge -Issues and Options*, found that only about 17% of projected reductions were directly linked to the Companies' involvement in the program. Wilkenfeld comments "The rest would most likely have taken place at the same time, irrespective [of the Greenhouse Challenge Program]."[54]

6. Discussion

Successive Australian Governments have met with considerable criticism with regard to climate change policy formulation and implementation. For example, in *Australia: State of the Environment 1996,* the State of the Environment Advisory Council commented "[T]o date, Australia has made only limited progress towards stabilising greenhouse gas emissions. . ." and "[F]urthermore, Australia has given little consideration to 'adaptive' responses to the enhanced greenhouse effect."[55]

There have been four major barriers to Australia's climate change policy formation and implementation. These are: a lack of political will to reduce greenhouse gas emissions, little serious implementation of ecologically sustainable development goals, weaknesses in policy development, and a failure to educate and engage the public. These barriers are further discussed in the following sections.

6.1 LACK OF POLITICAL WILL ON CLIMATE CHANGE

The Australian Government defends its lack of political will to act strongly on greenhouse on a number of grounds. It emphasises that Australia's economy depends very heavily on energy and greenhouse-intensive industry. Moreover, the government contends that the emissions from this sector are small, accounting for

52. *Id.*
53. *Id.*
54. G. Wilkenfeld, *Evaluating the Greenhouse Challenge - Issues and Options,* Report for the Greenhouse Challenge Office, George Wilkenfeld & Associates (1997).
55. State of the Environment Advisory Council, *Australia: State of the Environment 1996* (1996), at 30.

around 1.4 per cent of global emissions.[56] Finally, the Government argues that it is not interested in implementing response measures without commensurate actions by other major greenhouse gas producing countries.

Successive Australian Government's energy export policies are indicative of Australia's energy producer orientation. Both the previous Labor and the present Coalition Governments have very strongly encouraged exports of fossil fuels during the last decade. Under the Labor Government, in 1988, the Department of Primary Industries and Energy stated in a national policy paper, *Energy 2000*, that Australia needed "to continue to press overseas governments to restructure their industries and open up their markets to coal imports."[57] The current Coalition Government advocates the desirability of conventional energy exports even more strongly. For instance, with regard to the coal industry, the Liberal Party stated in its election platform that it wished to abolish all export controls on coal, and its policy would be "to enhance international competitiveness to maximise the returns to Australia."[58]

Also, the current Federal Minister for Resources and Energy, Senator Parer, has stated that there is enormous potential for the sale of Australian coal, natural gas and uranium given growing Asian demand.[59] He has further emphasised that it is very important for Australia to remain competitive against rivals in the long term with regard to exports of fossil fuel energy products.[60]

The Australian Government's unwillingness to reduce greenhouse emissions is also evinced in its financial cuts to energy efficiency and renewable energy programs. Since 1996, the Coalition Government has axed the Enterprise Energy Audit Programme, the National Energy Awards, and the Billion Trees Program as well as the Energy Research and Development Corporation and Renewable Energy Industry programs mentioned earlier. Also, it has annually cut A$3 million funding from the National Energy Efficiency Programme.[61] Some of this funding has been reinstated with the PM's Greenhouse Package, but certainly there has been inconsistency in policymaking in first cutting and then partially reinstating similar programs.

The Coalition Government also terminated the Development Import Finance Facility (DIFF) Scheme in 1996, which adversely affected Australia's renewable energy companies. The DIFF held a special position in assisting the

---

56. Department of Foreign Affairs and Trade, *supra* note 1, at 44.
57. Department of Primary Industries and Energy, *Energy 2000: A National Energy Policy Paper* (1988).
58. Liberal Party of Australia, *Coalition Resources Policy*, in *Federal Coalition Policies: Complete Resources for the 1996 Federal Election*. Internet, http://www.liberal.org.au/ (1996).
59. These comments were made before the 1998 Asian economic downturn.
60. D. McKenzie, *Japan seeks stronger energy links: Parer*, THE AUSTRALIAN, 14 October 1996, at 17.
61. P. Rogan, *Howard fails the leadership test on greenhouse*, Press Release, 27 June 1997.

export of Australia's renewable energy technology. The program was first introduced into Australia's aid portfolio in 1980, and accordingly, was a long-standing scheme.[62]

6.2 NO SERIOUS IMPLEMENTATION OF ECOLOGICALLY SUSTAINABLE DEVELOPMENT

The fact that successive Australian Governments have not seriously implemented ecologically sustainable development policy is another major factor. To date, Australia has produced only one comprehensive strategic national sustainability plan. This was the 1992 National Strategy for Ecologically Sustainable Development.[63] Implementation has been hampered due to the overemphasis on short term considerations of national economic development rather than longer term environmental imperatives.

To further economic well-being, the main efforts of successive Australia Governments have predominantly been focused on maintenance and expansion of export markets for energy, minerals and agricultural products. As a consequence, Australia's economy lacks structural diversity, although it is a developed country with highly trained human resources and potential for developing high technology and other industries.

Restructuring of Australia's energy industry to ensure ecologically sustainable development should be a fundamental priority of the federal government. Australia's energy industries should ideally be based on diversified sustainable energy systems. Although this type of reform is difficult, it will ultimately secure Australia's long term national economic interests and help to protect the global environment.

Furthermore, the Australian Government should increase assistance for research, development and commercialisation of the renewable energy industry beyond the initiatives in the *PM's Greenhouse Package.* Also, the Australian Government should develop overseas markets for renewable energy products and technologies through its aid programs and trade organisations. This could strengthen the renewable energy industry and help secure Australia's energy market for the long term. It is not wise for Australia to rely on fossil fuel exports when competing with other suppliers in the international energy markets. As it stands, Australia and other

62. Financing for DIFF projects came from two sources: one was provided by Australian Agency for International Development (AusAID); another was export credit financing by Export Finance and Insurance Corporation (EFIC). These combined funds were offered to recipient countries as a concessional loan. The objective of the DIFF was to promote sustainable socio-economic development in developing countries, as well as to promote Australian foreign policy and commercial interests. The scheme has encompassed a variety of sectors, including manufacturing, waste management, communications, transport, energy, mining, water supply, sanitation and rural electrification. *See* X. Yu, R Taplin & A. Gilmour, *Overseas Market Development: A Strategy For Australian Renewable Energy Industries,* 28 AUSTRALIAN GEOGRAPHER (1997), at 159-171.

63. Commonwealth of Australia, *supra* note 20.

nations continue to provide cheap fossil fuels, particularly coal to international markets. As a result of such efforts, the depletion of non-renewable resources continues, the progress of renewable energy and energy efficiency is retarded and global environmental pollution increases.

6.3 POOR POLICY IMPLEMENTATION

Reflecting on Australia's record on greenhouse issues, an important limiting factor has been policy implementation. Although Australia has some policies regarding energy efficiency, renewable energy and greenhouse gas emissions, these policies are largely restricted to policy documents or politician's statements. Hence, Australia has failed to develop renewable energy, promote energy efficiency and stabilise its greenhouse gas emissions. As David Bennett, Executive Director of the Australian Academy of the Humanities, pointed out with respect to climate change:

> As well as local and global responses Australia has the mechanisms in its courts, policies, and international obligations and it has options available at the local, national, and international levels. In each case it needs do no more than take advantage of existing options. Response is missing, not mechanisms and options.[64]

In the future, more concerted government efforts should be made towards policy implementation.

6.4 PUBLIC PARTICIPATION/COMMUNITY EDUCATION

Community engagement in climate change issues is not particularly strong in Australia. Climate change is viewed as a phenomenon that is difficult to understand by those that are not scientifically literate, and thus to be addressed in the domain of experts. It is often confused with the issue of ozone depletion by the Australian public. Only if community education programs on climate change are implemented will the public be adequately informed. In turn, public pressure for effective greenhouse policies could be the critical impetus for government action in the future.

64. D. Bennett, *Global Climate Change: An Environmental Viewpoint*, in THE CHALLENGE FOR AUSTRALIA ON GLOBAL CLIMATE CHANGE, National Academies Forum, 29-30 April 1997, Summary of Proceedings (1997), at 35, 43.

7. Conclusion

In this chapter, we have reviewed Australia's climate policy formation between 1995 and 1998. The focus has been to identify some major barriers to Australia's climate change policy formulation and implementation. Lack of political will, a failure to implement ecologically sustainable development policies, and a failure to galvanize the public thwart the effectiveness of Australia's efforts to address climate change. In the future, these barriers will need to be overcome or Australia will not be able to meet its international commitments.

6. THE AUSTRALIAN POSITION AT THE KYOTO CONFERENCE

X. YU AND R. TAPLIN
Graduate School of the Environment
Macquarie University
New South Wales, Sydney, Australia

1. Introduction

On 28 April 1997 on ABC Radio National, the Australian Prime Minister, John Howard, stated publicly that he believed that Australia should never have signed the United Nations Framework Convention on Climate Change (FCCC). This was the culmination of over a year of backpedaling by the Australian Liberal-National Party Government on the issue of climate change due to purported negative economic impacts. Climate change has been a difficult issue for Australia in international circles, in particular, because Australia has been viewed in the past as an exemplary international environmental citizen. From the 1970s onwards, Australia made significant contributions to protection of the global environment, including in the areas of nature conservation, whaling, ozone layer protection and conservation of Antarctica. Australia's current stance on climate change is a radical departure.

In the lead up to Kyoto, Australia consistently opposed the European Union's proposal to cut greenhouse gas emissions by 15% under the proposed Protocol. John Howard argued that "... the European Formula would cut Australian coal production by 20 per cent and aluminium smelting by 25 per cent."[1] He went on to suggest that "the [Australian] Government is not prepared to sign on to any agreement at Kyoto unless it is fair." He also threatened that Australia would "withdraw from any international agreement which is unfair to Australia."[2]

The Australian Government has been strongly criticised by many developed and developing countries alike. For example, the Cook Islands Prime Minister, Sir Geoffrey Henry, pointed out "that Australia's insistence on protecting its coal and energy intensive industries was self-serving."[3] During the South Pacific Forum Members leaders' meeting held in the Cook Islands in September 1997, the Cook Islands, Kiribati, Nauru, Tuvalu and Niue moved a resolution calling on developed countries to cut their greenhouse emission levels by 20 per cent below 1990 levels by 2005.[4] However, Mr. Howard strongly berated these Pacific Island governments for their alleged lack of

1. C. Skehan, *The Woman with a Global Mission*, THE SYDNEY MORNING HERALD, 23 September 1997, at 1.
2. C. Skehan, *Howard draws the line on emissions*, THE SYDNEY MORNING HERALD, 7 October 1997, at 7.
3. C. Skehan, *Greenhouse gas: Pacific nations warn PM*, THE SYDNEY MORNING HERALD, 16 September 1997, at 2.
4. *Id.*

A. Gillespie and W. C. G. Burns (eds.), Climate Change in the South Pacific: Impacts and Responses in Australia, New Zealand, and Small Island States, 113–119.

understanding of Australia's economic situation.

Australia's persistence in pushing for recognition of its purportedly unique circumstances concerning climate change was successful at Kyoto. While agreement was reached to cut emissions globally, Australia was permitted to *increase* output of greenhouse gases by 8% above 1990 levels by 2008-2012. Meg McDonald, the Australian Ambassador for the Environment who led Australia's delegation to Kyoto, summarised the outcome for Australia as "a major achievement that we . . . got much more significant cuts than anyone believed possible."[5]

The Australian Government's position for the Kyoto meeting is embodied in the document *Australia and Climate Change Negotiations: An Issues Paper* released in mid 1997.[6] However, it largely reflects an intense pursuit of economic self-interest and rejection of climate change as a global environmental issue. In the remainder of this chapter, issues related to the Australian position at the Kyoto conference are discussed.

2. Previous Conferences of the Parties (COPs)

Australia's position at COP1 (March-April 1995) was focused on two major issues. These were that strengthened commitments should not be limited to Annex I countries (developed nations and eastern European countries undergoing economic transition), and that any further commitments should be based on the principle of differentiation. In the latter context, Australia had contended during the negotiations over the formulation of the Berlin Mandate[7] that Annex I countries should have their differing economic structures taken into account and that equitable and appropriate contributions should be made according to the circumstances of different countries. Australia also argued that the negotiating process should include focussing on cutting emissions in developing countries because greenhouse emissions from developing countries, as a whole, exceed those of OECD countries and are projected to increase significantly in the next century.[8]

At COP2 (July 1996), Australia endorsed almost all aspects of the Ministerial Declaration, but disagreed with the statement that there should be a requirement for legally binding targets because "it committed Parties politically to legally binding targets without the nature and content of those targets being clear."[9] This was in opposition to the consensus on the Declaration forged by key participants in the negotiations, such as the United States, Japan and the European Union.

5. F. Gillies, *Career Path: Meg McDonald, Australian Ambassador for the Environment*, BUSINESS CLASS, March, 1998, at 16.
6. Commonwealth of Australia, *Australia and Climate Change Negotiations: An Issues Paper*, Department of Foreign Affairs and Trade, (1997), at xiii.
7. The Berlin Mandate was the Agreement arising from the First Conference of the Parties to the Framework Convention on Climate Change (COP1)
8. Commonwealth of Australia, *supra* note 6, at 140.
9. *Id.* at 33.

Another important Australian move at COP2 was to again raise the issue of differentiation among developed countries. Australia argued that any target to cut greenhouse gas emissions must reflect the individual circumstances that exist among Annex I countries, lest mitigation targets be unrealistic and unsustainable.[10]

3. Australia's Position at COP3

In *Australia and Climate Change Negotiations: An Issues Paper*, the Australian Government encapsulated its position for COP3 as:

> Australia's approach to the negotiations has been aimed at working towards the achievement of an international agreement that will contribute to achieving the ultimate objective of the Convention *whilst simultaneously ensuring that Australia's national interests are safeguarded.* This approach ensures that Australia's international responsibilities and objectives are consistent with the national interest.[11]

Australia's position at the COP3 addressed the issues of coverage; nature and level of the target; flexibility mechanisms; polices and measures; and developing countries' involvement.[12]

Coverage. Australia was in favour of a comprehensive approach that included all gases, sources and sinks. Australia believed that this approach would provide the Parties with the flexibility to maximise environmental benefit and minimise economic cost. Australia was concerned that if carbon dioxide emissions alone were targeted at Kyoto, the task for Australia would be more difficult and costly than for other nations. This is because Australia's energy emissions, and hence carbon dioxide emissions, have been projected to grow more rapidly than those of most other Annex I countries. The agreement to include six greenhouse gases[13] as emission sources and sinks for all sectors in the Protocol was accordingly to Australia's advantage. This is because Australia can make reductions in non-energy sectors, such as actions relating to land-use clearing, more easily than in the direct energy-use sectors.

Also, with respect to sinks, there was some debate in the initial stages of the meeting about the method of accounting for agriculture and forestry sector sinks. Australia was concerned that the accounting method ultimately adopted could result in less contribution of these sectors to reducing Australia's overall emission contributions. This point of contention was successfully resolved in Australia's favour.[14]

---

10. *Id.*
11. *Id.* at 149-150 (emphasis added).
12. *Id.* at 9.
13. Carbon dioxide, methane, nitrous oxide, hydrofluorocarbons, perfluorocarbons and sulphur hexaflouride.

Nature and level of the target, and policies and measures. Australia argued that its trade and competitive profile makes many specific greenhouse policies and measures inappropriate for Australia to adopt and implement. Australia thus argued for a flexible approach to policies and measures. Australia did not support any mandatory measure that could adversely impact competitiveness[15] and consistently opposed legally binding targets while the nature and content of commitments purportedly remained unclear. Australia proposed that targets should be different according to the circumstances of individual nations, with a focus on five economic indicators for the negotiations. The indicators were: projected population growth; GDP per capita growth; emission intensity of national economies; emission intensity of exports; and fossil fuel trade.[16] Michael Mugliston, Assistant Secretary, Australian Department of Foreign Affairs and Trade, said after Kyoto "[T]he acceptance of differentiation as a basis for the agreement was a major achievement for Australia . . .The Kyoto outcome is in line with key Australian objectives for the negotiations."[17]

Flexibility mechanisms, and developing country involvement. Emissions are growing rapidly in developing nations. Based on the assumption that more energy-intensive production processes may be adopted by developing nations in the future without any mitigation commitments from those states, Australia called for developing country involvement in the Kyoto Protocol. Accordingly, Australia supported the unsuccessful proposal of the United States, to establish a process to involve developing countries in meaningful efforts to reduce or limit emissions by a specified date,[18] and expressed its disappointment when this proposal was rejected.[19]

4. Discussion

As indicated above, the Australian position at the COP3 focused on three issues: legally binding targets; economic impacts; and developing country involvement. Further insights relating to Australia's interest in these issues follows.

---

14. E.W.R. Barlow, *Agriculture and forest industries beyond Kyoto*, in Greenhouse Beyond Kyoto: Issues, Opportunities and Challenges, conference organised by the Bureau of Resource Sciences, 31 March -1 April 1998.
15. Commonwealth of Australia, *supra* note 6, at 9.
16. *Id.*
17. M. Mugliston, *International climate change negotiations; implications of the Kyoto Protocol*, Greenhouse Beyond Kyoto conference, *supra* note 14.
18. Commonwealth of Australia, *supra* note 6, at 9.
19. Mugliston, *supra* note 17.

4.1 LEGALLY BINDING TARGETS

Australia argued that efforts to address global climate change need to be based on co-operation, not coercion.[20] In practice, however, legal instruments do appear to be necessary. Implementation of the FCCC to date has shown that in the absence of binding commitments little emission cutting occurs. The Convention was opened for signing in 1992 and came into force relatively quickly, yet greenhouse gas emissions from Australia (similarly to other Annex 1 countries) have not dropped. By 1995-96, for example, Australia's CO_2 emissions from the energy sector were 11 per cent above 1990 levels and are projected to rise by about 40 per cent above 1990 levels by 2010 without further measures.[21]

To date, Australia has made only limited progress towards stabilising greenhouse gas emissions. The response to this point has been based on voluntary agreements with industry and the introduction of 'no regrets' measures that have other benefits, or at least no net costs, apart from reducing emissions. The Australian State of the Environment Advisory Council (1996) has pointed out: "[I]n the absence of strong action by all sectors, Australia will not achieve even the more modest emission reduction foreshadowed in the Commonwealth's current policies and programs."[22]

4.2 ECONOMIC IMPACTS

The diplomatic team negotiating the Australian position for Kyoto relied on economic modelling research carried out by the Australian Bureau of Agricultural and Resource Economics (ABARE) to justify the need for Australia to be given a differentiated target. ABARE's models predict that meeting a target to return emissions to the 1990 level by 2010 would cost the Australian economy at least 0.9 to 1.5 per cent of GDP. In the worse case scenario, it would constitute an estimated 1.5-2.5 per cent of Gross National Expenditure.[23] Predicted impacts on the Australian economy included: a possible A$12 billion worth of investment going offshore; and A$68 billion of emission-intensive investments currently planned for the energy and energy-intensive sectors from 1997-2002 could be potentially rendered as non-viable.[24]

However, ABARE's modelling was strongly criticised in Australia. Hamilton has suggested that the models overestimate the costs and underestimate the benefits of reducing emissions.[25] He also has argued that a range of research shows that Aus-

20. Commonwealth of Australia, *supra* note 6, at 11.
21. Australian Bureau of Agricultural and Resource Economics, *Australian Energy Consumption and Production: Historical Trends and Projections 2009-10*, Australian Government Publication Service, http://www.abare.gov.au/ (1997).
22. State of the Environment Advisory Council, *Australia: State of the Environment 1996: Executive Summary*, Australian Government Publishing Service (1996), at 30.
23. Commonwealth of Australia, *supra* note 6, at 74.
24. Commonwealth of Australia, *supra* note 6, at 82-83.

tralia could reduce its emissions by 20-48 per cent below 1990 levels by 2010 without economic cost.[26] In June 1997, 131 Australian economists, including 16 professors of economics, also declared that Australia could reduce its emissions at no cost. They also warned that Australia was in danger of becoming locked into a fossil fuel-based economy just as the rest of the world is moving towards renewable energy options.[27]

These differing perspectives are not uncommon to economic modelling. This is because model results are dependent upon the selection of modelling approaches, assumptions, factors, methods of analysis and interpretation of the results. Moreover, many of the elements involved in the modelling process may involve political judgements.

4.3 DEVELOPING COUNTRY INVOLVEMENT

Engendering developing country involvement in emission reduction efforts was one of Australia's stronger stances at Kyoto. Australia argued that global carbon dioxide emissions will increase by 30-40 per cent under a moderate growth scenario and could be as high as 90 per cent above 1990 levels by 2020.[28] Much of this growth will come from developing countries, and in particular, Asian nations such as China and India.[29] However, the underlying issue for Australia, again, was national interest. If developing countries are not going to control the growth of their greenhouse gas emissions, the Australian Government is concerned that potential future investment in emission-intensive industries in Australia could go instead to developing countries, resulting in "carbon leakage."

However, Australia's assumption is not based on fact. Many developing countries have positively committed themselves towards reduction of greenhouse gas emissions. This is demonstrated in their renewable energy implementation initiatives. For example, in Asia, solar energy resources are being rapidly developed in Indonesia, the Philippines, and many other countries. In particular, the market for solar energy in India is growing at a rate of 30 per cent to 40 per cent per year. In 1996, more than 250,000 solar PV systems were installed for an aggregate capacity of 17 MW.[30] India was also expected to be the second-biggest wind energy producer in the world after the United States by 1997.[31] China has also made efforts to exploit its hydro, solar, wind, nuclear energy and improve its energy efficiency. For exam-

---

25. C. Hamilton, *Climate Change Policies in Australia: Background Paper*. The Australia Institute (1997), at 6.
26. *Id.* at 4.
27. M. Warby, P. Hartley & C. Hamilton, *Kyoto countdown: Monday Viewpoint*, THE SYDNEY MORNING HERALD, 30 June 1997, at 15; P. Cleary, *Our greenhouse stand may lead to trade bans*, THE SYDNEY MORNING HERALD, 25 June 1997, at 4.
28. Commonwealth of Australia, *supra* note 6, at 140.
29. *Id.* at 141.
30. J. Zubrzycki, *Solar power helps light Indian Dream*, THE AUSTRALIAN, 23 April 1996, at 62.
31. *Id.*

ple, a total of 48,284 hydro power stations have been built in China and 89 per cent of its rural population has electricity supplied by its rural hydro-electrification programmes.[32] Also, it is interesting to note that most Pacific Island states are also making efforts to develop renewable energy sources. This is despite the fact that these nations consume an extremely small amount of energy and produce a minuscule proportion of global greenhouse gas emissions.[33]

5. Conclusion

COP3 was a key stage in terms of implementation of the FCCC. Australia's negotiating stance was largely driven by overemphasised national economic interests. Paul Barratt, Australian Department of Defence Secretary, stated after Kyoto:

> Nobody else gives a damn about Australia's economic interests - if we don't look after ourselves, no-one will . . . for decades our closest friends have been reducing the economic opportunities of Australians . . .The same willingness to sacrifice our economic interests has been evident throughout the climate change negotiations.[34]

It is not surprising that this unvarnished emphasis on economic considerations to the almost total exclusion of environmental concerns has opened Australia to criticism. This is especially so because the impacts of emission cutting on Australia's economy remain a source of divergent opinion. Moreover, it appears that promotion of sustainable energy solutions to the greenhouse problem could actually benefit Australia's economy.

32. X. Yu & R. Taplin, *Policy Perspectives: Environmental Management And Renewable Energy In The Pacific Islands*, 51 J. ENVTL. MGMT. (1997), at 107-122.

33. X. R. Taplin & A. Gilmour, *Climate Convention Implementation: An Opportunity for Pacific Island Nations to Move Towards Sustainable Energy Systems*, 21 ENVTL. MGMT. (1997), at 493.

34. P. Barratt, *Implications of the Kyoto Protocol for Australian Government and Policymakers*, Greenhouse Beyond Kyoto conference, *supra* note 14.

7. THE IMPACTS OF CLIMATE CHANGE ON NEW ZEALAND

REID E. BASHER
National Institute of Water and Air
PO Box 14901
Kilburnie, Wellington, New Zealand

1. Climate and Culture

On Saturday, September 11, 1880, Wellington's *Evening Post* reported that a terrible railway accident had occurred at 11am that morning in the Rimutaka Ranges 80 kilometres north of the town, with at least three people killed and numerous others injured. But this was no ordinary railway accident; the train was blown off the rails by a powerful gust of wind, causing the wooden carriages to cascade down and smash into the ravine 30m below. Was the severe north-westerly gale experienced that day normal for the area, a notoriously windy gully called "Siberia," or was it something exceptional, or even some indication of a changing climate?

In 1880, these questions may not have been contemplated by many. But if they had been, they would not have been asked, or answered, with the insight and concern that we have 118 years later, in a world much richer in climatic data and knowledge and on the threshold of global climatic changes arising from the activities of humans. Yet, this small accident in the tiny, remote British colony of New Zealand so long ago contains some very instructive lessons for us today.

Firstly, the issue of climate *change* is predicated on the fact that the existing climate is a key determinant of the environment and human society, and always has had significant impacts. Secondly, climatic impacts are delivered in the form of day to day weather, usually in the form of extremes in weather, such as gales, floods, frosts, etc. Thirdly, many decades of historical data are required to know whether a climatic event is unusual or whether the climate is changing. Thus to deal with the impacts of *climate change*, we must always return to the building blocks of *existing climate*, *climatic variability*, *weather*, and *climate data*.

The accident also provides a way of illustrating some cultural views relevant to New Zealand's perceptions of the climate change issue. The indigenous people and immigrants of New Zealand all came from somewhere else, many thousands of kilometres away, and their oral and written histories contain many stories of the heartbreak, privation, injuries and death arising from their uprooting and arduous voyages to the new lands. Once here, they faced the rigours of an untamed and often untameable land, their endeavours in farming, commerce and family life and their lines of supply, at the mercy of the fickle weather and seas.

For example, on the windy day of the 1880 train accident, the newspaper also reported

A. Gillespie and W. C. G. Burns (eds.), Climate Change in the South Pacific: Impacts and Responses in Australia, New Zealand, and Small Island States, 121–142.

that the barque Waimea from London was unable to enter Wellington Harbour because of the "stiff NW wind blowing hard." Further down the page, in an item "Indications of Probable Weather", it was stated that warnings of strong winds had been telegraphed to places along the east coast.

It is often said that there is "a lot of weather" in New Zealand, lively weather with fast-changing mixtures of piercingly blue, still, sunny skies, of cloud and heavy rain, and blustery winds both warm and cold, sometimes all in one day, and possibly at any time of the year. To someone from the continental land masses of the Northern Hemisphere, this high daily variability and lack of great seasonal variation is a surprise, especially when a summer visit is punctuated by what seems like winter weather, and vice versa, but to the New Zealander it is just the natural result of living in the expanses of the Southern Hemisphere's weather-filled oceans. The mid-latitude westerly winds pulse and eddy, like swirls in a teacup, streaming around and over the mountain chains of this pair of mountainous islands, only temporarily interrupted by the settled weather of a slow-moving anticyclone.

In such a climate, few houses have central heating, or double glazed windows, or air conditioners, and most people spent a good deal of time outside, in gardening, walking, team sports, sailing, fishing, and other forms of recreation. New Zealanders live in their climate and know its variability and the often unpredictable nature of the weather. To some extent this closeness to the elements is also true of Australia and the South Pacific islands. So it might be said that if the climate were to vary a little, especially if it were to become a little warmer, would it matter much? Would it even be noticeable? Perhaps not, yet when an unpleasant summer or winter occurs it is usually no more than a degree Celsius lower than average; of more importance it seems is the combination of changes in wind, cloud and rain coupled with temperature.

Familiarity with the climate does not necessarily confer understanding, of course. Most people know that the El Niño phenomenon affects New Zealand's climate, but many confuse this with climate change, unwittingly interpreting the greater reporting of El Niños as evidence of a changing climate. Global climate change models at present cannot provide reliable predictions for El Niño under enhanced greenhouse gas concentrations, and in any case to date suggest relatively small changes in El Niño characteristics.

Unfortunately, the situation is complicated by two other factors; the increased frequency of El Niños over the past decade or two, and the fact that when an El Niño event occurs, it tends to temporally warm the global climate. The recent annual statements by the World Meteorological Organisation on record global mean temperatures and climate change usually comment on the role that the El Niño has had in boosting the year's temperature.

One salutary feature of living in a small, remote country is the recognition that most things happen somewhere else. A New Yorker might be forgiven for thinking nothing much of importance happens west of the Hudson River, or perhaps west of the Appalachians, but a New Zealander always knows that he or she must look overseas, and widely so, especially to Europe, North America and Asia, to know what is really going on in the world. Reminiscent of their ancestors, young New Zealanders leave in droves to undertake their OE (overseas experience). New Zealanders thus have a reasonably wide global perspective, and, familiar with the major environmental issues of the day, will readily support the appropriate "clean" side of the debate.

This is not difficult, of course, for an isolated maritime country which has ample renewable energy, relatively few polluting industries, and generally clean, clear skies and waterways. Products are consumed, but most of the more polluting industries that created them lie conveniently far away in the countries of Northern Hemisphere. The problem of excessive industrial production of atmospheric carbon dioxide is mainly a problem for the distant large countries of the north, with apparently negligible contribution from New Zealand. This may present the paradox of a clean low-population environment, with its inhabitants predisposed to environmental smugness, but with very little motivation to take environmental action.

2. Programmes, Politics, Personalities

After a decade or more of serious effort on the climate change problem, we are now in the position of having established through both national and international processes, an array of scientific assessments that set out the known facts and numerous uncertainties about the likely impact of enhanced greenhouse climate change, The three thick volumes of the Intergovernmental Panel on Climate Change (IPCC) Second Assessment Report[1] and its summary document[2] collate and summarise a vast amount of scientific knowledge relevant to the issue. The task involved several years of effort by thousands of scientists, many teams of lead authors, hundreds of reviewers and several stages of review by governments and experts.

1. IPCC, *Climate Change 1995: The Science of Climate Change. Contribution of Working Group I to the Second Assessment Report of the Intergovernmental Panel on Climate Change* (1996); *Climate Change 1995: Impacts, Adaptations, and Mitigation of Climate Change: Scientific-Technical Analyses. Contribution of Working Group II to the Second Assessment Report of the Intergovernmental Panel on Climate Change* (1996); *Climate Change 1995: Economic and Social Dimensions of Climate Change. Contribution of Working Group III to the Second Assessment Report of the Intergovernmental Panel on Climate Change* (1996).
2. IPCC, *Climate Change 1995: IPCC Second Assessment.* (Contains the *IPCC Second Assessment Synthesis of Scientific-Technical Information Relevant to Interpreting Article 2 of the UN Framework Convention on Climate Change)* (1996), together with the Summary for Policymakers of the three Working Group Reports, *supra* note 1.

The IPCC's role is to assess the available research - not create it. There are many gaps in our knowledge, and inevitably there are some grounds for criticism of both the process and the conclusions. Nevertheless, given the complexity and arduousness of the effort involved, it is unlikely that the process, and hence the conclusions, could have been greatly improved upon.

The now famous and compelling one-line conclusion of the Second Assessment was that "the balance of evidence suggests that there is a discernible human influence on global climate."[3] It is this final summary statement which encapsulates the concern of governments worldwide, New Zealand's included. It has also provided the impetus in the international political arena to limit global greenhouse gas emissions via the developing protocols of the Framework Convention on Climate Change (FCCC).

It may be asked how did New Zealand reach this stage in 1998 where the available science is well documented and its official political stance is firmly supportive of the FCCC? Many a book could be written about this, especially by those who were intimately involved. However, here the author will concentrate on a few key organisations, Propgrammes and personalities, and their roles of advancing and sometimes hindering this progress over the last decade.

Growing concern about climate change in the mid 1980s, and international action by the World Meteorological Organisation and the United Nations Environmental Programme, led senior government scientists, in particular Trevor Hatherton, President of the Royal Society of New Zealand and Director of the Department of Scientific and Industrial Research (DSIR)'s Geophysics Division, and John Hickman, Director of the New Zealand Meteorological Service, to consider what national action was required in New Zealand. Subsequently, in 1988, the New Zealand Climate Change Programme was convened by the Secretary for the Environment, Roger Blakeley.[4] This was a landmark step forward, with its IPCC-like assessment process of voluntary contributions of scientists coordinated through the work of three perfunctorily named working groups on Facts, Impacts and Policy. The Ministry for the Environment also initiated the Priority Research Contract Scheme, in which a set of specific research contracts concerning key issues of climate change was commissioned.

Behind the scenes, there was considerable jockeying for position, especially between those who saw the problem as solely a scientific one for scientists to deal with, and those who had a more applied and holistic view. The Royal Society took responsibility for the so-called Facts Working Group, under the convenorship of John Hickman, and organised a Workshop under the umbrella of the scientist-led International Geosphere Biosphere Propgramme (IGBP) to address the more basic underlying scientific questions concerning global change.[5] The Impacts Working

---

3. *Id.*
4. *See Climate Change in New Zealand*, Misc. Series 18, Royal Society of New Zealand (1988).
5. *Global Change in New Zealand*, Proceedings of conference 14-15 June 1990, Bulletin 29, Royal Society of New Zealand (1990).

Group was convened by Dr. Paul Mosley of the DSIR's Water Sciences Division, while the working group on policy implementation was convened by Judy Lawrence of the Ministry for the Environment (MfE).

The focus of this chapter will be on the work of the 36-member Impacts Working Group. The working group developed a network of over 300 individual contributors throughout the country and reported back in April 1990, in a summary document and comprehensive supporting report. The latter report provided a wide-ranging and detailed examination of what was known about the climate sensitivities and potential climate change impacts in New Zealand's natural and managed ecosystems, its industries and society, including indigenous Maori and its Pacific territory of Tokelau. A separate review of climate change information concerning the Pacific islands was conducted under the Priority Research Contract Scheme.[6] The Impacts Working Group report is an impressive collation and review, and as an information resource is unlikely to be bettered for a long time to come.

One feature of the report that is both a strength and a weakness is its extensive use of the regional scenarios of climate change prepared for the project. Based on an appraisal of climate model predictions and paleoclimatic evidence, these provided the principal starting point for analysing possible impacts. However, while the report's writers were generally careful to reflect the great uncertainties involved in scenarios by means of conditional language (*would* not *will*), the elaborate detail of the scenarios and of the many impacts based on them tends to convey an undue sense of confidence in the estimated impacts. It is all too easy to start treating scenarios as firm predictions.

The 1995 IPCC Second Assessment specifically addressed this issue by stating that while climate models provide useful predictions at global and continental scales, little confidence can be placed in predictions at the regional scale. A good example of the difficulty of predictions is the question of whether the predominant westerly winds across New Zealand will increase or decrease. This is important because these winds impinge on the mountain chains and control the marked rainfall - rainshadow pattern of the country. The outlook on this point is not yet clear. In the 1990 scenarios, the westerlies were expected to decrease, thus reducing the rainshadow effect, but in the recent IPCC regional impacts report[7] it is reported that new Australian research indicates possible increases in westerlies, depending on how the climate model deals with heat absorption by the Southern Ocean.

Irrespective of their detailed correctness, scenarios remain as very important tools to extend thinking and assessment, and to provide plausible indications of the possible size, range and complexity of climate changes. Furthermore, there is high

6. R.E. Basher et al., *Basic Studies For South Pacific Climate Change: Present Climate and Its Impacts, Data Resources and Scenario Possibilities*, Report to the New Zealand Climate Change Programme, New Zealand Meteorological Service (1990). This was revised and re-issued in 1992 as *Preliminary Studies for South Pacific Climate Change*.

7. R E. Basher et al., *Australasia*, in *The Regional Impacts of Climate Change*, Special report of IPCC Working Group II, (R.T. Watson, M.C. Zinyowera & R.H. Moss eds., 1997), at 105-148.

confidence in some components of scenarios, such as rising global temperature and rising global sea level. No one is predicting that rising greenhouse gas concentrations will cause these to reverse or fall.

The Labour Governments of 1984 - 1990 were governments of action and reforming zeal, with a strong focus on the machinery of government and economic rationalisation. The initiatives on climate change by the hard working and highly respected Minister for the Environment Geoffrey Palmer (later Sir Geoffrey) were just a small part of the environmental agenda, which included major reforms in the nation's environmental management.

Mr. Palmer's vision and drive in the creation of the Resource Management Act (RMA) are legendary, especially among the staff of his small and often stressed Ministry, which was responsible for the development of the Act. As an academic constitutional lawyer, Palmer wanted a sound scientific basis for the development of New Zealand's policy response on climate change, and as a politician he wanted it quickly. Mr. Palmer was well placed as Deputy Prime Minister, and then as Prime Minister after David Lange's resignation, to ensure that his objectives on climate change policy were pursued. In the foreword of the impacts summary he stated:

> Climate change is the most complex environmental crisis the world has ever faced. The causes are rooted deep in human industrial and agricultural development, and the results may affect every aspect of life as we know it. The international community is working fast towards agreements on how to control the greenhouse effect. At the same time, changes already under way pose various threats to our environment, economy and society. . . By examining the picture that emerges [from the impacts reports], we can try to identify potential threats and opportunities, in order to make the best of the situation . . . Many choices lie ahead of us, for which we all share some responsibility.[8]

While the National government that followed in 1990 continued to address the climate change issue, there were some differences in personality and approach. For most of the 1990s, the Minister for the Environment has been Simon Upton, a very able intellectual who has demonstrated an excellent grip on the climate change issue, and who has been able to bridge the gulf between the scientific and technical arena on the one hand, and the societal and political concerns of the community and the Cabinet on the other. In public speeches he can equally well chide the head-in-the-sand attitudes prevalent in industry and the more rabid pronouncements of some environmental organisations. He has been a strong supporter of scientific research and of the IPCC, recognising the critical need for a solid and expanding international scientific base for policy formation on climate change.

8. Ministry for the Environment, *Climate Change, A Review of Impacts on New Zealand*, New Zealand Climate Change Programme, Ministry for the Environment (1990).

It was Mr. Upton who in 1992 restructured New Zealand's government science departments along market lines. This was effectuated through creation of an independent "purchaser" agency, the Foundation for Research Science and Technology (FRST), and competing "provider" agencies, the ten or so corporatised but government-owned Crown Research Institutes (CRIs). By 1998, the pool of competition now extends to universities and industry research providers. The primary aim of the reform was to make science providers more responsive to society's changing needs, as determined by independent policy-setting and purchasing processes. Though not without shortcomings or critics, the science reforms have largely achieved their main objectives, and this success has helped Upton in Cabinet to maintain the science funding vote delivered to FRST.[9]

There are some ironies of the new "market" model for science. It would be very difficult now to speedily mount a major assessment like the previous 1988-1990 New Zealand Climate Change Propgramme without additional allocation of substantial public funds by Cabinet, since most of the nation's scientific funding and resources are locked up in tight multi-year research contracts with the FRST. The 1996 "Green Package," managed by MfE, demonstrated that it is not impossible to obtain new allocations, but the general rule is that Government departments requiring specialised science assessments and advice on topics like climate change must purchase them using their own budgetary resources.

This presents a hurdle for many departments, with the result that their use of external expert science inputs remains small, especially compared to PGSF science. For example, during the 1997/98 El Niño drought, it appears that government departments purchased barely a few thousand dollars worth of expert climatic assessment and prediction information, despite the fact that the impact on the economy has been estimated at around a billion dollars. Whatever advice and information was available to the community was mostly provided voluntarily by researchers.

A further case in point is the low level of funding to support New Zealand's participation in IPCC scientific work. The Ministry of Research Science and Technology (MoRST) funds the travel expenses of a few scientists to participate in IPCC meetings, but this is a minor cost compared to the substantial cost of the labour involved in the scientific assessment work, which is currently almost totally funded by the CRIs and universities concerned, or as free overtime by the scientists.

Thus, it is only through the *failure* of the market model that New Zealand secures its advice on the El Niño and its national inputs to the IPCC. The demand for IPCC inputs has increased with the commencement of the IPCC Third Assessment Report process (1998-2001) and the 1997 election of a New Zealand delegate (NIWA's Martin Manning) to the IPCC's standing committee, the IPCC Bureau. Where scientific information is a critical input to the nation's policy and decision

9. This may have encouraged Mr. Upton in his role as Minister of Health at the time to initiate similar but ultimately much less successful reforms in the public heath sector.

making, such as with climate change, the means of providing the information has to be more effective and formal than these ad hoc voluntary methods.

Another initiative of Mr. Upton was the establishment of the National Science Strategy Committee on Climate Change (NSSCCC), one of several similar committees set up to coordinate cross-cutting priority science themes following the 1992 science reforms. Given the structural fragmentation of entities in the science market—the departments, FRST, competing crown research institutes and universities, and private sector groups—it is clear that the NSSCCC has a critical role to play.

The NSSCCC Terms of Reference[10] charge it with developing a comprehensive strategy for climate change research, with emphasis on identifying research gaps, coordination of research efforts, identification of appropriate levels of funding and provision of advice to the Minister of Research Science and Technology (originally Simon Upton, later Maurice Williamson). The committee has 11 members, mainly scientists working on a voluntary basis, from CRIs, universities, government and the private sector, and is convened by environmental policy expert Judy Lawrence.

Annual reports[11] are a key output of the committee, providing impressively detailed assessments and recommendations on how well the current research efforts and annual science funding decisions match the recommended strategy. The committee also actively promotes linkages between science providers and users and it provides advice directly to the FRST, as well as to the Minister. Essentially, the committee has a departmental advisory role, yet once again, we see that a necessary and important component of the "science market" relies largely on voluntary inputs, and from a rather small number of committee members and their colleagues. Its small 1997-98 budget of $38,000 is mainly used to publish workshop reports and the annual report. Given the size of the climate change research effort that the committee strongly influences, and the inherent limitations of voluntary advice, it is clear that the resources and mechanisms of this committee need to be strengthened.

Fortunately there is a mood of change in science circles toward making science research more relevant and useful, with less zealous application of the "market" model. This follows five years during which the FRST emphasised research of a more fundamental nature (e.g., work publishable in international journals) and avoided work that appeared to be of direct benefit to specific groups in the community or that generated useful services. The major turnaround occurred in the 1997 bidding round, where a proposal's "relevance" became the key factor, this being assessed from its connectedness to the potential user groups and its likely contribution to the major outcomes sought by the Government, such as improved interna-

10. This is reprinted in the Annex of this chapter.

11. *See* for example, NSSCC, *Report of the National Science Strategy Committee for Climate Change 1995/96*, Royal Society of New Zealand (1996).

tional competitiveness and the protection of New Zealand's environment.

We have now reached the point where the contributions of climate researchers to IPCC work and to departmental policy formation are highly regarded as an indication of the real relevance of the science, rather than a costly drain on the resources of research projects or institute overheads.

It is unlikely that the National Cabinet fully shared Mr. Upton's views on climate change—the sources of National Party power in business and farming are not renowned for their identification with urban green concerns. In particular, Jim Bolger, the Prime Minister until replaced by Jenny Shipley in 1997 and originally a "back blocks" farmer, appeared to have little understanding of the nexus of environmental protection and sustainable land use and sustainable industry. For example, in a major speech on foreign policy on 13 March 1996 to the New Zealand Institute of International Affairs, Mr. Bolger made no mention of the actual or potential role of environmental issues in foreign policy, much less any mention of the climate change problem. When queried on this by the author, he indicated by the nature of his reply that he saw the demands of environmentalists as a potential threat to New Zealand's fragile economic recovery, and in particular he stated that he was not going to follow the example of other countries and make environmental promises he (or New Zealand) could not keep.

Environmental policy also does not appear to be a major interest of Mrs. Shipley, who is also a farmer by background. Nevertheless, continued support of the IPCC was explicitly noted in her budget speech in early 1998, and to date the Government's policies on climate change remain unaltered.

Ironically, part of the resistance by Mr. Bolger to making climate change commitments possibly may be laid at the door of Greenpeace. During the run up to the 1992 election, Greenpeace campaigner Kirsty Hamilton hounded the Prime Minister and other politicians throughout the country to agree to a commitment to limit greenhouse gas emissions. Ms Hamilton is an articulate young Scot, whose lilting but penetrating, persistent voice is the equal of any meeting hall politician. At one meeting along the trail, certain words must have escaped the lips of a tired Mr. Bolger and within an hour Greenpeace staff had let the world news channels know that New Zealand was going to limit its carbon dioxide levels to 20% below 1990 levels by the year 2000. Alas for Mr. Bolger, this was to prove totally unachievable by his subsequently elected government. His lack of judgement in making the promise probably still rankled him.

Worldwide, Greenpeace deserves considerable credit for raising the temperature of the climate change debate, increasing public awareness of its importance, and promoting necessary investment in research into the issue. On occasion, however, it has discredited itself and the climate change science community by taking liberties with the science involved. A memorable example was the ludicrous project that Greenpeace New Zealand undertook in the mid 1990s to seek from its members and from the wider public reports of evidence of climate change, such as changes in behaviour of garden growth, animal behaviour etc.[12]

Of more concern, however, are the group of scientists and lobbyists who contend that the IPCC conclusions were, at best, in error, and at worst, the result of a conspiracy among scientists hungry for research funds. These included a few local scientists who were familiar with the writings of similarly inclined individuals and lobbies overseas.

The realm of science is always redolent with tension between the end goal of solid reproducible scientific truths, and the necessary emotional commitment of individual scientists to the process of discovering and proving these truths. Progressively, the truths are established and the emotional content fades away. But in a complex matter like climate change, there are always questions that cannot be answered irrefutably, such as how the climate varied a million years ago, or how cloud patterns interact with the temperature of the atmosphere. Hence, there is always room for different positions that involve at least some degree of emotional content.

The anti-IPCC lobby has found a ready outlet for its views in New Zealand newspapers, which are always happy to fuel debate, irrespective of the factual foundations involved, and to allow a voice to the 'Davids' of the climate change issue against the 'Goliaths' of the science and environmental establishments. The sceptics' position has been imbued with a further degree of credibility by the support of the influential New Zealand Business Roundtable, a lobby group of chief executives of major corporations. The Roundtable has brought prominent foreign critics Patrick Michaels and Richard Lindzen to New Zealand to give public presentations and meet with senior members of the political and business communities.

There are, of course, grave concerns among large energy generating and consuming companies about the business impacts of any restrictions or additional costs on their operations. Moreover, it is fully legitimate, and desirable, for them to closely question the basis and implications of the science of climate change. Where able scientists such as Dr. Lindzen have criticisms, these deserve to be examined closely. To its credit, the NSSCCC hosted a public meeting at the Royal Society to hear Dr. Lindzen and to debate his views.

However, much of the criticism voiced locally has been a mixture of selected loose ends of research that contradict the main body of research, or as-yet unresolved scientific issues, together with innuendo about the motivations and judgements of the masses of scientists contributing to the PCC process. The arguments often extend to the ozone depletion issue, and there is usually a strong thread of denial, with prominent use of emotive language and words such as "myth," "dishonest,"

---

12. This project suffered from at least three glaring errors, first the absence of a valid survey technique, second the notorious fallibility of human perception of long term change in their surroundings, and third, the assumption that any observed changes were the result of human-induced climate change. The idea of actually analysing climate measurements did not seem to be part of the "study." Predictably, the anxious in the community reported all sorts of unhappy trends and these were reproduced in Greenpeace newsletters for its members to read and further worry about.

and "alarmist," replete with attacks on individuals and their motives.

Extraordinarily, such articles have even penetrated into serious journals such as the *New Zealand International Review*, the quarterly journal of the New Zealand Institute of International Affairs.[13] The publication of the original material in the Review reveals a surprising lack of awareness of international environmental issues among some in the foreign affairs community. More importantly, it also provided a salutary lesson to the climate science community on the continuing need for it to communicate its specialist knowledge and to obtain the understanding and support of decision-making peers in other spheres.

With such a barrage of conflicting information, ranging from some industry positions that climate change is simply a myth, to government statements that it is likely and that coordinated international action is essential, and to populist assertions that your dog and your garden vegetables are already affected, it is hardly surprising that the public and the business community remain confused and uncertain.

To help clarify the situation, in 1991, a report entitled *Climate Change, The Consensus and the Debate*[14] was commissioned by Judy Lawrence, then Manager of Global Issues at the Ministry for the Environment. Prepared under the leadership of NIWA's indefatigable David Wratt, this report still provides a generally valid perspective on the issues, seven years later, albeit with inevitable corrections and downward revisions of some of the data it uses. However, given the importance of a wide consensus of public and business understanding and support for major government's policies, it is clear there now needs to be a better, more systematic provision of information and advice to the community on climate change. High quality scientific research by itself is not enough.

3. IPCC, Impacts, Interpretations

Although the 1995 IPCC Second Assessment Report is the definitive source of information on impacts, it is an extremely difficult document to use if one wishes to obtain an overview for an individual country such as New Zealand. Hundreds of pages of text in numerous sector chapters have to be scoured to identify the range and importance of likely impacts, and more often than not there is no information specific to New Zealand.

13. P. Toynbee, *The Responsibilities Of The Mass Media In The Ecological Revolution*, 18(3) N.Z. INT'L REV. (1993), at 25-27; P. Toynbee, *Ozone Depletion Myths*, 18(4) N.Z. INT'L REV. (1993), at 16-19; R.E. Basher, *Who to Believe on Global Environmental Issues?*, 18(5) N.Z. INT'L REV. (1993), at 21-22; T. Clarkson, *Ozone Depletion: the Science Versus the Myths*, 18(5) N.Z. INT'L REV. (1993), at 2-4; J. Laird, *Mass Media and the Ecological Revolution Revisited*, 18(5) N.Z. INT'L REV. (1993), at 23-25. The correspondence of the following issue of the review is also of particular note.

14. D.S. Wratt et al., *Climate Change: The Consensus and the Debate*, New Zealand Climate Change Programme, Ministry for the Environment (1991).

In response to such criticisms from many countries, in late 1996 the IPCC initiated steps to prepare regionally-specific summaries of information on impacts and vulnerabilities. A division of the world into ten geopolitical regions was agreed. The resulting Special Report on the Regional Impacts of Climate Change (SR-RICC)[15] is largely based on the Second Assessment Report but it also contains later results and additional local studies that meet normal IPCC criteria for inclusion.

The regional impacts report was simultaneously reviewed by governments and experts in a two-stage review process, and was completed under a very tight schedule for release in time for the Third Conference of the Parties (COP-3) to the FCCC in Kyoto. Of particular note was the almost total lack of criticism during the review stages from business interests or from the usual critics of the IPCC. It seems that the focus of their criticism is climate system science and climate predictions, and not the likely impacts.

The Australasia chapter of the report[16] covers Australia, New Zealand, and their outlying islands. As a starting point for the consideration of impacts, the chapter contains a detailed but concise summary of the region's climate, current climate trends and climate scenarios. To encourage wider dissemination, the Australian Greenhouse Office is publishing a reduced version of the full report containing just the Australasia chapter along with the introductory and summary chapters and essential annexes.[17] In what follows, the points most relevant to New Zealand are noted and discussed. Interestingly, many of the basic ideas, along with a considerable amount of background detail, may be found in the earlier 1990 Impacts Working Group Report.[18]

When considering the assessment of the impacts, it is important to not over interpret the results of impact studies. The climate change predictions and scenarios upon which they are based evolve as scientific knowledge develops, and many aspects are marked by high levels of uncertainty.

A useful stance is to consider the predictions of climate changes in a hierarchy of uncertainty, with rising uncertainty as one moves down the list:

- Rising levels of carbon dioxide, a virtual certainty to continue in the future;
- Global air temperatures and sea levels, which are already generally rising and which most scientists believe will continue to rise;

---

15. *The Regional Impacts of Climate Change*, Special Report of IPCC Working Group II (R.T. Watson, M.C Zinyowera & R.H. Moss eds., 1998).
16. *See* note 6.
17. AGO, *Australasian Impacts of Climate Change: An Assessment of Vulnerability*, Extracts from Watson et. al., 1997, the IPCC Special Report on the Regional Impacts of Climate Change, Australian Greenhouse Office, (in press).
18. Ministry for the Environment, *Climate Change: Impacts on New Zealand*, Report of the Impacts Working Group, New Zealand Climate Change Programme, Ministry for the Environment (M. P. Mosley ed. 1990).

- Changes in broad weather patterns that are likely to occur but specific details are not yet defined;
- Increases in high rainfalls and drought, which many climate models indicate, are likely to occur, and changes in tropical cyclones and El Niño events are possible (but there is not clear evidence to date).

One should also not forget the possibility of surprises, of totally unforeseen climatic effects. The ozone hole and the dramatic fires in Southeast Asia during the 1997/98 El Niño, both of which are non-linear environmental responses to human influence and were not predicted in advance, serve as powerful reminders of the limits of our knowledge and predictive capabilities.

Thus, some features of climate change and associated impacts may be regarded with relatively high confidence while other features are very uncertain. Where the size and direction of a predicted effect is relatively uncertain, our knowledge will be confined to "sensitivities," where we only know that a system will change as the climate changes. Also, very often the available impacts knowledge is generic to the globe, rather than specific to New Zealand.

While the Australasia chapter of the special report is naturally dominated by the impacts on Australia, it still provides the best, most-current summary of impacts for New Zealand, and the situation of New Zealand is helpfully illuminated by the contrasts and similarities with the situation of its much larger and more populous continental neighbour. Geographically, New Zealand is very different, being cooler, more mountainous, and relatively well watered, but it shares some important features such as the uniqueness of flora and fauna, and the impacts of human influences, such as from agriculture, deforestation, introduced exotic plants and animals, and urban development.

Changes at fundamental biological levels are a highly likely result of changes in climate and carbon dioxide concentration. These includes changes in characteristics such as soil composition, water and nutrient cycling, plant productivity, and species interactions (competition, predation, parasitism, etc.) The exact detail of these basic changes are complex and rather uncertain, especially when taken in concert, and the report's authors had some difficulty in presenting a suitably clear picture of them. Such changes are important to agriculture and other managed land-use systems as well as to natural ecological systems.

Potential ecological problems highlighted in the report included the possibility of reduced nutritional status of plants, and of increased risks of fire occurrence and insect infestations. In aquatic systems, the predicted changes in evaporation, rainfall and rainfall intensity are expected to cause varied changes in runoff, riverflow and associated transport of nutrients, wastes, and sediments. Physical and ecological changes are inevitable in low-lying coastal areas affected by sea-level rise.

As is described in the Appendix 4 of this book, a variety of changes in ecosystems are therefore likely, but the complexity of ecosystem dynamics means that the net effects are difficult to predict. It is likely that not all species will be able to adapt; thus, a reduction of species diversity is highly likely. Those species, which are intrinsically less adaptable or are already under other ecological stresses, appear to be at greatest risk. Climate change can be viewed as another stress factor, adding to existing problems such as land degradation, weed infestations, and pest animals. Two key factors identified as likely to cause increased vulnerability are the relatively rapid rate of imposed climate change and the highly altered and fragmented landscapes in which the adaptations must take place.

Against this gloomy outlook for ecosystem vulnerability, it should be noted that New Zealand's lake ecosystems are very similar in composition and function despite the latitudinal range of 5°C in mean temperature, which suggests an insensitivity to projected temperature rises. New Zealand river ecosystems contain relatively simple generalist species with little diversity and have a resilience to existing hydrological variability, and perhaps to the predicted increases in rainfall variability.

Significant hydrological changes have been predicted. Paradoxically, a warmer world is predicted to have more frequent droughts and more frequent floods. In addition, if the prevailing westerly winds over New Zealand were to increase, as is indicated by the most recent research, the east coast rainshadow regions would experience an increased frequency of drought, and west coast areas would experience increased rainfall and possibly floods.

Any increase in the frequency of flooding would be very costly, especially in respect to urban areas, public infrastructure and major dams, both in respect to damage and for efforts to strengthen them. Flooding is already the largest source of natural hazard insurance cost in New Zealand. Higher temperatures are likely to reduce mountain snow fields and snow seasons (provided the snowfall itself does not increase), and will exacerbate the historic shrinkage of Southern Alps glaciers.

Agriculture (including forestry) will be affected by the basic bioclimatic changes noted above, such as CO_2-related plant growth increases and soil fertility changes, and there are likely to be changes in quality of grain and pasture nutrition, and possibly increased problems with weeds, pests, and diseases, and forest fires. Shifts in the suitability of districts for particular crops are likely—for example kiwifruit areas would shift southward as temperatures rise. Any increases in drought in east coast areas would affect the relative viability of cropping and pastoral farm systems in the region.

Of particular concern to New Zealand agriculture would be changes on a global scale, such as changes in global production that affect food commodity prices, or the introduction or establishment of new pests and diseases. However, New Zealand's mid-latitude climate facilitates the cultivation and adaptation of many different types of crops, and its farming industry is relatively well-informed and technologically capable. Thus, it is likely that agriculture will generally be able to adapt to climate change in the immediate future. Furthermore, it is likely that increases in productivity will occur, all other things being equal. With fisheries, there is little research available, and considerable uncertainty about impacts, but it appears that freshwater and near-shore fisheries are likely to be more affected than open-ocean fisheries.

Coastal systems are affected not only by sea level, but also weather systems, especially extreme storm events. The potential impacts of climate change related to weather events and rising sea levels include inundation, riverine flooding, saline intrusion, erosion, and wave damage. Coastal systems are naturally adaptive to the forces imposed upon them, but human land uses and infrastructure increasingly limit natural adaptation possibilities and pose great potential for considerable damage and cost. Coastal populations and capital investment at risk are already significant and both are growing rapidly.

Coastal areas and estuaries have particular cultural and economic importance to New Zealand's indigenous Maori, and any loss or damage to traditional land, sacred sites, or food-gathering areas over which Maori have traditional domain would have large ramifications. The Pacific atolls of Tokelau, which are under New Zealand administration, are at obvious risk to sea level rise, which, along with the possibility of environmental refugee fluxes from several other Pacific island states, adds an interesting extension to New Zealand's mid-latitude vulnerabilities.

Cities, towns, industry and business could be affected by a variety of direct impacts on water supply and drainage, air quality, energy production and distribution, in addition to possible extreme events such as floods. Relative to other influences, these impacts are likely to be small, but the total economic effect on large sectors may be significant. They may also trigger pre-existing weaknesses in systems. For example, it has been reported that higher than average summer temperatures, and hence greater demand for electricity, was a contributing factor in the cable failures which caused the loss of supply to Auckland's central business district for several weeks in February 1998.

Similarly, health impacts such as heat stress mortality, tropical diseases such as dengue, and respiratory problems may increase, and though small compared to the total burden of ill-health, may in the long run cause significant community impact and cost.

Much of our knowledge of impacts boils down to sensitivity studies, which stand as warnings, rather than predictions, of what might happen. The question then, is not whether these impacts are real or exaggerated (some may be) but how we can reduce the uncertainties in their estimation, and how we can respond to the possibilities in the interim, through the enhancement or development of adaptation capabilities and technologies.

Integrated studies that attempt to put together complete pictures for ecosystems, or watersheds, or economic regions, are in their infancy, but New Zealand has taken a lead in this work with the development of the CLIMPACTS project, led by Waikato University's Richard Warrick. This computer-based software package endeavours to model and geographically map the changes in a climate-dependent economic or environmental condition (say wheat productivity) resulting from a chosen climate scenario.

As described in the IPCC's Second Assessment Report[19] and the IPCC regional impacts study,[20] there are many adaptation methods already in routine use to deal with existing variability of climate and sea level, although the users of these methods generally do not think of them as related to climate change. These include statistical analyses of climate data to determine extreme climatic risks for engineering design purposes, El Niño climate prediction services, flood management works, agricultural crop choices, beach protection Propgrammes, public health measures and public information projects. Our ability to handle future change in climate will to a large extent depend on our current skills in dealing with existing climate variability.

In summary, while New Zealand may receive some benefit from an increase in agricultural productivity, many other negative impacts have been identified. There are many uncertainties in all of the estimates, arising especially from our inability to accurately predict future climatic conditions at the regional scale, and from our inability to understand how the multiple components of ecosystems and human communities interact. However, as a warning of what the total national impact potentially may be, a simple (albeit criticised) cross-sector economic costing study noted in the report has indicated that the costs for Australasia as a whole for a doubling of CO_2 may be as much as several percent per annum in GDP. If so, this would substantially depress projected GDP growth.

4. Regional Relationships

On the climate change issue, it could be said that New Zealand stands in the middle ground - a familiar position for a small country that likes to be innovative and to lead the development of consensus where possible. It lies between the big and regionally dominant Australia, with its large carbon-dioxide-producing coal and energy intensive industries, and the small and dependent Pacific islands, with their acute vulnerabilities, especially to sea level rise. New Zealand tries, with some success, to be a source of pragmatism and common sense and to provide a bridge between the incompatibilities of position that inevitably arise between these two disparate geopolitical entities.

As a set of smallish islands, and with many inhabitants of Pacific island origin or

19. *See* note 1.

20. Watson, *supra* note 13.

ancestry, including its indigenous Maori, it readily identifies with the Pacific islands and is a strong supporter of Pacific island involvement in the IPCC and FCCC processes. It also has constitutional responsibility for the Pacific islands of Tokelau, which lies north of Samoa.

However, as a mid-latitude country, New Zealand is inevitably set apart from the tropical concerns of Australia and the Pacific islands. In New Zealand, rising temperatures to some extent can be accommodated within its ranges of latitude and altitude, but in the flat expanses of northern Australia and on the tiny atolls of the Pacific Ocean there is no escape. Crops in Australia already may be near the upper level of adaptation, and hence may face declines, whereas in New Zealand warmer conditions are likely to bring improvements in crop options and crop yields. This is a fundamental difference between the two countries, arising mainly from latitude.

In terms of climate change research, New Zealand and Australia work closely together, with many Propgramme collaborations and personal connections. Australia's larger scientific establishment and its capabilities in climate system modelling and impacts analysis are of particular value to New Zealand's scientific community. In the preparation of the IPCC regional impacts report[21] it became clear that the expertise and knowledge of climate change in Australasia was world class (albeit with many knowledge gaps remaining).

Similarly, the two countries share broadly similar understanding on foreign policy, economic policy, and environmental policy, even if their political level concerns about climate change differ somewhat. There are a number of joint committees, e.g., the ministerial-level Agricultural Resource Management Committee of Australia and New Zealand, and a parallel committee on environmental matters. The Royal Society of New Zealand was invited to contribute an observer to participate in the Steering Committee of a major Australian review of climate change science.[22]

The situation in the South Pacific islands, however, is rather different. In most of these tiny countries, there is essentially no scientific establishment and the few scientists in the region focus on operational and policy responsibilities and have little opportunity or encouragement to do original research. Climate monitoring and advisory services are provided by the local meteorological services, often with the guidance and support of New Zealand and Australia. Although many climate impacts assessments have been done, most have been carried out or coordinated by foreigners, and the local expertise and resources needed to independently assess, interpret or extend these studies are very limited.

The South Pacific islands have small, weak, dependent economies and they look to

21. Basher, *supra* note 6.

22. SCCCS, *Climate Change Science: Current Understanding and Uncertainties*, Steering Committee of the Climate Change Study, Australian Academy of Technological Sciences and Engineering (1995); *see also*, *Climate Change: IPCC '95 and Beyond*, Proceedings of workshop, Misc. Series 32, Royal Society of New Zealand, Wellington (1996).

the larger foreign countries to deal with the serious climate change impacts that are expected to affect them. In this respect, many were bitterly disappointed by the aggressive stance of Australia's Prime Minister, John Howard, at the 1997 South Pacific Forum meeting in Rarotonga, where he defended Australia's policy to allow continued growth in its greenhouse gas emissions.

In climate and environmental matters, the South Pacific Islands have been successful in overcoming some disadvantages of their size by means of regional cooperation under the South Pacific Regional Environment Propgramme (SPREP) and under the Alliance of Small Island States (AOSIS). Australia and New Zealand are strong supporters of SPREP, which focuses on regional environmental issues such as marine pollution, biodiversity, and climate change. Most of its climate change work is done under the Pacific Island Climate Change Assistance Propgramme (PICCAP), a new multi-year Propgramme that is largely funded by the Global Environment Facility.

In contrast, Australia and New Zealand have no place in AOSIS, whose numerous member countries, with well-focussed and clearly articulated concerns on climate change form a powerful force for action. The interests of AOSIS nations are generally aligned against those of OECD countries, Australia and New Zealand included.

The socio-cultural contexts of the South Pacific islands are also very different than those of Australia and New Zealand, and may result in important differences in understanding and action on climate change. In particular, the Christian religion is a dominant force in most islands and local perceptions of climate change are strongly influenced by church teachings, especially by the moral teachings of the Bible. At the closing address of a regional meeting of Meteorological Service Directors called in the mid 1990s to discuss climate matters, an elderly Director expressed the view that climate change would not occur; that God would not allow this to happen to the God-fearing Pacific islanders, noting as supporting evidence the fact that birds could rest on power lines without being electrocuted. Similarly, the clergymen of the Cook Islands reportedly made it very clear to the people of Manihiki that their devastation by the El Niño-related cyclone of late 1997 was only to be expected for people who adopted modern ways and failed to attend church regularly.

SPREP's work Propgrammes make use of village meetings and other traditional social systems to communicate environmental information in the Pacific islands. However, it seems that the climate change community has not properly appreciated the differences in value systems and communication channels in the islands and the importance of closely involving key entities such as the churches in getting their message across.

5. Conclusion

Overall, it can be said that the impacts component of the climate change problem has been handled reasonably well by New Zealand. The main factors that stand out as critical to the successes are the leadership shown by numerous individuals, the encouragement of wide-ranging voluntary scientific input to the debate and to the assessment reports, and the increasingly informed climate-aware public.

The leadership shown by the two Ministers for the Environment, Geoffrey Palmer and Simon Upton, is especially commendable. Each developed a thorough understanding of the problem, moved quickly to support scientific research and to assess impacts and policy implications, and obtained from Cabinet the support and resources necessary to achieve their objectives.

Of course, much of the real thinking and initiatives over the years occurred at the working level, by Ministry officials led by Roger Blakeley, John Gilbert, Judy Lawrence and Ralph Chapman, and by others in the science community such as John Hickman, Paul Mosley, David Wratt, and members of the NSSCCC. New Zealand's small size, its small and largely Wellington-based science research and policy community, and the informal relationships that exist among these people, have been important factors in the rapid development of assessment and policy responses. This closely-knit character is much less possible in the more dispersed communities of both Australia and the South Pacific islands.

The voluntary networking developed for the New Zealand Climate Propgramme, and for New Zealand's contributions to the IPCC assessments, is a relatively new development for the science community. The networked involvement of climate scientists, agricultural scientists, hydrologists, coastal specialists, environmental managers, industry experts, social and medical researchers, has set in place a very effective model for how science can rapidly respond to societal concerns, and for enhancing collaboration among scientists, especially in these days of the Internet.

Currently, the New Zealand science funding system does little to encourage the development of such responsive networking and collaboration, owing to its high stress on competitive bidding between agencies, its narrow sector focus, and its tight contractual relationships, but there are signs that this situation may improve in future.

Although exasperating at times, the willingness of newspapers to print any view-point on climate change has certainly contributed to greater awareness of the issue by the public and to the greater dissemination of factual material. Additionally, the presentation of sceptical views has helped to reduce complacency among scientists and to sharpen their knowledge of the critical points. This open public process is like a mirror to the parallel open debate and evolution of expert thinking that goes on in the specialist forums of science, where scientists constantly propose and test ideas, and energetically criticise what they see as incorrect. Public debate also helps to remind scientists of the continuing need to communicate their knowledge to inform the policy process and society.

Three shortcomings have led to the suggested loss of the 3/10 marks. The first is the failure to develop and support an ongoing well integrated climate change research programme that follows up on the earlier impacts assessment work and provides systematic input to government policy making and to IPCC assessments. This problem lies not with the structural arrangements of the existing science policy, purchaser and provider organisations, but with the unduly ideological market-modelled rules governing their relationships, and in particular the reduced influence of policy departments over science resources. Of course it is not uncommon at times of zealous reform for the rules of a process to be elevated to a higher plane than the desired outcomes.

The second shortcoming is the failure to involve industry and business in science impacts research and assessment in a meaningful and constructive way, and the consequent ambivalence (and sometimes hostility) of these sectors toward climate change science and responses. The third shortcoming, which is somewhat outside the scope of this chapter but is closely related to the other two shortcomings, is the failure to develop a national political consensus and will to undertake significant national 'no-regrets' responses (both mitigation and adaptation) based on the science available.

On the important question of climate change impacts and vulnerabilities for New Zealand, it is clear that there are many potential impacts, some good, many bad, and most with a high degree of uncertainty. While there are reasons to believe that some natural systems may cope in the near future, and there may gains from the ready availability of suitable crops and likely increases in agricultural production, this has to be set against a general picture of vulnerability. This is particularly true in the context of ecosystems, hydrology, coastal zones, settlements, industry, health and potentially large economic impacts, for example from floods and in GDP costs of possibly several percent per annum.

Much remains to be done, especially to reduce the uncertainties in impacts estimates, and to develop adaptation methods to deal with present and future climate variability. Adaptation has been the Cinderella of the climate change debate to date, but is about to move to centre stage. It has to be remembered that greenhouse gas concentrations will continue to increase beyond the "doubling" that fixates much current thinking and most impacts estimates, and hence that some degree of climate change is inevitable, whether sooner or later.

Finally, there is always the potential for totally unexpected and unwelcome surprises, of the "ozone hole" variety, and for large externally imposed impacts arising from the confluence of global factors, including trade, politics and disease. In these respects, New Zealand may not be a polar opposite to Australia and the Pacific islands but a member of an unhappy threesome huddling together to defend themselves from northerly winds of not just climatic change, but also related economic and social change.

Annex 1

New Zealand National Science Strategy Committee for Climate Change

Terms of Reference (July 1996)

1. To develop a comprehensive strategy for climate change research which specifically includes the following:

   1.1 identification of priority research and important gaps in the overall climate change research agenda;

   1.2 coordination of research efforts between CRIs, universities, government dements and the private sector;

   1.3 identification of the overall level of funding appropriate for research topics by consultation with science funders, science providers and the end users; and

   1.4 report, through the Ministry of Research, Science and Technology, to the Minister of Research, Science and Technology by mid August each year.

2. To develop, in consultation with all interested parties, a portfolio of research which meets the objectives of the National Science Strategy.

3. To establish and maintain strong linkages with researchers and end users in the NSS topic area to actively promote knowledge of and access to information of common interest.

4. To provide advice to funding agencies on the priority and integration of research proposals in National Science Strategy topic areas.

5. To provide advice to the Ministry of Research, Science and Technology and the Foundation for Research, Science and Technology on priorities and funding levels for research closely related to development of government policy.

6. To provide advice on the establishment and maintenance of linkages within the international climate change programme, with particular emphasis on joint/complimentary activity with overseas funding agencies within the World Climate Research Programme and the International Geosphere-Biosphere Programme, and on international meetings and conferences of importance to New Zealand.

7. To monitor the progress of implementation of the climate change research strategy and its priorities and provide advice to the Ministry of Research, Science and Technology and the Foundation for Research, Science and Technology on progress; and to modify the strategy, if necessary, in the light of significant new information.

8. To provide advice to the Ministry of Research, Science and Technology on strategies and mechanisms that could be used to strengthen the effectiveness and accountability of climate change research.

8. NEW ZEALAND CLIMATE POLICY BETWEEN 1990 AND 1996: A GREENPEACE PERSPECTIVE

KIRSTY HAMILTON
Greenpeace International
Keizersgracht -176, 106DW Amsterdam, Netherlands

1. Kyoto: A Point To Look Back From

The unspoken tragedy of the United Nations Framework Convention on Climate Change's (FCCC) Third Conference of the Parties (COP3) in Kyoto was the implications for the Pacific. The demands of New Zealand's neighbouring nations, along with other small island states, for strong international reduction targets for CO_2 emissions were categorically disregarded in the thick of international *realpolitik.* This allowed New Zealand and Australia to walk away from Kyoto without any greenhouse gas reduction targets at all. Ironically, both countries in the early years of the decade had stated clearly, along with Pacific Island leaders at the South Pacific Forum, that climate change was the greatest threat to the South Pacific region.[1] Seven years later, New Zealand and Australia agreed to do less than the United States, the 15 members of the European Union, 10 Central and Eastern European countries, Canada and Japan.

During negotiations in Kyoto, New Zealand was subjected to international opprobrium because of its efforts to secure international agreement on a formula for net emissions accounting. New Zealand also spearheaded efforts to extract commitments from developing nations to reduce their greenhouse gas emissions, even before industrialised nations had finalized their own commitments. Both positions were highly controversial.[2] While few would disagree that at some point developing countries will need to find mechanisms for decreasing their emissions, the critical issue in Kyoto was that New Zealand's comments came well before the industrialised world had expressed a sincere commitment to adopt emissions reduction targets and set out to achieve them. There was distinct fear that this could lead to a collapse of the negotiations.[3]

---

1. For example, in the 24th South Pacific Forum Communique, 10-11 August, Nauru stated: "The Forum reaffirmed that global warming and sea level rise are among the most significant threats to the Pacific region and the survival of some island states. The Forum therefore reiterated its strong support for the Framework Convention on Climate Change and urged all states to sign and ratify the Convention as soon as possible."
2. For example, the World Energy Council's analysis of Kyoto commented on the "strong EU opposition" to New Zealand's argument that carbon sinks be permitted "on the grounds that carbon sinks involve great uncertainties of measurement and verification." WEC's appraisal of COP3 also notes that "uproar ensued" following New Zealand's intervention in the process for negotiating future developing country commitments. *See* World Energy Council, *Climate Change 1997 COP-3 to the UN Framework Convention on Climate Change and the Kyoto Protocol.* Report No.8 (1997).

A. Gillespie and W. C. G. Burns (eds.), Climate Change in the South Pacific: Impacts and Responses in Australia, New Zealand, and Small Island States, 143–163.

New Zealand adopted this 'brokering' role because its domestic policy predicament appears to have left it little alternative. Hiding in the folds was the sorry fact that New Zealand's net and gross emissions were rising rapidly and that policies that could, and should have, been put in place to reverse this trend were never implemented.[4]

This was particularly disappointing because most of the ingredients for a world class national response were present in New Zealand at the end of the 1980s. However, by 1997 New Zealand had gone from being in a prime position to serve as a positive catalyst in the debate, to at best sitting on the fence on key issues, and at worst becoming an impediment to the formulation of an effective international response to climate change.

2. 1990-1992: Science, Impacts And Initial Government Policy Response

2.1 POLICY FORMATION

1990 was a baseline year for the development of climate change policy in New Zealand, during which a number of very significant national and international reports were released. This was also the year that both the Labour and then the succeeding National governments adopted climate change policies and greenhouse gas reduction targets for the first time.

Internationally, the first report of the Intergovernmental Panel on Climate Change (IPCC)[5] appeared. This was an historic effort in collating information and analysis of the world's top experts on climate change. It detailed the best available science of climate change, the possible impacts, including the vulnerability of the South Pacific coral atoll nations to inundation and evacuation. It also concluded that

3. On this point the World Energy Council, *id.*, commented also that: "It was, in short, widely felt that some industry interests in the USA wished to scuttle the chances of reaching agreement in Kyoto by causing a stand-off between developed and developing country Parties. If so, the ploy very nearly succeeded in Kyoto, and may still do so, if US ratification of the Protocol falls on this point."
4. New Zealand's first National Communication to the FCCC projected an emissions increase of 18 to 22% by the year 2000 for CO_2 emissions, escalating to 35 to 40% by the year 2005. Following a revision of carbon absorption in the 1990 baseline year, it was reported that New Zealand would have a projected net emissions increase of 73%; this was further revised and a 60.7% increase in net emissions by the year 2000 was reported by the Ministry of the Environment. *Discussion Document of the Working Group on CO_2 Policy - Climate Change and CO_2 Policy - A Durable Response* (1996), at 33.
5. In August 1990, the IPCC presented the summary documents of its historic First Assessment Report, stating that to maintain CO_2 concentrations at their 1990 levels in the atmosphere would require greenhouse gas emissions reductions of more than 60 percent. IPCC reported that concentrations of CO_2 in 1990 were already around 25% higher than pre-industrial levels due to fossil fuel combustion and deforestation. Intergovernmental Panel on Climate Change (IPCC), *First Assessment Report* (1990), Overview.

these nations faced "very significant costs" for coastal protection.[6] The IPCC also highlighted the necessity to start negotiations for a framework convention on climate change.

The IPCC report presented both the scientific rationale for action globally and nationally and illuminated the risks of climate change. It also provided the international context for New Zealand's own policy debate and demonstrated there were feasible policy solutions available to governments.

In early 1990, two major New Zealand reports on the science[7] and impacts[8] of climate change were released following a lengthy collaboration of experts, government departments and others. This was a process started in 1988 with the establishment of the New Zealand Climate Change Programme (NZCCP), coordinated by the Ministry for the Environment.[9] The report on climate change science provided an assessment of the scientific basis for climate predictions, and suggested future scenarios for climate change.[10]

The 'impacts' reports pointed to the possibly potentially serious threats to local climate,[11] Maori[12] indigenous biodiversity,[13] agriculture,[14] and associated costs - including, for example, the invasion of new pests which may be able to reside in New Zealand under changing climatic conditions.[15] Despite the magnitude of the

---

6. Intergovernmental Panel on Climate Change, *Climate Change the IPCC Impacts Assessment; Summary*, Australian Government Publishing Service, Canberra (1996), at 6-12.
7. The New Zealand Climate Committee of the Royal Society of New Zealand. This was known as the 'facts' report (1990).
8. Ministry for the Environment, *Climate Change: Impacts on New Zealand. Implications for the Environment, Economy and Society* (1990). *See also* its sister summary report Ministry for the Environment, *Climate Change: A Review of Impacts on New Zealand* (1990).
9. Under the NZCCP four working groups were established: 'Facts' on climate science; 'Impacts' 'policy' on government responses and a Maori working group to advise on matters relevant to Maoridom. This programme is described in the preface of the reports the different working groups released.
10. This is described in the summary document on impacts, supra note 8, at 9.
11. "The greatest immediate impacts could arise from changes in the severity and frequency of climatic 'extreme events' such as floods and storms. Their impact might be disastrous at any time but if they became more widespread and frequent the consequences would be great." *Impacts, supra* note 8, at 12.
12. "Modifications to the environment due to climatic change have the potential to significantly impact on Maori and their tribal resources . . . .The preservation and promotion of rangatiratanga, as affirmed in the Treaty of Waitangi, is essential in examining potential impacts that climate change may have on traditional tribal resources." *Impacts, supra* note 8, at 10.
13. "The adverse effects of climatic change could be most severe for those parts of the environment and society which are least able to adjust. These include native ecosystems and species which are already under threat . . . ." *Impacts, supra* note 8, at 12.
14. On trade, this early report acknowledged the vulnerability that arises from having a 'narrow menu' of exports - in 1989 cattle, sheep and forest products accounted for all but 5% of export earnings. "Shifts of less than 5% in any one of several beef-producing regions are equivalent to New Zealand's entire [1989] exports. Shifts of around 6% in dairy production in Europe or Eastern Europe/ USSR are also equivalent to our entire output." *Impacts, supra* note 8, at 18.

possible implications for New Zealand, these were recognised as far less severe than those imposed upon New Zealand's South Pacific Neighbours.

For New Zealand, there is a further international dimension. Although the impacts of climate change on New Zealand itself may be significant, it is clear that many of our island neighbours in the Pacific stand to suffer much greater consequences. Given the small land area of many of these islands and their vulnerability to destructive tropical cyclones, the effects of climate change could be devastating.[16]

Around the same time, the Ministry for the Environment launched the first public consultation document on climate policy in May 1990.[17] All options - from carbon taxes to energy efficiency regulations- were put on the table and public responses sought.

Policy options put forward for consideration included a National Policy Statement under the then Resource Management Bill; taxes on greenhouse gas emissions; tradeable emission permits; subsidies for emissions reductions or sink creation; mandatory offsets for emissions increases; mandatory or voluntary appliance standards and a series of options for competitive and tariff restructuring and other legislative options to encourage energy efficiency in the electricity industry. Interestingly, the report raised the Maori gas take-or-pay contracts, the inefficiency of synthetic petrol manufacturing and the institutional arrangements governing oil and gas exploration that did not prioritise gas over oil. A range of instruments were also suggested for the transport sector, including fleet fuel efficiency goals, rebates for old vehicles, business taxation provisions and allowances to encourage fuel efficiency and integrated transport planning.[18]

The report was comprehensive and recognised the Treaty of Waitangi and equity implications of climate change policy, calling for "the policy options [to] be examined and developed by both parties to the Treaty."[19] There was little evidence of these issues being taken seriously in the years that followed.[20]

---

15. More extreme weather and the potential introduction of new pests added an additional layer of risk and potentially crippling costs. The 1990 report put the cost of eradication and lost production from a fruitfly outbreak at over half a billion dollars to the horticulture sector. *Impacts*, *supra* note 8.

16. Ministry for the Environment, *New Zealand Climate Change Programme, Responding to Climate Change A Discussion of Options for New Zealand* (1990), at 20; *see also* chapter by William C. G. Burns in this volume.

17. *Id.*

18. The transport sector remained virtually absent from policy debate to cut CO_2 emissions from this point on.

19. Ministry of the Environment, *supra* note 16, at 41.

20. For example, in the 1996 report of the Working Group, *supra* note 4. Treaty issues in relation to policies such as tradable emissions certificates are only briefly mentioned, despite these market instruments having been discussed for two or three years. *Id.* at 96.

Following public input on the New Zealand reports described above, in August 1990 Prime Minister and Minister for the Environment, Rt. Hon Geoffrey Palmer, announced Labour's climate policy, in what were the final months of the Labour government. The government pledged to cut CO_2 emissions by 20% from 1990 levels by the year 2005. The 20% target was in accordance with the call made at an earlier international meeting in Toronto in 1988.[21] Palmer also announced that the Labour government would assess the potential for cutting CO_2 emissions even more. The 'policy options' report stated:

> Through the Australia and New Zealand Environment Council (ANZEC), the environment ministers of Australia and New Zealand have agreed to investigate ways of achieving limitations on emissions of greenhouse gases - looking at targets of up to 40% of the 1988 levels of emission, as well as considering what the impacts of implementing such targets might be.[22]

Later that year, the National Party, in reaction to public pressure, adopted its formal climate policy only two days before the October 1990 election.[23] Environmentalists had pointed out an immediate concern - two days after the election government officials would be attending the Ministerial level international meeting, the Second World Climate Conference, without a climate policy should the National Party come to power. Accordingly, it too adopted a 20% reduction target for CO_2 emissions, but moved the target date up five years.[24]

2.2 THE POSITIVE OUTLOOK

Two weeks after the 1990 elections in New Zealand, the Second World Climate Conference Ministerial Declaration called for the negotiations on a framework convention on climate change to 'begin without delay' so that it could be signed at the United Nations Conference on Environment and Development (UNCED).[25] From this point the debate within the New Zealand government had to turn towards implementation. New Zealand was in an ideal situation for putting a

---

21. At this meeting, many governments voluntarily adopted a "call for action." They advocated a 20% cut in CO_2 emissions by 2005 over 1988 levels as a global goal. Additionally, on reviewing the science available at the time, delegates concluded that "Humanity is conducting an unintended, uncontrolled, globally pervasive experiment whose ultimate consequences could be second only to a global nuclear war," Conference Statement, *The Changing Atmosphere: Implications for Global Security* (1988), at vii.
22. Ministry of Environment, *supra* note 8, at 21.
23. Press statement by Deputy Leader of the Opposition, Hon. Don McKinnon, Spokesperson for Foreign Affairs, 25 October 1990.
24. This was the target that Hon Rob Storey, Opposition Spokesperson on Environment and then Environment Minister, had established as a National Party goal, although prior to the election and not as formal Party policy.
25. Noted in *Climate Change: Science Impacts and Policy, Proceedings of the Second World Climate Conference* (1991), From the Ministerial Declaration, 7 November 1990.

strong and effective policy in place. The public was already sensitised to energy issues[26] and effective policies. Policymakers early on also recognized that there were 'no regrets' measures that could be taken as part of a climate policy that made sense anyway. For example, energy efficiency to reduce demand for electricity provided a means of avoiding the projected economic costs of new power stations (subsequently estimated at between $15 - 25 billion dollars by 2010) associated with meeting energy demand growth forecasts.[27]

The specific feasible alternatives were energy efficiency, renewable energy and new transport policies. Indeed, because of such options environmentalists and others were concerned that New Zealand's initial 20% reduction target was not ambitious enough.

An important overlay to the discussion on energy efficiency in New Zealand at this time was a report by the International Energy Agency in 1990. This demonstrated that, comparatively, New Zealand had the worst record in the OECD in terms of increasing intensity of energy use, which is a measure of the amount of energy used to provide a unit of economic wealth, translated into GDP.[28] This could partly be explained by the energy intensive so-called 'Think Big' projects, such as the Synthetic petrol plant in the 1980s, however this was not sufficient to explain why the trend continued. Moreover, New Zealand's energy intensity was dramatically out of step with the trends in industrialised countries. Indeed, it had risen by over 30% between 1973 and 1987, in contrast to the 20% reduction for the OECD as a whole.

Juxtaposed with this poor international record were a series of reports within New Zealand that showed that there was significant potential for energy savings, carbon dioxide reduction[29] and even job creation associated with such policies.[30] Adding to these studies was information brought to light in the Wanganui River flows case[31] being heard by the legal disputes Planning Tribunal. In the context of displacing the need for greater hydro flows on the Wanganui River the debate over

---

26. Energy and environment issues were raised in the 1960s, for example, surrounding the debate over raising the water level of Lake Manapouri for hydro-electricity purposes. This generated such a degree of public concern that it became a major issue in the 1972 election. In the 1980s public debate around energy intensive developments known as the 'Think Big' projects also raised environmental and energy issues and again were an important factor in the 1984 election.
27. The Electricity Supply Association estimates were quoted in the Parliamentary Commissioner for the Environment's Report, *Sustainable Energy Management in New Zealand* (1992).
28. For details of this area, *see* OECD, *Environmental Performance Reviews: New Zealand* (1990), at 113-130. For the specific document, *see* International Energy Agency, *Energy Policies and Programmes of IEA Countries: New Zealand* (1990).
29. *See* for example the Energy Management Group, *Energy Management and the Greenhouse Effect*, Ministry of Commerce (1991).
30. New Zealand Institute for Economic Research, *Job Creation and Energy Management Programmes* (1988).
31. Whanganui Rivers case: Sheppard J (1990), ECNZ vs Whanganui River, Maori Trust Board and Manawatu-Wanganui Regional Council, W70/90.

the potential for New Zealand's energy efficiency came fully into the public eye. Specifically, under the Official Information Act, Electricorp - the State Owned Enterprise which was then the monopoly supplier of electricity in New Zealand - was forced to release an internal report showing energy savings of 57% of electricity demand were not only feasible but cost-effective.[32]

The ability to make huge gains in energy efficiency were also demonstrated in the context of renewable energy. Resource mapping at the end of the 1980s had shown that New Zealand possessed world class wind resources. For example, a New Zealand Energy Research and Development Committee (NZRCD) report in 1987 demonstrated that there was the potential for at least twelve 250MW wind farms, capable of meeting 40% of New Zealand's energy demand at that time.[33] Additionally, supply curves from the Electricity Supply Association in the early 1990s confirmed that wind was close to being economically viable. However, at that time in New Zealand there were no commercial windfarms - not even the single 'test' wind turbine in Wellington. This was in marked contrast to countries such as Denmark that already had an active export industry in wind turbines.[34]

2.3 THE ANATOMY OF THE DEBATE

Against this background of opportunities, the first signs of what were to become systematic efforts over the next five years by the traditional industry lobby to protect its fossil fuel and energy intensive interests started appearing. In December 1990 Professor Fred Singer was invited to New Zealand. He was the first of several well-known US climate science "sceptics" who came to New Zealand. Others included Richard Lindzen, Patrick Michaels and Robert Balling. Hosted by organisations including the Business Round Table, the Centre for Independent Studies, the Energy Foundation, NZ Coal Research Association and Electricorp in 1990, 1991, 1992, 1995 and 1996, they set about trying to debunk the science and some of the proposed responses to climatic change. These efforts mirrored similar strategies in the United States.[35]

Industry also had a high degree of influence due to the high priority given by the National government to the protection of national economic interests and a strong Treasury-driven leaning towards almost pure market-orientated policies. This

32. Electricorp, *Potential Savings in Electricity Consumption Through Increased Energy Efficiency in New Zealand* (1989).

33. New Zealand Energy Research and Development Committee, *Wind Energy Resource Survey of New Zealand, National Resource Assessment, Summary and Final Report* (1987).

34. The domestic market for Danish wind energy was established in the 1980s when the government's 1981 National Energy Plan set a goal for 1000 megawatts of wind energy by the year 2000, by 1993 there was 500 MW installed in Denmark, enough to supply 3% of its electricity needs. By 1998, the wind power capacity reached 1,200MW. According to Environment and Energy Minister Svend Auken, the wind energy industry is one of the fastest expanding industries in Denmark with 60% of the world market, exporting turbines to 50 countries. This is an industry that could have been fostered in New Zealand. References for these facts are contained in Greenpeace International, *Danish Wind Energy - An Industrial Success Story* (1998).

drive shaped the response to climatic change. Additionally, industry tapped into fears that climate and other environmental policies would make New Zealand industry non-competitive overseas, contending that regulations would distort the 'free' market and thus impose higher costs on home companies. As a consequence, they argued that increased energy prices (flowing from policies such as carbon taxation) would result in industries packing up and leaving (even though New Zealand's electricity prices were amongst the lowest in the world).

The first sign of the failure to take climate change into account in the energy policy context was a change in the oil and gas taxation regime to make oil exploration cheaper and less risky for oil companies in New Zealand. A second sign of the conflict between the new market orientated considerations and emerging environmental concerns was the 1991 electricity sector 'reform.' This reform presented a clear opportunity in the electricity industry to avoid two concerns. The first was the projected rise in greenhouse gas emissions. Indeed, the Ministry for the Environment had pointed out in 1990 that one large (1000MW) new coal fired power station alone could have increased New Zealand's national CO_2 emissions by 30%.[36] The second concern was the large economic costs of meeting demand growth by traditional means and the consequent opportunity to lock in new incentives for conservation, efficiency and renewables.

In July of 1991 the cabinet declared that "whenever policy development within government may have significant effects on greenhouse gas emissions this should be included in the terms of reference for the work."[37] However, climate considerations, as well as related policies, such as energy efficiency, were given short shrift during this period.[38] This was not surprising, because while climate change policy was managed by the Ministry for the Environment in consultation with other government departments,[39] electricity sector restructuring was firmly in the hands of

35. For example, a 1992 press conference in the US by the Global Climate Coalition included Fred Singer and Patrick Michaels. Michaels and Singer attended the international FCCC negotiations right up to COP3. More information on the sceptics can be obtained in Greenpeace International, *The Scourge of the Sceptics: Prepared for the Second Conference of Parties (COP2), Geneva* (1996). These visits prompted the Ministry for the Environment to produce a special report entitled *The Consensus and the Debate* to put some perspective on their wild statements.

36. Ministry for the Environment, *Responding to Climate Change A Discussion of Options for New Zealand* (1990), at 29.

37. Cabinet paper document, ENV (91) 37. This is referenced in Greenpeace New Zealand, *Power and Energy* (1991).

38. The policy decision to reduce emissions by 20% by the year 2000 meant that even new power station proposals using gas would have a significant impact on the ability to meet targets. The 400MW 'efficient' combined cycle gas fired power station proposed for Stratford had the capacity to increase New Zealand's emissions by over 5 percent. This was one of the reasons that the government finally decided to initiate a national legal inquiry under the Resource Management Act into the project.

39. This meant that much behind the scenes negotiation went on with the Ministry of Commerce and the Treasury on one side and Ministry for the Environment often on the other before documents were made public or reached Cabinet. This was revealed by research of interdepartmental documents obtained under the Official Information Act throughout the 1990's.

Treasury and Commerce, with its singular focus on economic efficiency. This was despite the fact that the success of similar experiments with electricity privatisation in England had already proven to be "patchy and unimpressive" by the early 1990s.[40]

Furthermore, rising concerns that the Maui gas field would run out before the 'contract' date of 2009 raised doubts about how much cheap gas was available. This in turn led to speculation that the 1000MW baseload power station at Huntly (a dual fuel gas/coal station) would be switched from gas to coal in order to free up gas supplies for the new plant. Such a fuel switch would have increased CO_2 emissions from Huntly power station by 73 percent.[41]

The market mantra in New Zealand soon reached ridiculous proportions. For example, a letter from Treasury to the Ministry for the Environment in 1991 stated:

> Because this [cost-effective energy efficiency measures] increases people's income and changes the relative price of goods and services it could increase the demand for energy and have the perverse effect of increasing CO_2 emissions.[42]

This implied that Treasury would advocate continuing to pay to waste energy just in case people used more than they needed. Moreover, it meant that Treasury was prepared to ignore evidence emerging from economically successful countries like Japan, then recognised as a world leader in energy efficiency, having reduced its energy intensity by approximately 30% between 1973 and 1987.

Despite this fact, the New Zealand electricity debate was dominated by considerations of economic efficiency and the establishment of competitive markets by carving up and corporatising as much as possible - at the retail and supply ends of the industry. This was an agenda that started in the late 1980s but came into its own under the National government. The previously elected Power Boards were corporatised to form companies with commercial directors; a protracted and frustrating series of consultations on the establishment of a Wholesale Electricity Market (WEM) were started; plans to chop up Electricorp into 'competitive' sections (which would trade electricity on the WEM) were under discussion; and the elec-

---

40. UK government House of Commons Energy Committee reports *Energy Efficiency* (March 1991) and *Consequences of Privatisation* (February 1992) were both critical of the outcomes of the electricity privatisation process in the UK from the standpoint of energy efficiency. The text from the 1992 report is worth quoting more fully: "In our 1988 report on electricity privatisation we assigned the highest priority to the promotion of energy efficiency . . . . We later called for Public Electricity Suppliers to be under statutory obligation to promote more efficient electricity use . . . .No such provision was incorporated into the Electricity Act . . . . So far the effect of greater competition on the efficient use of electricity appears to be patchy and unimpressive." *See also* Greenpeace New Zealand, *Power and Energy* (1992).

41. Electricorp still applied for resource consents to Hamilton Regional Council under the RMA, 1991, which would have allowed Huntly to fully operate on coal.

42. Quoted in Climate Commentary section of Greenpeace, *Power and Energy* (September 1991). The original letter was from John Wilson, Treasury, to the Ministry for the Environment.

tricity transmission and retail tariff structure was under scrutiny. It was a veritable spaghetti bowl of options - some crucial, others pointless and potentially destructive, but the main problem was the lack of a clear vision. New Zealanders were left to rely on the assertion of the Minister and his officials that competition and 'free' markets were an objective in themselves and would deliver energy efficiency and CO_2 reductions. However, although the Energy Minister at the time, John Luxton, acknowledged that the omission of environmental external costs (e.g., CO_2 emissions) from energy pricing or investment decisions constituted a market failure, he was not prepared to solve this as part of, or even in parallel to, the electricity restructuring debate. Indeed, a letter from the Ministry of Commerce to the Minister of Energy in September of 1991 stated that despite reform of the market, "one conclusion is expected to be that barriers to energy efficiency will remain after energy reform."[43]

Meanwhile, environmentalists had approached the Parliamentary Commissioner for the Environment about the government's approach to energy and environment. In April 1992 she released a comprehensive study, with recommendations, covering the electricity restructuring process and the environment.[44] One of her recommendations was that Minister Luxton draft a National Sustainable Energy policy for New Zealand before passing the Energy Sector Reform Bill. The report also recommended specific goals for energy efficiency improvements and the need to address energy related greenhouse-gas emissions, as well as a strategy for a transition to renewable energy. The Energy Minister ignored the report.

3. 1992: The Earth Summit & The Power Crisis At Home

In 1992, the world was beginning to grapple with the international dimensions of environment and development at the Earth Summit, and in particular the signing of the Framework Convention on Climate Change (FCCC). Simultaneously, it seemed that the domestic commitment to climate change was waning rapidly in New Zealand.

It was revealed that at the end of the previous year a Cabinet Committee[45] had concluded that it would be 'difficult and costly' to meet the CO_2 target, pegging the cost at $6 billion, only half of which would be cost-effective. Official Information requests revealed that the basis for this conclusion was little more than 'back of the envelope' calculations by the Ministry of Commerce.[46] Using exactly the same assumptions, an analysis commissioned for Greenpeace found there would be a net saving of $1.8 billion dollars. Government calculations had failed to consider the

43. The letter was obtained under the Official Information Act and was quoted in Greenpeace, *Power and Energy* (January 1992).
44. Parliamentary Commissioner, *supra* note 27.
45. Cabinet Paper, *Strategic Issues and Options*, November 26, 1991.
46. A three page letter was sent to Greenpeace describing the calculations.

benefits of policies, and had only focused on the costs of their implementation.[47] The Ministry of Commerce calculations both fuelled and reflected the increasing lack of commitment to deal with climate change. The Cabinet was being advised that independent domestic action would have little effect on global warming and could impose economic and social costs on New Zealand if other countries did not take similar action.[48]

By May of 1992, there were fears that the National government was about to abandon its 20% target in the lead up to the Rio Earth Summit in June. Environment Minister Rob Storey was quoted as saying New Zealand would have to virtually halve car use within eight years to achieve the target. Economic impacts of policies by then were the foremost consideration. He stated:

> Having seen the preliminary work that has now been done on the potential for achieving it [CO_2 reduction target] without economic damage, quite frankly it doesn't lead me to believe that we can change people's behaviour to the extent of a 40 per cent reduction before the year 2000.[49]

By this time, the forecast increase in emissions, 20% between 1990 and 2000, was announced, evidently to make the objective look impossible and to make a policy U-turn more palatable.

When the government finally did announce its CO_2 Reduction Action Plan in July 1992, it was confirmed that there was no intention of adhering to its policy to "aim for" a 20% cut in CO_2 emissions by the year 2000. Few concrete actions were included. Official Information Act requests covering interdepartmental correspondence in the lead up to the decision revealed that Treasury tried to block measures further. This was done by actively opposing requests for further funding, and even suggesting that the threat of the 'call-in' provisions given to the Minister for the Environment under the Resource Management Act "may impact adversely on the government's policy for encouraging foreign direct investment."[50]

This elite policy debate - and even the Rio Earth Summit - was soon overshadowed by the major event of the year - the hydro electricity shortage - or the "power crisis" as it became known. Low water levels in the main hydro 'storage lakes' in the South Island created a shortage of electricity - and the threat of electricity rationing

---

47. K.H.P. Kammler, *Cost and Benefit of Achieving Government's CO_2 Target* (1992) (paper prepared for Greenpeace).

48. *The Dominion*, 23 March 1992.

49. *Dioxide Target Looks Elusive*, NEW ZEALAND HERALD, 12 May 1992. *See also*, *Emission goal hard to meet – Storey*, THE DOMINION, 12 May 1992.

50. Reported in Greenpeace, *Power and Energy* (September 1992). The original papers pertaining to the 5 July announcement were obtained by Greenpeace under the Official Information Act. Note that the 'call-in' powers under the Resource Management Act over energy projects were one element of the CO_2 plan. It meant the Minister for the Environment could require a legal inquiry into projects deemed to be of national significance under the terms of the Act.

before the end of winter. New Zealanders were urged to save as much electricity as possible and Electricorp was forced to fire up the mothballed Marsden A power station.[51]

Amid much speculation of the real causes of the power crisis, it was clear that Electricorp was not concerned with trying to reduce New Zealanders long term consumption patterns. Rather, its 'sell like hell' strategy was designed to increase, not decrease, electricity use. The reason was clear - it had no commercial incentive to do otherwise. At the height of the crisis in July, Electricorp wrote to the Minister of State Owned Enterprises complaining about the "extremely high" level of energy savings being achieved by New Zealanders - a staggering 25% one week compared to normal usage - was "a matter for some concern . . . and is adversely affecting Corporation revenues . . . "[52]

From the standpoint of climate policy, the power crisis was a huge missed opportunity to capitalise on public attention, antagonism towards Electricorp and support for energy efficiency. Such support could have helped implement the sort of recommendations given by the Parliamentary Commissioner for the Environment. An inquiry into the power crisis, with a narrow terms of reference, was set up and a report was released in January of 1993.

Despite the inquiry, there were no substantive changes (apart from a name change by Electricorp to ECNZ) in the operation of Electricorp, or at the supply end. The one report ECNZ produced on the opportunity for energy efficiency in New Zealand[53] concluded under one scenario that reduced energy consumption and co-generation could delay the need for a new power station in New Zealand until at least 2005, and cut CO_2 emissions by 4%. However, less than three months later, ECNZ went on to make plans for a 400MW combined cycle gas power station to be built at Stratford in Taranaki. The proposed station had the capacity to increase New Zealand's CO_2 emissions by 1.5 million tonnes a year - a 5 to 6% rise. This proposal was highly controversial, and resulted in an unprecedented legal inquiry initiated by the Minister for the Environment after numerous requests for an examination of the situation under the "call-in" option of the Resource Management Act.[54] In Decem-

51. Greenpeace calculated that one month's fuel cost for the rekindled Marsden power station could have supplied a quarter of New Zealand's households with a free hot water heater insulation cylinder - a key area of domestic energy loss. *See* Media Statement, Mike Moore, 5 June 1992. In releasing a five point policy plan for energy efficiency he noted "Greenpeace, for example, estimates that the $7 million monthly fuel bill for Marsden A would be enough to provide hot water insulation free for one quarter of New Zealand's homes."

52. Confidential letter to State Owned Enterprises Minister, Maurice McTigue, 10 July 1992, obtained under the Official Information Act by Greenpeace.

53. ECNZ, *The Developing Market for Energy Efficiency* (1992).

54. Section 140 (2) provides a number of 'relevant factors' such as: (a) widespread public concern regarding its likely effect on the environment (including global environment); (b) involves significant use of natural and physical resources; (d) is likely to affect more than one region; (e) affects New Zealand's international obligations to the global environment; and (g) contributes to significant or irreversible changes to the environment (including the global environment).

ber, the Cabinet decided that such an inquiry should be conducted.

4. 1993: Election Year and Comalco

Climate change was not a key issue in the November 1993 election, however the issue of government accountability was. Not only had the government ignored its climate policy, it was seemingly insouciant about the spate of new power station proposals from the newly commercialised Electricity Supply Companies and ECNZ. This was despite a Greenpeace commissioned Heylen poll in September of that year that found that 73% of New Zealanders wanted the government to help them save energy rather than building new power stations. The controversy surrounding supply versus demand side approaches was echoed by the International Energy Agency in its 1993 annual chapter on New Zealand which noted that one issue confronting New Zealand will be balancing construction of power plants and energy saving measures. The IEA also showed graphically that New Zealand's rising energy intensity was still dramatically out of step with trends in the rest of the OECD. This is itself was not surprising when operators such as ECNZ were still planning for 3% annual growth in electricity demand.[55]

Epitomising their approach to core issues in the first three-year term in the lead up to the election, National government Cabinet Ministers and ECNZ, in conjunction with Comalco, were negotiating deals behind closed doors over renewed electricity supply contracts between ECNZ and Comalco. Environmentalists and others were using Official Information Act emergency procedures - supported by the Ombudsman - to get the papers made public. The government refused, using the hung Parliament to stall disclosure of the relevant information until after the deals were finally signed.

By according Comalco contracts for around 15% of New Zealand's electricity needs- still at cheap prices - for up to three decades (out to year 2012 with options for an additional period to 2022), the government exacerbated the pressure for new power stations, and the prospect for escalating CO_2 emissions. These strategic, long term issues should have been debated in public.

By the time of the 1993 election, the National government's official Party policy on climate change had been dramatically modified to a call for stabilisation of net CO_2 emissions at 1990 levels by 2000. The 20% cut in CO_2 emissions below 1990 levels by the year 2000 was retained as an "ultimate objective, subject to various conditions."[56] By mid-1994, alternative packages of complementary policy instruments were expected to be reported on.

55. The 3% figure appeared in statistics obtained under the Official Information Act by Greenpeace. *NZ's Energy Record Appalling - And Worsening*, THE INDEPENDENT, 1 October 1993.

56. The 'interim objective' "to reduce net carbon dioxide emissions to 1990 levels by the year 2000 and keep them at that level beyond then," together with the "ultimate goal," was adopted in May in Cabinet paper CAB (93) M 18/29.

In according such a low priority to climate change the government was blithely ignoring national and international warnings emerging from sectors previously uninvolved in the debate, such as the insurance industry. In August 1993, the Chief Executive of the New Zealand Insurance Council had told a government Business and the Environment Seminar that it was in the interests of all businesses to be "actively lobbying governments for reduction in greenhouse gas emissions."[57] This followed international expressions of concern from an industry reeling from a string of billion dollar losses from weather catastrophes. For example, Munich Re, the major German reinsurer, had stated as early as 1991 that:

> . . . mankind is about to change the climate significantly and possibly irreversibly, without having any idea of the consequences that will have . . . In the future, we must anticipate a further, probably dramatic increase in catastrophe losses from currently US$20 billion to a magnitude of US$100 billion in overall economic terms.[58]

5. The 1994 Stratford Inquiry & The Net Approach

The most prominent and important event in climate policy in 1994 was the legal assessment of the proposed 400MW combined cycle gas power station that ECNZ intended to build in Stratford. The CO_2 emissions from the operating station would have increased New Zealand's national emissions by over 5 percent. Due to the national implications and the transboundary nature of the emissions, Greenpeace had written to the Minister for the Environment, Simon Upton. Its letter requested that he exercise his 'call-in' authority under the Resource Management Act and hold a national inquiry (which he agreed to) into the CO_2 discharge permit application.

This was a unique test case. The July 1994 hearings, before a three-person legal Board of Inquiry, was the first national, public, non-government hearings concerning energy and climate change. Over 30 expert witnesses testified at the hearings, including from ECNZ, the Supply Authorities, Taranaki Regional Council, environmental groups and other energy companies.

Immediately prior to the hearings, Greenpeace, via the Ombudsman, secured the full Cabinet paper leading to the decision to opt for 'call-in.'[59] It became clear that

57. Presentation to a Ministry for the Environment Business and the Environment conference, August 1993, by David Sargeant, Chief Executive of the NZIC. Sargeant also described during an interview how rising 'catastrophe' premiums overseas, due to the spate of super-expensive weather related insurance losses, had spiked earthquake premiums in New Zealand, as earthquake insurance comes out of the same category of catastrophe insurance. This had meant a 100% rise in reinsurance rates for one New Zealand company in 1992 and a further 50% increase 1993.

58. Munich Re, *Windstorm* (1991).

59. Paper from the Office of the Minister for the Environment to the Chair of the Cabinet, *Proposed Taranaki Combined Cycle Power Station: Government's Action Under the Resource Management Act 1991* (1994).

the government was well aware of the difficulties it faced if this power station went ahead. The cabinet paper stated, among other things:

> 11. We consider that the proposal would have significant implications for the Government's ability to meet its climate change policy objectives and is relevant to New Zealand's international obligations in relation to the global environment.
>
> 12. Increasing New Zealand's CO_2 emissions by up to 5% from one source would necessitate greater reductions elsewhere if the Government wishes to meet its policy objectives on CO_2. This would require stronger CO_2 policies in all sectors (including electricity) and greater consequent adjustment costs. This is likely to be a problem in the post-2000 period as the Government attempts to hold CO_2 emissions to their 1990 level. It may also be a problem before 2000 if once-displaced generation is brought back into use by then.
>
> 13. In the event that Stratford is build, ECNZ would have a strong incentive to maximise its use of both the new Stratford station and Huntly. ECNZ will have an incentive to seek new markets for electricity in order to maximise the return on its generation assets.
>
> 17. ECNZ may claim that its project is being unfairly singled out to meet climate change objectives. However, if the Stratford proposal is not consistent with the Government's climate change policies and its emissions are significant, ECNZ will be increasing the adjustment costs to the rest of the economy.
>
> 20. Doing nothing on the Stratford proposal could also affect New Zealand's credibility in seeking to have the international community accept our forestry actions as contributing to meeting our international obligations under the Framework Convention on Climate Change.

It is instructive that the Ministry of Commerce and the Treasury opposed the use of the call-in option.[60] They preferred to accept ECNZ's contention that there would be no increase in CO_2 before 2000, and therefore the government's FCCC target would not be affected. They also referred to the Ministry of Commerce's demand forecasts, ignoring the government's and ECNZ's own reports demonstrating the great potential of energy efficiency programs, and instead concluded that "new electricity generating capacity will be required in any event before 2000."[61]

---

60. Appendix C to the Cabinet paper, *id*, expresses the views of these two departments.

61. ECNZ, *supra* note 53.

The final report by the Board, with recommendations to the Minister for the Environment, was submitted in February 1995.[62] The recommendation was that the Minister grant resource consents for CO_2 emissions but subject to full mitigation, by tree planting, to create a permanent sink. The Board also recommended that the government develop a National Policy Statement (under the RMA) covering CO_2 emissions from energy and transport.

The Board recognised that the Stratford CO_2 emissions were "likely to result in or contribute to significant changes to the global environment." This was viewed as antithetic to New Zealand's international obligations as stipulated in the FCCC to limit greenhouse gas emissions. The Inquiry also stated that ECNZ had not established the case that the Stratford proposal was the only viable option for meeting New Zealand's electricity needs, as opposed to non-CO_2 producing alternatives, and they disputed ECNZ's electricity demand forecasts. In other words, the majority of ECNZ's key points did not stand up to independent scrutiny. Importantly, the Board stated: "If we had reached the view that there was no jurisdiction for the Minister to impose the mitigation condition we would have recommended that the consent be refused."[63] This was, in Greenpeace's view, in effect, an acknowledgment that any increase in atmospheric CO_2 emissions ran counter to sustainability and the Purpose and Principles of the RMA. However, to circumvent the full implications of this conclusion, the Board sanctioned the planting of increased numbers of trees to sequester an equivalent quantity of atmospheric carbon dioxide. This was the so-called 'net' approach, a central and controversial part of the government's policy approach to climate change.

5.1 NET EMISSIONS AND CO_2 POLICY

Greenpeace and others have strenuously opposed the use of the 'net' emissions approach since it was first mooted. Analysis by Greenpeace's Legal Department and other New Zealand legal academics argued that this approach was not consistent with the FCCC.[64] Overt criticism of this approach was also made on scientific grounds. For example, the Climate Committee of the Royal Society of New Zealand sent a letter to the Minister prior to his decision, with the assessment of one of its leading experts, which stated that:

---

62. *Proposed Taranaki Power Station - Air Discharge Effects. Report and Recommendation of the Board of Inquiry pursuant to Section 148 of the Resource Management Act, 1991,* Chairman D, Williams. QC (1995).

63. *Id.* at 227.

64. Greenpeace International Legal Department, *Anthropogenic Carbon Dioxide Emissions and Sinks Under International Law* (1994). *Legal Opinion For the Australia and New Zealand Solar Energy Society on the Meaning and Effect of Article 4(2)(b) of the United Nations Framework Convention on Climate Change* 1992. Note that in this context the precise role that forestry can play in terms of net emissions was one of the very controversial aspects - at the end of 1997 - of the Kyoto negotiations, and indeed prompted a call for a special report to be completed by the IPCC on methodologies and other matters at the Bonn session of negotiations, mid-1998.

> More generally the claim, sometimes made, that 'a molecule of CO_2 absorbed by a tree is equal to a molecule of CO_2 not emitted in the first place' is too simplistic and would be roundly criticised by overseas scientists. This equation is only valid if one can guarantee that the carbon stored by afforestation will remain locked out of the atmosphere for geological timescales . . . There does not appear to be any scientific basis by which the integrity of forests as carbon stores can be guaranteed for centuries.[65]

In the final instance, the Minister decided to grant the CO_2 discharge consents, but subject to full mitigation once the overall CO_2 emissions from the electricity sector increased following the start-up of the plant. In other words, he accepted ECNZ's argument that Stratford would push more polluting stations further down the merit order (i.e., the hierarchy of power station usage to meet demand). He allowed tree planting and other contracted CO_2 emissions reductions from energy and end-use efficiency improvements outside the electricity generation system. This would effectively allow the plant to operate for, perhaps, a number of years with no obligation or penalty to reflect the environmental damage caused by its CO_2 emissions. Other key implementation issues were also overlooked. For example, how the operators could be held genuinely liable in perpetuity for the maintenance of stored carbon, or baseline issues for contracted emissions reductions elsewhere - in other words how to assess the actual reduction over what it would otherwise been.

The fact was that the net approach was already at the core of government policy and this was formalised further with the Stratford Inquiry. Nationally, the role of carbon sequestration was to account for 80% of meeting emissions commitments, with the remaining 20% being met through direct emissions reductions.[66] The Ministry for the Environment and the Ministry of Foreign Affairs and Trade both argued for a 60:40 mix, but were overruled.[67] Moreover, despite being unable to implement much talked about economic instruments to induce market-led reductions of CO_2 emissions, the taxation regime for forestry was changed to create further incentives.

---

65. M. Manning, *Comment on the Board of Inquiry report on the Taranaki Combined Cycle Power Plant*, sent with a letter from V. R. Moore, Chief Executive Officer of the Royal Society of New Zealand, 20 March 1995. Obtained by Greenpeace under the Official Information Act.
66. Ministry for the Environment, *Climate Change: The New Zealand Response* (1994).
67. Based on the interdepartmental correspondence in the lead up to the July 1994 CO_2 policy decision, obtained by Greenpeace under the Official Information Act; quoted in Greenpeace, *ECO, Royal Forest and Bird Protection Society* (1995). *The New Zealand Effect: Why Sinks Won't Save the Climate, Report for the Expert Review of New Zealand's First National Communication under the Framework Convention on Climate Change.* Note that in this report to the Expert Review Panel Greenpeace recommended that SBSTA request an IPCC review of issues surrounding carbon offsets. This was agreed at the 1998 June negotiating session in Bonn, although the final terms of reference for the review remain to be resolved.

The net approach was also very convenient for New Zealand's first National Communication to the FCCC (in September 1994), which laid out its greenhouse gas national inventory and policy. New Zealand's emissions, under a Business as Usual forecast, were forecast to rise 18-22% between 1990 and year 2000, rising to 35-40% higher by 2005. The outlined emissions reduction policies in the report would cut year 2000 total emissions by only 0.5% - 4.0%, leaving a significant increase over 1990 levels. However, the government argued that overall net emissions would be 50-59% below 1990 levels due to the level of forest planting. This approach reduced pressure to even initiate emissions reductions in the energy and transportation sectors, and instead turned into a directive to ensure this approach was accepted and adopted internationally.

The Minister remained silent on the secondary but strategically important Board of Inquiry recommendation for a National Policy Statement.[68] Thus, the inconsistency of approach among Regional Councils in terms of CO_2 discharge applications continued. This became obvious when Mercury Energy applied for CO_2 resource consents for its proposed 110MW gas power station to be built at Penrose, Auckland. The Auckland Regional Council went on to grant consents with review clauses but no mitigation conditions at all. The issue was evaded by delegating it to a Working Group on CO_2 Policy established in mid 1995. Throughout this time, industry maintained a behind-the-scenes opposition to the National Policy Statement.

6. The 1995 Berlin Climate Summit

The Stratford Inquiry released its results just before the Berlin "Climate Summit" - the First Conference of the Parties (COP1) under the FCCC, which Environment Minister Simon Upton attended. At this meeting, New Zealand was closely associated with the notorious 'JUSCANZ'[69] group of countries, which were widely associated with blocking efforts being made by the European Union and a broad group of developing countries to establish binding emissions reduction targets for industrialised countries.

The net approach was a "critical principle for New Zealand" and it was a "a major objective" to build it into the FCCC framework.[70] During his Ministerial Statement to the COP, Mr. Upton announced New Zealand would achieve a net emissions reduction of 50-59% compared to 1990 levels, in line with estimates submitted in the first National Communication. The cabinet-approved brief to the New Zealand delegation revealed that a core part of its mission at the meeting was to "continue its efforts to lock-in the net approach to FCCC commitments, includ-

68. The New Zealand government was itself painted in a corner. Without the ability to implement its agreed carbon taxes in 1997, it appeared that the government could not even signal its commitment to domestic action, and instead puffed up the rhetoric of 'global least cost' action.

69. Japan, United States, Canada, Australia and New Zealand.

70. Ministry of Foreign Affairs and Trade, *New Zealand Delegation Brief to the First Conference of Parties* (1995).

ing through further representations in selected capitals by the Minister for the Environment."[71] This was despite the recognition within Cabinet that there was a risk to future (international) emissions reductions[72] and "NZ's current reputation as a good international citizen on environment issues, and our international reputation as an environmentally aware country."[73] Accordingly, it was not surprising that when the Minister put the official position forward:

> The reactions ranged from fascination and support to sceptical curiosity. In the case of the small Pacific Island states there is some outright hostility which the Prime Minister will have to bear in mind at future meetings of the [South Pacific] Forum.[74]

He further acknowledged the domestic risks of this approach because the National Party was the "only major party" supporting the net approach in New Zealand. This "rais[ed] significant uncertainties for the policy beyond the next election." This was especially so given that New Zealand had to "accommodate the second highest growth rate in CO_2 emissions within the OECD."[75] Moreover, the bubble on sinks was about to burst a few months later, as new measurements for carbon sequestration in the 1990 base year revealed a dramatic change in fortune, with net emissions set to rise by over 60%.[76]

Far from causing a change in policy, in August 1995, the New Zealand government delegated this and other controversial issues to an appointed Working Group on CO_2 Policy, thus avoiding making any final decisions for nearly a year.[77] The

71. MFAT, *supra* note 70
72. The Treasury noted under a subheading 'Benefits of limiting gross emissions' that "It also remains unclear what the consequences are of not providing an 'adequate' response in reducing gross emissions." This quote is in a fax from The Treasury to Officials working on the July 1994 Cabinet paper.
73. Ministry of Commerce, *Briefing to the Minister for the Environment*, 28 June 1995. Even the Ministry of Commerce agreed: "The main point is that we probably cannot get away with doing nothing on sources for the reasons MFAT raise in their paper."
74. Report from the Minister of the Environment to the Cabinet Committee on Enterprise, Industry and Environment, *Reviewing the Government's Climate Change Policy*, 26 June 1995.
75. *Id.* By 1995, Cabinet ministers were fully informed that New Zealand's projected emissions growth were among the highest in the OECD. Indeed the collation of national emissions reporting to the FCCC showed that New Zealand was projecting a 15.7% increase in gross emissions by the year 2000, and higher projections were noted subsequently. *See* UNFCCC, *Projected Anthropogenic Emissions Of CO_2, Excluding Land Use Change And Forestry*, Document FCCC/SB/1996/1 Table B.1(Provisional) (1996). There were some concerns that projections were overinflated, making it easier to achieve the 20% gross emissions reduction, compared to the 80% from tree planting.
76. Report by Forestry Research Institute, *Carbon Sequestration by Plantation Forests; Calculations revised as at June 1995* (1995). Note that the news release on the report by Hon. Simon Upton, Minister for the Environment and Hon John Falloon, Minister of Agriculture, 29 August 1995, did not mention the substantial revision of figures, although the Working Group on CO_2 Policy did so in its Discussion Document, *Climate Change and CO_2 Policy, A Durable Response*, at p. 57.
77. The Working Group reported in June 1996. Working Group, *supra* note 4.

Working Group's job was to analyse and report on the net approach, the focus of the National Policy Statement under the Resource Management Act, and the question of the best economic instrument for New Zealand (broadly another iteration of the carbon taxes versus tradeable emissions permits discussion, which were issues raised in the first 'policy options' document in May 1990).

The conclusions of the Working Group report left the net approach in place and the National Policy Statement was deemed unnecessary if an effective economic instrument was put in place. At the end of 1996, the government failed to implement the one economic instrument - a low level carbon charge - that it had been talking about since 1990. At this point, it deferred the issue once more until after the Kyoto COP.

7. Conclusion

The background to New Zealand's current climate policy position reflects a major missed opportunity: to mobilize a public that was concerned about the environment and that supported solutions to rising energy-related CO_2 emissions. However, the public was never presented with real on-the-ground choices. The power crisis and the energy sector 'reform' did not produce genuine change. Transport policy never took off, and the government's energy efficiency authority did not have the mandate nor budget to move the nation in a new direction.

The government constrained itself by fixating on a theoretically pure 'market' approach. As such, it failed to address the larger question of where New Zealand was going to position itself in a world increasingly taking climate change seriously. An excessive reliance on fossil fuels could ultimately turn the nation's energy policy on its head.

New Zealand has superb wind resources, yet had only one test turbine in operation by the middle of this decade. By contrast, Denmark now commands 60% of the burgeoning international wind export market. The Danish government established clear goals for renewable energy development and use, as well as short-term incentives. New Zealand's government appears loathe to take any actions that could render its markets 'impure.' The fact that there was no representative from the 'solutions' end of the private sector on the government's Working Group on CO_2 Policy - appointed in 1995 to advise the government on key aspects of its policy - tells a story in itself. Yet three of the four members of the group, chosen by government, were associated with the energy sector (ECNZ), the forestry industry and petroleum. The government could have chosen to break up the lobbying monopoly of the industries with as much enthusiasm as it did in breaking up the electricity industry.

In choosing to ignore its relations with the Pacific in its approach to climate change, the New Zealand government has not only displayed a marked lack of statesmanship and moral leadership, but it has also failed to lay the foundation for a new energy economy. Rather, the New Zealand government dug itself ever deeper into a hole in the first half of this decade, and is now casting around in the international arena for a way out. It is doing this by seeking to weave into the text of new international agreements sufficient 'flexibility' to obviate the need for domestic action. This is unfortunate, because New Zealand's maneuverings may prove to be a Pyrrhic victory if it ultimately fails to prepare itself for a world in which the pressure grows stronger to reduce emissions and the need for a diversified energy base becomes more critical.

9. NEW ZEALAND AND THE CLIMATE CHANGE DEBATE: 1995-1998

ALEXANDER GILLESPIE
Senior Lecturer in Law, Waikato University
Te Whare Wananga o Waikato, Hamilton, New Zealand

1. Introduction

New Zealand (NZ) has the fourth highest per capita emissions (on a cumulative basis) on the planet. However, NZ does not have to make the average 5% cuts in greenhouse gases that most other Annex I countries have to make following Kyoto. Moreover, with the current way the Kyoto Protocol is structured, it is possible that positive economic impacts will result for NZ. These benefits will be achieved with minimum disturbance of the market as it currently operates. Although this may be 'good' for NZ, it may also be highly counterproductive to achieving a solution to what may be the biggest environmental problem to face the planet in the next century.

2. The Scientific Consensus on Climate Change

Since the beginning of the 1990s, the global warming debate has been heating up. Within the last few years there have been a growing number of reports, all reiterating a similar theme - the planet could be in for a large climatic shock. The mid-1990s were very hot years for the planet. 1997 was another record year (although El Niño kept NZ cool).[1] 1998 has seen heatwave records all over the country, manifesting itself in buckled railway tracks, dried up rivers and very bad droughts.[2] These record-breaking temperatures are not direct evidence of climatic change.[3] However, they may be the portent of a nearly incomprehensible environmental disaster, which if not properly addressed, could leave a disastrous legacy for the next generation.[4] This is the background by which all analysis of success in climate negotiations must be judged.

---

1. NZPA, *El-Nino Blows Cool On NZ*, NZ HERALD, 19 January 1998.
2. K. Perry, *Heatwave Sets National Record*, NZ HERALD, 3 March 1998.
3. *See* Chs. 1 & 2 of A. GILLESPIE, BURNING ISSUES: THE FAILURE OF THE NEW ZEALAND RESPONSE TO CLIMATIC CHANGE (1997).
4. *See* S. Retallack, *Kyoto: Our Last Chance*, 27(6) ECOLOGIST (1997), at 229-235.

A. Gillespie and W. C. G. Burns (eds.), Climate Change in the South Pacific: Impacts and Responses in Australia, New Zealand, and Small Island States, **165–187.**

At the end of 1995 the Intergovernmental Panel on Climate Change (IPCC), released its fourth report.[5] The IPCC expects that increasing concentrations of carbon dioxide, methane, nitrous oxide and halocarbons will probably begin to affect the climate by a process of warming within the next 100 years. Specifically, it suggests as a "best guess" that the average temperature will be two degrees (Celsius) warmer by 2100 than it is now.[6] This increase represents a warming greater than in the previous 10,000 years.[7] Accompanying the temperature increases may be rises in sea levels of around 50 centimetres.[8]

Although these projections are ostensibly small, the implications may be cataclysmic, as climatic change is expected to occur at a rate that outpaces the speed by which certain ecosystems grow, reproduce and re-establish themselves. Accordingly, complete ecosystems may end up either migrating or collapsing in the face of temperature changes. This is expected to lead not only to a reduction in global biodiversity, but also to a disruption of crop productivity and possibly more famine[9] and the disappearance of up to one-third of the world's forests.[10] Aquatic and coastal ecosystems such as wetlands, salt water marshes, coral reefs and sandy beaches are also at great risk. Between one-third and one-half of existing mountain glacier mass could disappear over the next 100 years. Desertification is expected to become more extreme. There may also be an intensification of the global hydrological cycle, which may have major impacts upon regional water resources.[11] The coastal populations of the world are facing a particularly precarious future. Estimated land losses from sea level rises range from 17.5% for Bangladesh up to 80% for the Majuro Atoll in the Marshall Islands.[12]

Climate change may also have a direct impact upon human health, with temperature extremes both in terms of intensity and duration. This may result in increased mortality and illness. Additionally, the incidence of vector-borne diseases such as malaria, dengue and yellow fever is expected to increase in existing regions and spread to new areas.[13]

---

5. Intergovernmental Panel on Climate Change, *Climate Change 1995; Impacts, Adaptations, and Mitigation*. United Nations Environmental Program & World Meteorological Organisation (1996). *The Science of Climate Change: Summary for Policy Makers*. United Nations Environmental Program & World Health Organisation. Geneva. Together, these two publications are known as the 'Synthesis' Report.
6. *Id.* at 44-45.
7. This '10,000 year' statement originated at the "Toronto Summit." *See* United Nations Environmental Program & World Meteorological Organisation, *Proceedings of the World Conference on the Changing Atmosphere: Implications For Global Security* (1998), at vii. The statement was subsequently adopted by the IPCC in its Synthesis Report, paragraph 2.7.
8. *Science of Climate Change*, *supra* note 5, at 45-47.
9. *Climate Change 1995*, *supra* note 5, at 10-12.
10. *Id.* at 6-9.
11. *Id.* at 9-10.
12. *Id.* at 9-10 & 13-14.
13. *See* World Health Organisation & United Nations Environmental Program, *Climate Change and Human Health* (1996); *Climate Change 1995*, *supra* note 5, at 13-14.

The likely impacts of climatic change on NZ remain unclear. It is generally expected that the effects of climate change will not be as severe in NZ as in many other countries.[14] While some consequences in NZ may be obvious, such as a 50cm increase in water levels around NZ's 1,100 kilometre coastline,[15] other effects may not be. For example, in some areas increased crop growth may occur due to the warmer temperatures, whereas in others, certain crops may no longer be viable.[16] The kiwifruit industry situated in the Bay of Plenty may be an example of the latter. NZ may also expect incursions of undesirable subtropical grasses, pests and diseases.[17] It is possible that all of these effects could drastically and adversely alter NZ's agricultural and tourist-based economy.[18]

Despite these threats, NZ's climate change response has tended to move between the sublime and the ridiculous. For example, in early 1996 it was reported that the "Long range climate forecasters" in NZ were "optimistic" about the effects of climate change. There were two principal reasons given for this. The first was that warmer temperatures could lower winter electricity demand whilst melting snow would increase hydro-lake inflows, thus enhancing electricity production. The second reason was that as less mountainous countries lose their snow, NZ could gain a strong tourist drawcard.[19]

3. From Rio to Kyoto

As the science of climatic change has become more precise, scientists have pinpointed the required reductions in emissions of the climatic gases to avoid serious impacts have been pinpointed. Specifically, the IPCC has been telling the community of nations since 1992 that at least a 60% reduction in carbon dioxide emissions (the main greenhouse gas) is required just to stabilise its build-up in the atmosphere.[20] This 60% reduction yardstick is substantially more than the wealthy countries of the international community have agreed to reduce their carbon dioxide emissions. The international community was initially guided by the 1992 Framework Convention on Climate Change[21] (FCCC) under which the industrialised countries of the world agreed to "aim" at stabilising their carbon

---

14. M. Manning, *Climate Change -What is Happening and Why?*, AIC Conferences; *Greenhouse Gases and Carbon Dioxide Policy in New Zealand* (1996), at Ch. XIII.
15. *See* M. McGlone, T. Clarkson & B. Fitzharris, Unsettled Outlook: New Zealand in a Greenhouse World (1990), at 77-86.
16. J. Salinger, Greenhouse New Zealand. Our Climate: Past, Present & Future (1991), at Chs. 6-11. See also McGlone, *supra* note 15, at 86-97.
17. National Science Strategy Committee, *Report on Climate Change, Royal Society of New Zealand* (1995) 28-29, 33; New Zealand Climate Change Programme, *Climate Change: Impacts Upon New Zealand,* Ministry for the Environment (1990), at 108-133.
18. National Science Strategy Committee, *supra* note 17, at 26-27.
19. W. Gamble, *Long-Range Climate Forecasters Optimistic*, New Zealand Herald, 12 March 1996, at 1.
20. Intergovernmental Panel on Climate Change, *Climate Change 1994: Radiative Forcing Of Climate Change & An Evaluation of the IPCC 1992 Emission Scenario* (1995), at 12-14 & 19-24

dioxide emissions at 1990 levels.[22] NZ signed and ratified this agreement,[23] but suggested at the time of signing that it intended to go even further by reducing its carbon dioxide emissions by 20% below 1990 levels by the year 2000.[24]

Unfortunately, even the goal of stabilising carbon dioxide emissions at 1990 levels has proved too onerous for a number of countries, who are now on track for a "carbon-blowout."[25] NZ has been among the more notable failures in reducing its carbon dioxide emissions to 1990 levels.[26] Among the other developed countries of the world, only Portugal and Finland are projected to have a higher growth rate in carbon emissions than NZ by the year 2000.[27]

Since 1987, NZ's emissions have risen from 23 million tonnes per year, to around 27.4 in 1994.[28] These are projected to reach 31.5 million tonnes by the year 2000.[29] Thus, by the year 2000, NZ's gross emissions will have risen above their 1990 levels by between 22-25%. If the focus is on the net approach (where a total is arrived at after the gross emissions have been added or subtracted to by whatever amounts of carbon dioxide have been absorbed or released by a country's sinks - usually forests) then the increase in total net emissions is projected to be 61% higher than they were in 1990.[30]

---

21. *United Nations Framework Convention on Climate Change*. UNCED. A/ AC.237 / 18 (Part II) / Add.1. 15 May 1992.
22. *Id.* at Art. 4 (2)(c).
23. New Zealand ratified the FCCC on 18 September 1993. It was the 34th country to do so.
24. This was accepted as a formal target in Ministry for the Environment, Climate Change: The New Zealand Response. New Zealand's First National Communication Under the Framework Convention on Climate Change (1994), at 7.
25. *See Second Compilation & Synthesis Report of National Communications*, Doc. FCCC/CP/1996/12. *See also* C. Flavin, *Facing Up To the Risks of Climate Change*, in STATE OF THE WORLD: 1996 (L. Brown ed., 1996), at 21, 29-32.
26. *Working Group on Carbon Dioxide Policy, Climate Change and CO2 Policy: A Durable Response*, Ministry for the Environment (1996), at 59, 62.
27. Ministry of Foreign Affairs & Trade, *Third Meeting of the Ad-Hoc Group on the Berlin Mandate and the Second Meetings of the Subsidiary Bodies: New Zealand Delegation Brief*, Ministry of Foreign Affairs (1996), at 2-3.
28. NZPA, Emissions of Carbon Dioxide Rise by 7%, NEW ZEALAND HERALD, 12 Jan. 1996.
29. Working Group, *supra* note 26, at 59.
30. Delegation Brief, *supra* note 27, at 2. This blow-out from the perspective of the net approach is due to the high rate of absorption when the counting started in 1990, as well as a failure to reach the necessary planting rates needed to contain the already sequested carbon and the growth in gross carbon dioxide emissions.

In the light of such failures by NZ and other nations to take substantive measures to reduce emissions, the international community reassembled in Berlin in 1995, Geneva in 1996 and Kyoto in 1997 in attempts to establish more stringent reduction targets. At the Berlin meeting, the Alliance of Small Island States (AOSIS) proposed a 20% reduction in carbon dioxide emissions by 2000 and specific directives on energy efficiency.[31] NZ voted against the adoption of these proposals.[32] Nevertheless, NZ did support the Berlin Mandate,[33] by which the signatories promised to commit themselves to further specific reductions at the conference of the parties in 1997 at Kyoto. However, at Kyoto it was successful in exempting itself from the average 5% cut in emissions agreed to by most other Annex I countries.[34]

4. New Zealand's Net Response

Between 1994 and 1995, NZ's climate policy coalesced.[35] Leading up to the 1992 FCCC, the NZ delegations concurred with the necessity to reduce greenhouse emissions and followed the broad trend of international opinion.[36] In 1994, this changed as the government adopted two basic precepts for its climate change response. The first is minimal interference in the market. The second, a corollary of this, is that if action is required, it must be done at with the "lowest" or "least" ("total lowest cost to society")[37] possible economic cost to society.[38]

---

31. AOSIS Protocol. INC. A/AC.237/L.23. 27 September 1994. Article 3 (1)(a) of the Draft Protocol suggested that each of the Annex 1 parties shall, 'Reduce its level of anthropogenic emissions of carbon dioxide by at least 20 % by the year 2005.' The Protocol is included in this volume as Appendix 3.
32. *See* NZPA, *Greenhouse Gas Policy Under Fire*, N.Z. HERALD, 1 Apr. 1995, at 5; *New Zealand Supporting Island Nations*, N.Z. HERALD, 6 Apr. (1995), at 9; *New Zealand Takes Lonely Stand At Climate Summit*, N.Z. HERALD, 7 Apr. 1995, at 15. The specific justifications for this can be found in the Delegation document, Ministry for the Environment, *New Zealand Intervention, 1995. February 8: Agenda Item 7 B: Review of the Adequacy of the Commitments* (1995).
33. *The Berlin Mandate: Decision*. 1/CP. 1 (1995).
34. The NZ target is a 0% increase from the 1992 FCCC target by the end of the first commitment period (2008-2012). That compares with an average for developed countries of around -5%, and a target for the US of -7%. The precise targets vary among countries on a 'differentiated' basis, ranging from Iceland, at the upper limit, which received a target of 110% of its 1990 level; Australia - 108%, NZ - 100%, Canada and Japan - 94%, US - 93%, and the EU - 92%. The cumulative reduction, should (according to Article 3) be in the order of 5%. *See* Ministry of Environment, *Kyoto Outcome: A Summary*, 1 Dec. 1997, at 17.
35. This is not to suggest that New Zealand had not formulated a response to climate change before 1994. Indeed, prior to this time three important reports were issued by the New Zealand Climate Change Programme, operating through the Ministry for the Environment. However, it was not until 1994 that a specific method to reduce carbon emissions was formally introduced. *See* Ministry for the Environment, *Exploring the Options for Reducing Net Emissions of Carbon Dioxide: Consultation Document*. Ministry for the Environment (1994).
36. *See* G. Palmer, *Environmental Politics: A Greenprint For New Zealand* (1990), at 59-75.
37. Minister for the Environment, *Upton Announces Climate Change Position*, Press Release, 2 Dec. 1997.

The lowest cost option for NZ with regard to carbon dioxide emissions is to plant more (pine) forests, which sequester carbon from the atmosphere.[39] In theory, if more trees are planted than carbon dioxide is emitted into the atmosphere then NZ should be able to achieve a net reduction in its emissions. This 'net approach' is in contrast to an approach where less emissions are produced at the source through, for example, approaches such as either energy efficiency or by reducing demand for carbon dioxide intensive processes.[40] This is known as gross reductions. NZ rejected the gross approach, and instead, in 1994 proposed that the control of carbon dioxide emissions would be achieved domestically with more tree plantations making up 80% of the response, and voluntary agreements with industry and the promotion of energy efficiency making up the residual 20%.[41] In 1997, sequestration as a response mechanism was reduced to 60%.

The net approach was definitively embraced in the 1995 Stratford Inquiry[42] into the proposed gas-fired power station at Stratford. This Inquiry was sparked by the challenge of a consortium of "Green Groups" who argued that the proposed power station, which would increase NZ's gross emissions of carbon dioxide by 5%, was inconsistent with NZ's obligation under the FCCC to stabilise its carbon emission rates at 1990 levels.[43] The Inquiry concluded that the building of the power station was in accordance with the FCCC, provided that enough trees were planted to offset the growth in emissions.[44] Accordingly, it was accepted that the net approach

---

38. S. Upton, *Address to NZ National Committee*, in National Committee, World Energy Council, *Impacts of Climate Change Policy on New Zealand Incorporated* (1994), at 2, 3-4. *See also* The New Zealand Response, *supra* note 24, at 7.
39. R.K. Dixon et al., *Conservation and Sequestration of Carbon: The Potential of Forest and Agroforest Management Practices*, GLOBAL ENVIRONMENTAL CHANGE (June, 1993), at 159, 168-169, 172. A. Chisholm & A. Moran, *A Perspective on the Potential Economic Impacts of Climate Change Policy in New Zealand*, in *Impacts of Climate Change Policy on New Zealand Incorporated.* (New Zealand National Committee of World Energy Council ed. 1994), at 1, 7-9. The argument that the planting of carbon sinks was not a 'costly option' was accepted by the Board of Inquiry, *Proposed Taranki Power Station: Air Discharge Effects* (The 'Stratford Inquiry') (1995). *See* paragraphs 9.55, 9.57-9.62, 9.62.
40. This net approach was recognised by the INC at its first session. *See* Decision 1/1 which refers to 'appropriate commitments . . . for limiting and reducing net emissions of carbon dioxide and other gases.' [emphasis added] *Report of the Inter governmental Negotiating Committee For A Framework Convention on Climate Change on the Work of its First Session.* U.N. GAOR. INC/FCCC, 1st Sess. U.N. Doc. A/AC.237/6 (1991), at 24. The approach was favoured by certain countries as it took the pressure off to reduce specific gases, such as carbon dioxide. Theoretically, this should allow countries some 'breathing time' as they can concentrate on other ways to reduce greenhouse emissions. *See* W.A. Nitze, *A Failure of Presidential Leadership*, in NEGOTIATING CLIMATE CHANGE: THE INSIDE STORY OF THE RIO CONVENTION (I.M. Mintzer ed., 1994), at 187-189.
41. *New Zealand Response*, *supra* note 24, at 9-10.
42. *The Stratford Inquiry*, *supra* note 39.
43. *Stratford Inquiry*; *Greenpeace and Eco Submission*, 1994, in possession of the author.
44. Paragraph 12.2 (23) concluded that 'Mitigation . . . of the adverse environmental effects of allowing the discharge of CO2 could be carried out by ongoing planting of a sufficient number of trees to act as a permanent carbon sink. The imposition of a mitigation condition to this effect would be reasonable and practicable in the circumstance.' *Stratford Inquiry*, *supra* note 39.

was consistent with the objectives of the FCCC, as the FCCC did not expressly preclude this strategy.[45]

Advocacy of the net approach as a legitimate strategy to meet FCCC obligations has formed the basis of NZ's climate policy, and was again re-emphasised by the 1996 Working Group on Carbon Dioxide.[46] These two documents have given NZ the mandate to push (or at least, justify) the net approach at all relevant international meetings as the foremost way (for NZ) to reduce greenhouse emissions.

The adoption and overt advocacy of the net approach resulted in NZ taking an "isolated stand"[47] at some of the international meetings in the run-up to Kyoto, with many countries reacting with "open hostility."[48] At Kyoto, NZ played a dominant role in the railroading of the agenda which, according to the *Guardian International* resulted in "angry exchanges about sinks . . . the inclusion of such schemes . . . was regarded by some participants as an attempt to create a loophole for evasion."[49] NZ's unwavering advocacy of this approach led to portrayal of NZ's stance as "parochial self-interest, combined with sleight of hand."[50]

In the run-up to the Kyoto Conference, there was no agreement on whether or how carbon sinks should be included in the Protocol.[51] NZ pressed strongly for the inclusion of sinks,[52] characterizing their inclusion in any emissions regime at various times to be "very important"[53] "vital," and "most important."[54] Within the domestic frame of reference, the impetus came from the assertion that without inclusion of sinks there could be "a major loss to the NZ economy"[55] as the sinks issue involved "big winners and losers."[56] NZ forged the way on sinks to such an extent that the Ministry for the Environment noted that "had NZ not been there sinks might well have dropped off the table . . . NZ played a lead role in the sinks negotiations."[57]

---

45. Thus, 'The FCCC in effect urges a net approach since it is worded in terms of both reducing the emissions and enhancing absorption.' *Stratford Inquiry*, *supra* note 39, para. 4.44 (i). *See also* paras. 4.30- 4.34, 9.2, 9.14-9.21, 10.8, 12.4 (4).
46. Working Group, *supra* note 26, at 66-71.
47. *Supra* note 32.
48. Cabinet Committee on Enterprise, Industry & Environment, *Reviewing The Government's Climate Change Policy*, 26 June 1995.
49. P. Brown, *Kyoto Fails Test on Climate Change*, GUARDIAN INT'L, 14 December 1997.
50. *Kyoto*, 3(2) ECO NEWSLETTER (1997).
51. Associate Minister of Foreign Affairs and Trade, Late Paper to CIE Climate Change Negotiations, 20 November 1997, at 3.
52. Draft Supplementary, Note to Minister Re: AGBM8, 31 October 1997, at 1.
53. *Kyoto Outcome*, *supra* note 34, at 2.
54. *Late Paper, supra* note 51, at 2 & 5.
55. *Kyoto Outcome*, *supra* note 34, at 2.
56. Office of the Minister of the Environment, *Report on Overseas Travel: Third Conference of the Parties to the Framework Convention on Climate Change*, 8-10 Dec. 1997, at 2.
57. *Kyoto Outcome*, *supra* note 34, at 2.

The fervour of NZ's support for the sinks approach at Kyoto was not surprising given the structure of its mitigation policy when compared to other countries. That is, working from the first and second national communications of Annex I parties under the FCCC the aggregate removals from this category represent 6.7% and 8% of total greenhouse gas (GHG) emissions in carbon dioxide equivalent and total carbon dioxide emissions of the reporting Annex I Parties.[58] However, the amounts varied widely amongst the parties. For example, whereas the average was 8% reductions by sinks of national carbon dioxide emissions,[59] for NZ the rate was the highest of any Annex I country at 81%.[60]

This is not to infer that the NZ approach was met with universal hostility. Indeed, Australia, Canada, Norway, Iceland, the United States, and the Russian Federation all supported NZ's drive to include sinks within the Kyoto Protocol. However, opposed to them were the European Union, Japan, Denmark, the United Kingdom, France, Kenya, the Marshall Islands, Nauru, Peru, Uzbekistan, China, Brazil, and delegations speaking on behalf of the Alliance of Small Island States.[61] These countries all wished to defer any decisions upon what role sinks should play in the FCCC until many of the uncertainties had been satisfactorily resolved.[62] They argued that it was inequitable to adopt an approach that only a few countries can geographically and demographically adopt. Moreover, they contended that sequestering carbon only postpones the inevitable, as the carbon that is stored will ultimately be released.[63]

58. United Nations, *Methodological Issues: Synthesis of Information From National Communications on Sources and Sinks in the Land-Use Change and Forestry Sector*, Technical Paper. FCCC/TP/1997/5. 20 Nov. 1997, at 5-6.

59. Only in the UK and Australia were these areas not a sink. That is, they were a source of emissions. For Australia, they added 24% and for the UK it was a 3% increase. *See id.*, Table 1.

60. *Id.* By 1997, the projected removal rate over the period 1990-2020 had fallen to 63% of the response of dealing with the gross carbon dioxide emissions from energy sources and industrial processes, *Climate Change: The New Zealand Response II*, Second National Communication under the FCCC (1997), at 9, 78.

61. *See* International Energy Agency/OECD, *Climate Change Initiatives: 1994 Update, OECD Countries* (1995), at 19-23.

62. Technical Paper, *supra* note 58, at 4-11. The specific results to the questionnaire on sinks can be found in United Nations, *Response From Parties on Issues Relating to Sinks*, FCCC/AGBM/1997/MISC.4. & MISC.4.Add 1. & MISC.4.Add 2, Ministry for the Environment (1997), Draft Supplementary Note to Minister Re: AGBM8., 31 Oct. 1997, at 1. Minister for the Environment, *supra* note 37; *The Third Conference of the Parties to the United Nations Framework on Climate Change*, EARTH NEGOTIATIONS BULL. (ENB) 12 (67):1; 12(71):1; 12 (76):1,2,3; 8 ECO NEWSLETTER, 9 December 1997, at 3; Ministry of Foreign Affairs and Trade, *The United Nations Framework Convention on Climate Change: The Third Session of the Conference of the Parties: The New Zealand Delegation Report* (1997), at 4.

63. *See* NATIONAL ACADEMY OF SCIENCES, POLICY IMPLICATIONS OF GREENHOUSE WARMING: MITIGATION, ADOPTION AND THE SCIENCE BASIS (1992), at 76-77.

It is very important to recognise that this was not an argument against the possible role of sinks as part of an *appropriate* mechanism in *context* for combating global warming.[64] However, this is the view that NZ advanced. It contended at Kyoto that if sinks were not included in the Protocol, then "taking it from the perspective of the atmosphere"[65] an emissions loophole of about a billion tonnes of carbon dioxide per year would be created.[66] It was also suggested that if they were not addressed "a crucial part of the Berlin Mandate will not be met."[67]

Admittedly, the FCCC does refer to commitments by the Parties with regard to "anthropogenic emissions by sources and removals by sinks" in the context of inventories and mitigation actions.[68] The Berlin Mandate utilised similar words.[69] However, the Mandate did not specify how sinks were to be incorporated into reduction targets. Moreover, any failure to include them into the Kyoto Protocol would not have implied (as NZ suggested) that they could be ignored. Indeed, both the European Community and Japan explicitly rejected the idea that excluding sinks from the Protocol would reduce any incentive to foster and protect them. Thus, they maintained that the protection and enhancement of sinks should be done regardless of obligations under the FCCC.[70]

5. The Difficulties of the Net Approach

As brought out at Kyoto, the net approach is troubling for three reasons: scientific uncertainties; definition problems (exactly what the term "sink" encompasses); and policy concerns.

Scientific uncertainties have been highlighted in both domestic and international contexts. In NZ, the 1995 Report of Climate Change by the NZ National Science Strategy Committee urged caution in the government's pursuit of this net approach because "a number of research questions" have not yet been satisfactorily addressed.[71] Internationally, the IPCC Second Assess-

---

64. Indeed, paragraph 21 of the 1989 *Noordwijk Ministerial Conference on Climate Change* calls for an additional 12 million hectares of forests to be planted by the year 2000. This document is reprinted in 5 AM. U. J. INT'L L. & POL'Y (1990), at 592. For the general recognition of the importance of sinks, *See* Principle 2(b) and Principle 4 of the *Non-Legally Binding Authoritative Statement of Principles For a Global Consensus on the Management, Conservation and Sustainable Development of all Types of Forests*. UNCED. A/CONF. 151/26 (Vol. III) 14 August 1992.

65. *See After Kyoto*, ENERGY WISE NEWS, 13 April 1998.

66. *NZ Intervention*, December 2, Committee of the Whole, FCCC, COP 3. Kyoto. 1, *Draft Supplementary Note to Minister Re: AGBM8*, 31 October (1997), at 1.

67. *Statement on Behalf of New Zealand*, the Honourable Simon Upton, Minister for the Environment, Plenary, FCCC. COP 3, 8 December 1997, at 2.

68. *Id.* at arts. 4.1 & 4.2

69. *Id.* at art. II 2(a). "by sources and removals by sinks . . . "

70. *Response From Parties, supra* note 62, FCCC/AGBM/1997/MISC.4, at 12. *See also* the United Kingdom (On behalf of the EU), United Nations, Response From the Parties; *see also Issues Relating to Sinks*, FCCC/AGBM/1997/MISC.4. Add 1. (1997), at 22.

71. *Report on Climate Change, supra* note 17, at 35.

ment Report[72] clearly sets out the methodological problems and scientific uncertainties surrounding the measurement of sinks.[73] The IPCC Revised Guidelines in 1996 also recognised that major uncertainties exist relating to emissions factors and activity data for sinks. In his report to the seventh session of the Subsidiary Body For Scientific And Technological Advice in October 1997, Professor Bolin, on behalf of the IPCC, reiterated the problems relating to sinks. He specifically pointed out in relation to terrestrial ecosystems, "the error margin for the determination of sources and sinks are quite large" and that "because of our limited understanding and lack of observations simplified methods have been proposed by the IPCC and have been adopted by the FCCC for the assessment of sources and sinks by countries." As these were "very approximate," Professor Bolin highlighted the importance of analysing "their possible shortcomings" in the context of the IPCC's work.[74]

According to the Secretariat, there are multiple concerns in this area.[75] As such, it was not surprising that the rate of uncertainty in this area was pegged at 60% by the IPCC.[76] This figure was higher than NZ's current rate of uncertainty (25%)[77] but less than Australia's (about 80%).[78] A useful indicator of the uncertainties associated with calculating sinks in this area is the substantial variation between the first and second national communications for the Annex I signatories.[79] These ranged from 430% for the United Kingdom, down to 3% for France (NZ had a 16% percent rework between reports).[80]

72. IPCC, *Second Assessment Report*, Vol. 2. (1995).

73. For example, it is difficult to obtain accurate estimates of carbon sequestration rates for different species of trees, accurate information on the type of land being planted; the effects that mass planting will have on volatile hydrocarbons and carbon monoxide. Information is also required to improve the understanding of the carbon exchanges between soil, forest and atmosphere; as well as decomposition and storage times of carbon deposited after soil erosion. There is also a need to develop sequestration models for indigenous as well as plantation forests. *Report on Climate Change, supra* note 17, at 35-36. These problems have also been reflected on the international stage. For example, Borine and Ripert stated, "It is in fact quite impossible, in the present state of scientific knowledge, to aggregate on a common basis the figures of reductions of emissions of the different greenhouse gases . . . it is also difficult at this stage, to identify and to quantify all the sinks. Commitments on net emissions could therefore be considered impossible to implement." D. Borine & J. Ripert, *Exercising Common But Differentiated Responsibilities*, Mintzer, *supra* note 40, at 77, 88. *See also* K. Brown & N. Adger, *Estimating National Greenhouse Gas Emissions Under the Climate Change Convention, Global Environmental Change*, June, 1995, at 145-158.

74. Noted by the Marshall Islands, *Response From Parties, supra* note 62.

75. United Nations, *Compilation of Responses From Parties on Issues Related to Sinks: Comments from Parties & Note By Secretariat*, FCCC/AGBM/1997/INF.2. 29 November (1997), 3, 4-11.

76. IPCC, *Guidelines for National Greenhouse Gas Inventories* (1996), at A1-1, Table A1-1.

77. New Zealand, *Response From Parties, supra* note 62, FCCC/AGBM/1997/ MISC.4, at 40.

78. Australia, *Response From Parties, supra* note 62, FCCC/AGBM/1997/MISC.4. Add 1, at 7.

79. The Technical Paper on sinks, *supra* note 58, noted that "None of the problems with comparability of CO_2 emission estimates from this sector identified in the compilation and synthesis of first national communications appear to be resolved. The information provided did not shed additional light on various assumptions related to the definitions of anthropogenic activities and their treatment for emissions reporting purposes." FCCC/SBI/1997/19, Annex, para 29.

The 1998 Bonn meeting of the Subsidiary Body for Scientific and Technological Advice suggested several sources of technical uncertainties:

[1] Differing interpretations of source and sink categories or other definitions, assumptions or units;

[2] Use of simplified data formats and average values (especially emission sequestration factors);

[3] Uncertainties introduced by changing national models for estimating activities, or random errors in reporting; and

[4] Inherent uncertainty in the scientific understanding of the basic processes leading to emissions and removals.[81]

In addition to the technical difficulties are a number of problems with definitions referring to sinks within the Protocol. This problem, manifest since the first national communications, has reflected the absence of a common reporting framework for emissions from the sub-categories of land-use change and forestry category.[82] That is, many countries have differing ideas about what does -- or should -- constitute sinks.[83] Although some of the issues in this area have been addressed within the definitions of the Protocol,[84] a number of important difficulties remain unresolved. The Protocol acknowledges this, stating: "Uncertainties, transparencies in reporting, verifiability and methodological work"[85] all pertaining to the sinks question still have to be worked out in the short-term future of the Convention.[86]

5.1 EQUITY AND OTHER RELATED CONCERNS

What the nation of Nauru called "the broader arena of socio-economic concerns"[87] related to the sinks issue is yet to be fully resolved. Moreover, the linkage with emissions trading may exacerbate this. Specifically, carbon sequestration, if unchecked, may create a powerful incentive to begin or accelerate the felling of old growth forests, with attendant destruction of biodiversity and expulsion of indigenous peoples - all in efforts to secure space so that sig-

---

80. Marshall Islands, *Response From Parties, supra* note 62, FCCC/AGBM/1997/MISC.4, at 18.

81. *See* SBSTA, *Methodological Issues: Issues Related to Land Use Change and Forestry*. FCCC/SBSTA/1998/INF.1 (1998).

82. This category originally (1995) included changes in forests and other woods biomass stocks; forests and grasslands conversion and abandonment of managed lands. In 1996 the IPCC added changes in mineral soil carbon stocks.

83. With such considerations in mind, the Technical Paper on sinks concluded, "It is clear that further methodological work is necessary in order to ensure that the estimation and reporting of GHG inventory data for land-use change and forestry are consistent, transparent and comparable." Technical Paper, supra note 58, at 10. Very similar conclusions were reached with the compilation and synthesis of the second national communications, "[F]urther research and methodological work is needed to ensure that estimation and reporting is done in a consistent, transparent, and comparable manner." FCCC/SBI/1997/19, Annex, para 29.

natories or their financially poorer partners can take credit for planting fast growing mono-culture forests to fix carbon. The promise of financial benefits for poor countries from sinks may ultimately turn developing countries in favor of incorporating them into the emissions reduction regime. Accordingly, the inclusion of sinks in Qualified Emission Limitation and Reduction Objectives (QELROs) may ultimately end up running counter to the objectives of other international treaties. As such, measures designed to benefit the climate may "do greater harm to the environment at large."[88]

Equity concerns also enter into the debate through the consideration of appropriate actions between responding countries. Equity, by definition, is about fairness and

84. The specific language agreed upon in Article 3 (3) stipulated, "The net changes in greenhouse gas emissions from sources and removals by sinks resulting from *direct human-induced land use change and forestry activities, limited to afforestation, reforestation, and deforestation since 1990,* measured as verifiable changes in stocks in each commitment period shall be used to meet commitments in this Article of each party included in Annex I." (emphasis added) *Use Change and Forestry.* FCCC/SBSTA/1998/INF.1. The definitional concerns with these are such that depending of which approaches are taken, forests could either be classified as sources or sinks. *See* J. Greenough, et al., *Influence of Methodology and Assumptions on Reported National Carbon Flux Inventories*, 2 *Mitigation and Adaptation Strategies for Global Change* (1997), at 267-283.

At Kyoto, there was serious concern over the parameters of this article. "Sink" is defined in Article 1 of the FCCC to mean "any process, activity or mechanism which removes a greenhouse gas, an aerosol or precursor of a greenhouse gas into the atmosphere." The Protocol does not cover "all" forests, but is limited to *direct human-induced land use change and forestry activities, limited to afforestation, reforestation, and deforestation.* The question of whether this includes natural/indigenous forests was addressed within the IPCC Guidelines which state that natural, *undisturbed forests, where still in equilibrium should not be considered* either as an anthropogenic source or sink, and should therefore be excluded from the national inventory calculations. Of course, the question that arises is - what about undisturbed forests that are not in equilibrium?

As it stands, the protection and enhancement of carbon sinks and reservoirs, including commercially planted forests and indigenous forests are a central part of NZ's policy. However, the scientific research on the role of indigenous forests and sinks has been "slow" and it is likely that "it will be some years before the source/sink situation of our indigenous forests with respect to CO_2 can be clarified." It is still "not known whether this reservoir is expanding or shrinking, i.e., whether it is a sink or source." Second National Communication, *supra* note 60, at 9. *See also* Royal Society of NZ, *National Science Strategy Committee for Climate Change* (1996), at 7, 17. This last consideration is of particular concern, as the net approach has, to date, only been built upon the sequestration by planted forests and has largely ignored the sequestration of carbon (or emissions) by indigenous forests. The distinct problem is that no-one is certain how much carbon is being held in these forest, which make up around 95% of New Zealand's forest cover, or how much is being released through residual clearances by humans or by non-anthropogenic natural phenomena, from fires to opossums. These losses or net sequestration could throw off New Zealand's national net estimates of carbon dioxide by up to 300% either way. Working Group, *supra* note 26, at 58-59 & 61-61.

85. Kyoto Protocol, at art. 3(4).

86. *Id.* at arts 3(4) & 5.

87. Nauru Response, United Nations, Response From Parties on Issues Relating to Sinks. FCCC/AGBM/1997/MISC.4 (1997), at 27.

88. Nauru, the Marshall Islands and Kenya all made similar points on this issue. *See Response From Parties, id.*, MISC.4.pp.28. & MISC. Add 1, at 19.

one of the best ways to assess fairness is to determine what would happen if every other nation pursued the same policies (sequestration as the primary response mechanism) as NZ does. This policy is not only geographically impossible (domestically[89] and internationally) for many countries (and impractical as a suitable global response method)[90] but it also touches upon the debate with regard to historical causation[91] and per-capita emission differences between countries. It is also deeply problematic, because although a ton of (weighted) greenhouse gas absorbed from planting trees may be the same for the atmosphere as a weighted ton of greenhouse gas prevented from entering the atmosphere from methane from rice paddies, the social and political context of these reductions is different.

The question must be asked, is it fair that a population of 3.5 million people produce more carbon than much smaller, yet vastly more populated countries than NZ, and then claim that they are doing their bit by planting trees? The answer, from the perspective of international equity, must be no, as the concern is not just about how much these wealthy 3.5 million people emit, but how disproportionate

---

89. For example, it was estimated in 1994 that planting rates of around 100,000 extra hectares each year would be required to offset our carbon dioxide emissions. This planting rate never transpired, topping off at approximately 70,000 hectares per year. This is insufficient to maintain the high levels of sequestration achieved in 1990 (the base year of the FCCC obligations) and gross carbon dioxide emissions that are substantially higher than predicted. The result has been a substantial abyss between what was projected and what has actually been achieved. *The New Zealand Response, supra* note 24, at 40; Working Group, *supra* note 26, at 61-62.

90. For example, just to keep pace with global carbon dioxide emissions (about 3.2 billion tonnes per year), would necessitate planting trees in an area the size of India annually. Even if all the available land was reforested, - approximately 4 million square kilometres (about half the size of Australia) even then, only 10% of the estimated emissions from fossil fuel burning world wide would be achieved by sequestration. This is especially the case in a number of developing countries where there are other demands for the land that may be accorded higher priority. *See* S.H. SCHNEIDER, GLOBAL WARMING (1989), at 188-189; W.N. ADGER & K. BROWN, LAND USE AND THE CAUSES OF GLOBAL WARMING (1994), at 189-195 & 227-230.

91. The balance of equities is clearly tipped against New Zealand given its rather prolific historical record of anthropogenic releases of greenhouse gases through deforestation, agricultural and industrial practices. Because greenhouse gases such as carbon dioxide have life spans up to 200 years, the gases that New Zealand released over the past few centuries may still be having an effect upon the climate system. *See* IPCC, 1994; *supra* note 20, at 32-34. The current emissions total that is extrapolated from the 1990 base year of the FCCC, only reflects a "snap-shot" of any given year, as opposed to, for example, a 150 year period. The importance of historical contributions is recognised in paragraph 3 of the Preamble, which states "that the largest share of historical and current global emissions of greenhouse gases has originated in developed countries, that per capita emissions from developing countries are still relatively low and that the share of global emissions originating in developing countries will grow to meet their social and development needs." Additionally, paragraph 6 of the preamble identifies "the differentiated responsibilities and respective capabilities" of the parties. This differentiation of response measures is operationalised in Article 4, which distinguishes between the commitments of developed and developing countries. Thus, as Justin Ford-Robertson noted, 'Trees planted now, even if they sequestrate CO_2 only serve to credit the historical carbon debt that New Zealand has incurred due to earlier deforestation and consequent release of carbon to the atmosphere.' *The Carbon Balance of Plantation Forestry in New Zealand. A Report for Greenpeace New Zealand* (1993), at 23.

such emissions are on a per-capita basis, as opposed to a sovereign (country-by-country) basis.[92] Thus, the fact that NZ only produces 0.2% of the global greenhouse emissions[93] should be assessed from an equitable perspective also. This is necessary because although NZ's per-capita carbon dioxide emissions are commensurate with other OECD countries,[94] (an average of 8.12 tonnes of carbon dioxide emission per person per year), this is still radically greater than the global average of 2.59 tonnes per person.[95] Moreover, on an "all greenhouse gas" basis NZ has the fourth highest per capita emissions on the planet.[96] Understandably, NZ has tried to avoid discussion of per-capita emission comparisons.[97]

Such uncertainties and concerns should have spoken loudly against including sinks from the Protocol. QELROs must be subject to the highest degree of certainty. Conversely, if uncertainty was to be injected in the Protocol by the inclusion of sinks, they should have been limited in their application. This could have been achieved in two ways.

First, it would have made sense to restrict the use of sinks to verifiable changes in stock up to a set percentage of the QELROs. A logical standard would have been to limit sequestration as part of national response strategies to the global average (8%). If anything, it may have been more sensible to set the ceiling below the 8% global average due to scientific uncertainties in this area.[98] NZ argued against placing any limits on the amounts of sinks in a QELRO,[99] and the Protocol ultimately reflected this position.

The second limitation should have been to restrict sink credits to within national boundaries. However, as it stands the drive is to make sinks part of the international emissions trading market, and this objective has already been encompassed in Article 6 of the Protocol. This too, fosters NZ's drive to make "sink credits . . . an integral part of the international emission trading market."[100] Exactly how far emissions trading will progress is a matter of intense debate. Nevertheless, the

92. *See* V. Bhaskar, *Distributive Justice and the Case of Global Warming*, in THE NORTH, THE SOUTH AND THE ENVIRONMENT: ECOLOGICAL CONSTRAINTS AND THE GLOBAL ECONOMY (V. Bhaskar & A. Glyn eds., 1995), at 102, 103, 105, 115-116; C. Dasgupta, *The Climate Change Negotiations*, in Mintzer, *supra* note 40, at 129, 133-136; A. Rahman, *A View From the Ground Up, id.* at 239, 263.

93. Energy Foundation of New Zealand, *Global Warming: An Alternative Perspective* (1993), at 20.

94. *Climate Change Initiatives. supra* note 61, at 19-26.

95. WORLD RESOURCES INSTITUTE & THE UNITED NATIONS ENVIRONMENTAL PROGRAMME, WORLD RESOURCES 1994-1995: A GUIDE TO THE GLOBAL ENVIRONMENT (1995), at 201-203.

96. Ministry for the Environment, *supra* note 53, at 6.

97. "Whether our per-capita . . . emissions are comparatively high is irrelevant . . . ," R. Chapman & L. Gray, *Responding to Global Warming: A Critique of Al Gillespie's Burning Issues*, N.Z. J. ENVTL. L., at 4 (draft).

98. Marshall Islands, United Nations, *Response From Parties on Issues Relating to Sinks*, FCCC/AGBM/1997/MISC.4 (1997), at 22.

99. New Zealand, *Response From Parties, id.* at 40.

100. S. Upton, *New Zealand's Climate Change Policy: Speech to the Energy Foundation of NZ*, 31 October 1997, at 8.

flow-over, in terms of scientific certainty (mechanisms such as Joint Implementation and emissions trading require high standards of reporting that sinks may not be able to meet) and political considerations in the area of sinks may now be becoming apparent.

6. Methane

As disproportionate as NZ's carbon dioxide emissions are on a global per-capita basis setting, they pale in comparison with its methane emissions. On an individual gas basis, NZ's per-capita methane emission rate is about 10 times higher than the global average.[101] Currently,[102] methane emissions amount to 44.9% of NZ's total greenhouse emissions.[103] NZ has the highest ratio of (cumulative) non-carbon dioxide greenhouse gas emissions to carbon dioxide emissions of any of the Annex I parties to the FCCC.[104]

The methane figure is primarily attributable to the ratio of NZ's relatively small human population and its vast number of ruminants,[105] which account for around 71% of NZ's total methane emissions of around 2.2 million tonnes per year.[106] Although this figure is substantially less than NZ's carbon dioxide emissions in absolute terms, it is NZ's most potent greenhouse gas due to its greater heat trapping strength than comparable gases, such as carbon dioxide.[107] NZ's other prominent[108] non-carbon dioxide greenhouse gas is nitrous oxide.[109] This accounts for 18.9% of NZ's total greenhouse gas emissions.[110]

Thankfully for NZ, the international focus has shifted from a focus solely on carbon dioxide to a comprehensive approach.[111] Under the Kyoto Protocol, six dif-

101. Ministry for the Environment, 6 *The State of New Zealand's Environment: 1997* (1997), at 32; M. Ulyatt, *Is Emission Control Technology Advanced Enough to Control or Monitor Emissions Other Than Carbon Dioxide?* in *Greenhouse Gases and CO_2 Policy, supra* note 14, at Chs. 2 & 18.

102. This will change over a very long period of time, with carbon dioxide emissions ultimately expected to eclipse methane.

103. Livestock digestive processes account for 39.9%; animal waste 0.5%, landfills 3.5%; wastewater 0.1%, and energy 1.0%.

104. *See* Ministry for the Environment, *Climate Change: More Than Just Carbon Dioxide*, (1998), at vii.

105. Ulyatt, *supra* note 101.

106. *Id.*

107. IPCC, *supra* note 20, at 32-34.

108. NZ's remaining non-CO_2 greenhouse gases are SF_6 (sulphur hexafluoride), HFCs (hydrofluorocarbons) & PFCs (perfluorcarbons). These make up less than 0.5% of the total greenhouse gas emissions for NZ.

109. Around 94% of this originates in the agricultural sector, from animal waste, soil processes and nitrogen fertiliser.

110. Ministry for the Environment, supra note 103, at ix.

111. "A comprehensive approach, covering all gases in a basket, and sink." NZ Intervention. 1997, 2 Dec. 1997, Committee of the Whole. FCCC, COP 3. Kyoto, at 1.

ferent greenhouse gases are cumulatively targeted for reductions. The proposal to list gases separately, the China/G-77 preference,[112] which was still on the table at the end of the first working week of Kyoto,[113] was ultimately rejected in favor of a "flexible basket" of gases. Within this basket, gases are aggregated and rated according to their global warming potential (GWP)[114] and parties can then, in seeking to make overall reductions (made up of a cumulative target) choose to make reductions in whichever gas or gases they choose.[115]

If the global community had decided to focus upon greenhouse gases separately, (such as methane) then NZ could have been left "very exposed"[116] with respect to its disproportionate emissions from the agricultural sector. As it now stands, NZ should be able to avoid confronting this problem by focusing on the reduction of other greenhouse gases. Accordingly, it is not surprising that NZ argued from the outset for the comprehensive approach[117] because it should benefit by ". . . lowering the costs of meeting a given target."[118]

The adoption of a comprehensive approach is also considered beneficial because emissions of methane and nitrous oxide were below their 1990 levels in 1996, with methane down by approximately 7%.[119] The methane reduction was due to prior elimination of agricultural subsidies, lower prices for beef and sheep meat and the increase in pastureland being used for planted forests. Consequently, it has been suggested that NZ's methane emissions will continue to decline, and by 2020 should be "well below their 1990 levels."[120] Accordingly, as the MfE noted, a "reduction in these gases can be offset against carbon dioxide emissions . . . so that, in aggregate, less action than otherwise will be needed for NZ to meet its legally binding target."[121]

---

112. *See* The Third Conference, *supra* note 56; Late Paper, supra note 51. *See* also P. McCully, *Discord in the Greenhouse: How the WRI Is Attempting to Shift the Blame for Global Warming*, 21 ECOLOGIST (1991), at 213. M. Redclift, *Throwing Stones in the Greenhouse*, GLOBAL ENVIRONMENTAL CHANGE (June, 1992), at 90-92

113. The Third Conference, *supra* note 21; ENB, *supra* note 62, at 1-2.

114. Due to uncertainties with these, Article 5 of the Kyoto Protocol suggests that GWPs shall be those agreed to by the IPCC and the COP, and these may be revised in the future.

115. Late Paper, *supra* note 51, at 4. This approach originated with the World Research Institute, GREENHOUSE WARMING: NEGOTIATING A GLOBAL REGIME (1991). *See also* WORLD RESOURCES INSTITUTE, WORLD RESOURCES 1990-1991 (1990), at 14-18. *See also*, A.L. Hammond, et al., *Calculating National Accountability for Climate Change*, 33(1) ENVIRONMENT (1991), at 11-35.

116. *Report on Climate Change, supra* note 17, at 35.

117. *New Zealand Delegation Brief, supra* note 27, at paras. 15 & 16. *See also* D. Taylor, *New Zealand and the Climate Change Convention: Where to Now?*, in NEW ZEALAND AND THE CLIMATE CHANGE CONVENTION: WHERE TO NOW (New Zealand National Committee of the World Energy Council ed., 1995).

118. Minister for the Environment, *Points to Note Regarding Attached Reports on Modelling Impacts For NZ of Reducing Greenhouse Gas Emissions (1997)*, at 1.

119. Chapman and Gray, *supra* note 97, at 3.

120. Chapman and Gray, *supra* note 97, at 4.

Predictions of falling non-carbon dioxide greenhouse gas emissions may prove to be incorrect. This is because they are largely based upon the assumption that the total number of NZ's main methane producers -- ruminant livestock -- "will continue to decrease."[122] Similar claims that "NZ's biggest contributing gas (methane) happens to hold some excellent emission reduction prospects"[123] have been advanced elsewhere. Such contentions may be difficult to defend due to scientific uncertainty and market unpredictability.

The claim that methane emissions dropped about 7% between 1990 and 1991 is difficult to reconcile with earlier MfE emissions projections, which suggested that the (scientific and methodological) uncertainties[124] related to methane are in the range of 22%.[125] The 1996 IPCC Guidelines assessed the uncertainty for methane emissions from biomass burning at 100%, for animals at 25% and animal waste at 25%.[126] Thus, the uncertainty for future methane emissions is at least 3 times larger than the drop in emissions between 1990-1996. The MfE suggested an uncertainty of up to 59% for nitrous oxide,[127] whereas the IPCC placed the uncertainties for nitrous oxide between 50% and 100%.[128] These uncertainties increase to "very high levels" (due to data, process and methodological uncertainties)[129] the further into the future the projections are cast (i.e., from 2000-2010). According to the National Science Strategy Committee for Climate Change (1997) such uncertainties in emission figures for non-carbon dioxide greenhouse gases, "have implications for the robustness of policy."[130]

---

121. Ministry for the Environment, *supra* note 103, at 7. This produces "a good deal . . . for many developed countries." S. Upton, *Address to the Royal Institute of International Affairs: Climate Change, Addressing the Real Issues*, 19 Sept. 1997, at 4. At this point it needs to be asked whether one country should be allowed to claim a benefit i.e., increased emission space to be filled by rising CO_2 emissions, because of a fortuitous occurrence that creates less emissions than predicted? Should such a fortuitous outcome be an excuse to reduce need to control emissions in another field ?

122. Chapman & Gray, *supra* note 97, at 3-4.

123. Upton, *supra* note 100, at 2.

124. This uncertainty is particularly pronounced in the context of GWPs for non-CO2 gases. The National Science Strategy Committee for Climate Change noted in 1997, "there is a level of uncertainty in global warming potential's, typically of the order of -/+ 35%." This is particularly true in terms of methane and nitrous oxide. Royal Society of New Zealand, *National Science Strategy Committee for Climate Change* (1997), at 15. The Minister for the Environment has also noted that "the measurement uncertainty for non-CO2 emissions and net CO2 removals by sinks is higher than for energy emissions. In reality, due to uncertainty, emissions projections could be significantly higher or lower than the central estimate figures presented." Late Paper, *supra* note 51, at Annex A.

125. Ministry for the Environment, *supra* note 103, at viii.

126. IPCC, *supra* note 76, at A1-1, Table A1-1.

127. Ministry for the Environment, *supra* note 103, at viii.

128. IPCC, *supra* note 76, at A1-1, Table A1-1.

129. Ministry for the Environment, *supra* note 103, at 14 & 17, Royal Society of New Zealand, *supra* note 84, at 17.

130. Royal Society of New Zealand, *supra* note 124, at 24.

The connection of reduced ruminant numbers and falling methane emissions is also problematic. It has already been noted by the MfE in 1998 that "the degree of uncertainty in estimating livestock emissions beyond 2,000 is high; livestock numbers are driven by world prices, which cannot be predicted accurately."[131] This conclusion was probably supported by its 1997 report, which noted, "the increase in cattle numbers, particularly dairy cows in the mid-1990s, has partially reversed this trend [declining methane emissions in the order of 10%] so that overall methane reduction between 1990 and 1995 was 3.5%."[132] It is with the long-term predictions that projections of declining methane emissions are most questionable. For example, the recent OECD Environmental Performance Review of NZ came to the opposite conclusion. It suggested, "Recent economic developments suggest that continued agricultural expansion is likely."[133]

Such scientific and market uncertainties should give rise to inquiry about the technological prospects for reducing methane. Unfortunately, the most successful technologies in this area, such as dosing animals with anti-methane producing compounds, is problematic in NZ. First, the successes in the Northern Hemisphere may not be so easily replicated due to less intensive farming practices.[134] As the Second NZ national report under the FCCC noted, "possible measures, such as feed additives to reduce emissions of methane from ruminants would be difficult and prohibitively expensive to implement under NZ's extensive pastoral systems." Additionally, reductions through improved agricultural productivity, such as improving the reproductive performance of animals "are not quantifiable."[135]

Finally, it should also be noted that the above mentioned OECD report concluded that fertiliser use (which has a link to the greenhouse gas nitrous oxide) in the mid 1990s was at its highest level since 1980, and sales were exceeding the levels during the period fertiliser was subsidised.[136] Despite this, to quote NZ's second national report under the FCCC, "[A]s yet, few rules have been introduced to restrict fertiliser application."[137]

7. Advanced Myopia in the Climate Change Response

NZ's climate policy is currently stuck between a rock and a rising sea. With a steadfast focus upon problematic approaches such as sinks, the cumulative approach and international market mechanisms, NZ has refused to adopt a more direct approach to reduce emissions. This is despite obvious anomalies in current approaches. There are two particularly striking examples of this.

131. Ministry for the Environment, *supra* note 103, at 13.

132. *Id.*

133. OECD, *Environmental Performance Reviews: New Zealand* (1996), at 134.

134. Ministry for the Environment, *supra* note 103, at 21-23 & 26.

135. Second National Communication, *supra* note 60, at 64.

136. OECD, *supra* note 133, at 134.

137. Second National Communication, *supra* note 60, at 67.

7.1 TRANSPORT

The transport sector in NZ is responsible for around 40% of its total carbon dioxide emissions,[138] making it the single largest contributor of carbon dioxide in the country.[139] It is also largely responsible for the higher than expected gross carbon dioxide emissions revealed in 1996.[140] This growth, in part, reflects the addition of 100,000 cars[141] on NZ roads since 1992, as well as increases in road commercial transportation, as opposed to less greenhouse gas intensive mass transportation systems.[142] Currently, NZ has 69 vehicles for every 100 people. The global average is 11 vehicles per 100 people. NZ possesses the second highest car ownership ratio in the world. The number of licensed vehicles in NZ has grown at more than twice the rate of the human population since 1972. Moreover, the light vehicle fleet is expected to expand by a further 40% by 2021.[143]

This problem is exacerbated by the aging (and overall inefficiency) of NZ's vehicle fleet.[144] Finally, despite leading the international community in support of alternative (and less greenhouse inducing) fuels such as Compressed Natural Gas in the 1980s, the government has now withdrawn support for such programs.[145]

NZ is also notable among OECD countries for not imposing greenhouse emission controls of vehicles, nor does it require either existing or new vehicles to have catalytic converters.[146] This is in contrast to comparable countries such as the United

---

138. *The New Zealand Response, supra* note 24, places the figure at 34.2%. However, when the industrial processes of creating the fuel are added, this rises to approximately 40%. *See* the Working Group, *supra* note 26, at 42. For a detailed account of aggregate greenhouse emissions caused by New Zealand's transport infrastructure, *see* Ministry of Transport, *Greenhouse Gas Emissions From New Zealand Transport* (1995). This report puts the transportation contribution, when including the fuel creation process, at 45%.
139. Second National Communication, *supra* note 60, at 8; State of the Environment, *supra* note 101, at 31.
140. Indeed, 77% of the growth of this figure was attributed to increases in emissions from domestic transport. *See* Working Group, *supra* note 26, at 103.
141. Statistics New Zealand, *The New Zealand Official Yearbook: 1996* (1996).
142. *See* Energy Efficiency & Conservation Authority, *Transport Sector Energy Use: Highlights*, 2 ENERGY WISE MONITORING Q. (1995), at 1-2.
143. Ministry of Transport, *supra* note 138, at 1. 6 *State of the Environment, supra* note 101, at 13. New Zealand has one car for every two people. The USA has the highest rate, with one car for every 1.7 persons. *See* D. REDSHAW & K. DAWBER, SUSTAINABLE ENERGY: OPTIONS FOR NEW ZEALAND (1997), at 87.
144. *Greenhouse Gas Emissions From New Zealand Transport*, supra note 138, at 19.
145. NZPA, *The Decline of Compressed Natural Gas As A Transportation Fuel*, NEW ZEALAND HERALD, 13 March 1996.
146. OECD, *supra* note 133, at 123; *Greenhouse Gas Emissions From New Zealand Transport; supra* note 138, at 18. NZ does not require vehicles to be equipped with pollution control devices or to meet emission standards. The only laws targeting air pollution from motor vehicles are the Petroleum Products Specifications Regulations 1995, issued pursuant to the Ministry of Energy Abolition Act 1989, which ban the sale of leaded petrol and Traffic Regulation Number 28, issued pursuant to the Transport Act 1962, which makes it an offence to emit smoke from a vehicle to such an extent that it obstructs the visibility of other drivers.

States, the European Union, Japan and Australia,[147] which have emission regulations. Of course, catalytic converters do not reduce greenhouse gas emissions, but are important for reducing localised air pollution,[148] including carbon monoxide, dioxins, fine particulate matter, butadiene & benzene.[149] The failure of the government to confront pollution associated with vehicles, is akin to its failure to confront the carbon dioxide side of this equation. However, such technical solutions, although salutary, will ultimately be eclipsed by the sheer increases in vehicles projected over the next few decades.[150] Accordingly, long-term solutions to such problems reside in measures such as public transport. However, support for such initiatives is wavering. Central government funding for public transport was $50.43 million in 1990/91. Eight years later, following cuts of 40% between 1992-1996,[151] the figure has declined to $41.87 million.[152]

7.2 ENERGY EFFICIENCY

NZ is notoriously inefficient in its production of energy, with its total primary energy supply, as a proportion of GDP, substantially higher than the OECD average.[153] This is hardly surprising, given the fact that overall OECD energy intensity is declining (that is, other OECD countries are getting more results for their energy expenditure) whereas NZ's is increasing.[154] Thus, as the OECD review noted, "[I]n contrast with most OECD counties, NZ's energy intensity increased markedly in the 1980s and early 1990s, rising more than 30% since 1982 . . ."[155]

---

147. Ministry of Transport, *Vehicle Fleet Emissions Control Strategy for Local Air Quality Management* (1997), at 20-24; Ministry of Transport, Full Report: Environmental Externalities (1996) 119-123.

148. *See State of the Environment, supra* note 101, at 10-11.

149. Ministry of Transport, *supra* note 138, at 159-164.

150. Thus, as the Ministry for the Environment noted in 1997, "[I]t is becoming clear that behaviour change, public transport systems and strategic urban planning have as great a role to play as technical solutions." *State of the Environment, supra* note 101, at 14.

151. OECD, *supra* note 133, at 123. This was reduced due to the regional petrol tax.

152. Personal communication, Transfund NZ, 9 June, 1998. The figures are also available in Transfund's National Roading Programme Annual Reports.

153. Parliamentary Commissioner for the Environment, *Report to the House of Representatives: Sustainable Energy Management in New Zealand: Improvements Required in Government Policy* (1992), at 4. Between 1979 and 1990, New Zealand's energy intensity increased by 32.8%, at an annual rate of increase of 2.4%. From 1990 to 1993, energy intensity grew at a rate of 1.7% per year. Between 1993 and 1994, consumer energy intensity fell by 2.91%. This was the first reversal in energy intensity since 1979. *See also* Energy Efficiency & Conservation Authority, *1994/95 Annual Report* (1996), at 17.

154. *Id.* at 6. For analysis that rejects the conclusion that New Zealand's energy production is more inefficient that other OECD countries, *see* Energy Efficiency & Conservation Authority, *1993/94 Annual Report and 1994/95 Business Plan* (1995), at 12-13.

155. OECD, *supra* note 133, at 113.

Unfortunately, the NZ government has done little to engender energy efficiency beyond the provision of public information and some research.[156] It has rarely directly subsidised costs associated with energy efficiency programs to make them more attractive to the consumer.[157] Accordingly, the government's energy efficiency strategy is only expected to reduce gross emissions by around 1% by the year 2000. However, it is arguable whether even this small saving will transpire. According to the Second National Communication by NZ under the FCCC, "It is difficult to accurately predict how much carbon dioxide these indirect programs will save . . ."[158]

The same logic applies to new forms of renewable energy in NZ. Despite growing evidence that solar and wind power could effectuate huge reductions in NZ's carbon dioxide emissions,[159] the government is doing very little to encourage development of alternative energy resources.

8. The Core Problem

Despite the clear benefits of increased energy efficiencies and renewable technologies, these options remain fundamentally restrained by an overall pricing system which makes nearly all forms of NZ's energy comparatively cheap.[160] Accordingly, there is little economic incentive to reduce or conserve.

The government has done little to ensure that greenhouse producing sources and gases are priced progressively, so as to internalise the environmental costs[161] or

---

156. 1993/94 *Annual Report, supra* note 154, at 20-24. Indeed, as the *1994/95 Annual Report* emphasised, 'Above all, the Authority is a facilitator.' *Id.* at 2. The only dictate to facilitate energy efficiency is in the 1992 *Building Regulations*, Clause H1. However, this clause says little more than that energy efficiency should be encouraged, without laying down any specifics of how this is to be done.

157. This is not to suggest that the EECA provides no assistance in this context. Indeed, in the period of 1994-1995, $2.9 million was advanced to public sector bodies for improved energy efficiency. However, given the vast market for energy efficiency in New Zealand, *see* Ministry of Commerce, *Renewable Energy Opportunities in New Zealand* (1993); this amount is minuscule and its scope (i.e., only certain public bodies) is far too limited. This is especially so when compared with the trends in other countries that are actively promoting alternative forms of renewable energy and increased energy efficiencies. *See* C. Flavin, *Harnessing the Sun and the Wind*, in STATE OF THE WORLD: 1995 (L.R. Brown ed., 1995), at 58-76.

158. *The New Zealand Response II, supra* note 60, at 9, 72.

159. Ministry of Commerce, *supra* note 157.

160. *See* Ministry of Commerce, *Energy Data File: 1996* (1996).

161. World Commission on Environment and Development, *Our Common Future* (1987), at 168-169, 196, 198, 201. A. Williamson, *Technology and Market Issues, in Sustainable Energy For New Zealand: How Do We Make It Happen ?* (1994), at 37, 38; G. Bertram, *Economics and Finance Issues, id.* at 49, 52-55; J. Peet, et al., *Climate Change and Energy: Challenge and Choice*, in *Climate Change: The New Zealand Response: A Workshop*, (Ministry for the Environment ed., 1988), at 157, 161. The Parliamentary Commissioner for the Environment, *supra* note 153, was very direct on this point, at 8, 23-26, 41, 44. The same conclusions were recognised by the Stratford Inquiry; *supra* note 39, at paras. 8.63, 8.64, 8.68, 8.74-8.76, 8.81-8.83, 8.85-8.87, 8.91, 12.2 (17).

encourage the development of cleaner energy resources. In this regard, the OECD Environmental review was quite explicit and is worth quoting at length. It stated:

> The country has made relatively little use of prices, taxation or subsidies to encourage conservation and energy efficiency . . . The Government has offered no significant financial inducements or incentives to invest in or consume energy produced from renewable resources . . . NZ's low energy prices, compared to other OECD countries encourage relatively high consumption . . . a further issue is the need for electricity prices to fully reflect costs, including environmental costs . . . because this is not currently the case, prices are lower than they would otherwise be . . .[162]

The NZ government has continually resisted such considerations, as they necessitate market intervention. This is not a pleasant option for a government striving to deregulate many sectors of the economy. Any suggestions of imposing extra costs or burdens upon free markets, from commercial road transportation to electricity generation,[163] are overtly unpopular, and the government has gone to great lengths to avoid them. For example, the government has continually deferred serious consideration of a carbon tax over the last four years, merely promising to draft yet another "public consultation document."[164]

If NZ seriously intends to reduce its greenhouse emissions it is necessary to either increase the price of the processes which produce them to reflect externalities, or subsidise alternatives which reduce these emissions. These points are merely indicative of the nationally[165] and internationally accepted principle that polluters should pay the full price (internalise the cost) of the pollution they produce. However, NZ has resisted implementation of this principle in the context of greenhouse gas emissions. For example, it can be argued that the inclusion of sinks in the Kyoto Protocol runs counter to the Polluter Pays Principle because this shifts the burden of pollution control to a sector that is *not the major source* of the pollution. Given this stance, it shall be interesting to see how the government intends to apply the polluter pays principle to the agricultural sector in particular, which currently contributes 58% of NZ's total greenhouse gas emissions (on a carbon dioxide equivalent basis).[166]

---

162. OECD, *supra* note 133, at 113, 116, 122, 126-130.

163. The legislation creating competitive gas and electricity markets "do not in themselves ensure that energy efficiency will be promoted, but, instead, set the scene for this to happen where commercially viable prospects exist." As such, energy efficiency and the pursuit of new renewables are "secondary objectives." *The New Zealand Response II, supra* note 60, at 49-50.

164. New Zealand Treasury, *The Design of a Possible Low-Level Carbon Charge for New Zealand* (1997); B. Orsman, *Carbon Tax Deferral Is Surrender, Govt Told*, N.Z. HERALD, 13 March 1997, Second National Communication, *supra* note 60, at 47.

165. Ministry for the Environment, *Environment 2010 Strategy* (1995), at Principle 4.

166. Ministry for the Environment, *supra* note 103, at viii. The earlier report of the Ministry for the Environment, (1997) placed this figure at "approximately 60%." Second National Communication, supra note 60, at 7.

Rather than addressing greenhouse gas emissions directly, the government has focused on 'least cost options.' It has been so successful in pressing this approach that responding to climate change may actually prove to be economically beneficial for NZ. [167] This is attributable to NZ's successful effort to have sinks included in the Kyoto Protocol, differentiated emission reduction targets, which result in more lenient obligations for NZ vis-a-vis other industrialised nations,[168] the possibility of emissions trading,[169] the adoption of the comprehensive approach, and the possibility of a drop in the number of ruminants in the nation. Therefore, the nation with the fourth highest per-capita emissions on the planet[170] is free to blithely take a 'business as usual' approach to perhaps the greatest environment threat in the next century.

9. Conclusion

The honourable Simon Upton was, in a sense, correct when he suggested that the Kyoto Protocol was a "successful outcome for NZ."[171] This success is a direct reflection of his achievements in negotiations which he forthrightly characterised as "self-interested hard-ball from beginning to end."[172] However, in terms of the much larger picture, NZ's "success" is dubious. That is, a prolific producer of greenhouse gas emissions may now be poised to obtain a net financial benefit from processes that are scientifically uncertain, and may make no further emission reductions than it initially promised in 1992. Successes of this nature may ultimately come back to haunt NZ and the world community in the decades to come.

---

167. That is, the Kyoto outcome "may even create positive welfare impacts for NZ." Late Paper, *supra* note 51. Annex B. Irrespective of the possible benefits, the inclusion of such approaches could reduce the costs of reducing greenhouse gases by up to 70-80%. Minister for the Environment, *supra* note 37, at 4. Kyoto Outcome, *supra* note 34. The Ministry for the Environment even went so far as to suggest that, "even if we do not achieve an outcome involving 'sinks the way we want them' the costs for NZ of achieving a stabilisation or a -5% target may not be intolerable" [emphasis in original], Ministry for the Environment, Climate Change Negotiations Paper: Comment, 24 November 1997, at 2. *Cf. New Zealand Institute of Economic Research, Potential Macroeconomic Impacts of Trade in Carbon Removal: Report to the Ministry for the Environment* (1997), at 2.

168. *See supra* note 1.

169. *See* Peter Alsop's chapter in this book.

170. Ministry for the Environment, *supra* note 103, at 6.

171. *See* R. Chapman, *Successful Outcome at Kyoto*, MINISTRY FOR THE ENVIRONMENT NEWSLETTER, 8 March 1998.

172. S. Upton, *Reflections on the Kyoto Conference, Press Release*, 29 December 1997.

10. THE ROLE OF CARBON SEQUESTRATION AS A RESPONSE STRATEGY TO GLOBAL WARMING, WITH A PARTICULAR FOCUS ON NEW ZEALAND

J.B.FORD-ROBERTSON, J.P.MACLAREN AND S.J.WAKELIN
Forest Research Institute
Private Bag 3020, Rotura, New Zealand

1. Sequestration

Plants remove carbon from the air by the process of photosynthesis and retain this in the form of wood. Half the dry weight of wood is elemental carbon. Land that supports a forest cover will hold considerably more carbon than land that does not, even if the forest is managed for production and parts are being felled and replanted at any particular time. The high carbon-density (tonnes of carbon per hectare) of forests is their primary contribution to the mitigation of global warming.

Many people readily acknowledge *deforestation* as a cause of increased atmospheric carbon dioxide. The conversion of a high carbon-density landscape (such as tropical rainforest) to one of lower carbon-density (such as pasture) results in a transfer of carbon from the earth's surface to the atmosphere. Indeed, it has been calculated[1] that global deforestation between 1850 and 1990 has resulted in cumulative emissions of 122±40 GtC, compared to a contribution of approximately 230 GtC from fossil fuels. In other words, deforestation has been responsible for perhaps one-third or more of the CO_2 emissions over the last 140 years.

Many commentators fail to note that *reforestation* is merely the inverse of deforestation. Restoration of all of the forests that have been lost would result in a net transfer of carbon from the atmosphere back to the land's surface of approximately 122 Gt, compensating for the emissions caused by earlier forest removal. Just as forest clearance puts carbon into the air, so the establishment of forests takes carbon out of the air.

1.1 NATURAL FORESTS AND PRODUCTION FORESTS

The New Zealand Working Group on CO_2 Policy[2] drew a clear distinction between "permanent" forests and "rotation" forests. This distinction is one of perception rather than substance in the context of carbon sequestration. In a "permanent," or natural forest, trees die from natural causes and release their carbon to the atmosphere through the

1. J.T. Houghton et al., *Climate Change 1994. Radiative Forcing Of Climate Change And An Evaluation Of The Ipcc Is92 Emission Scenarios* (1995).
2. W.J. Falconer, et al., *Climate Change and CO_2 Policy: a Durable Response. Discussion Document of the Working Group on CO_2 Policy*, Ministry for the Environment (1996).

A. Gillespie and W. C. G. Burns (eds.), Climate Change in the South Pacific: Impacts and Responses in Australia, New Zealand, and Small Island States, 189–207.

normal process of decay. The carbon-density of the estate is maintained by the recruitment of younger trees. Over a long period, such a forest can be considered steady-state, in that it is not a carbon 'sink'[3] or source, but continues to be a carbon reservoir. In a shorter time-span, the stock of carbon in the reservoir will fluctuate as a result of natural perturbations, but will remain within distinct bounds. The weight of biomass carbon per hectare will not dip below a minimum level, nor attain amounts above a certain maximum level.

It is essential to understand that forests are not necessarily carbon sinks. Within decades or centuries, the biosphere reaches a saturation point where there can be no further net gain in carbon. The long-term benefit of a forest is merely that this land-use can retain high levels of carbon in perpetuity. Carbon that is held in the form of biomass is not available as carbon dioxide to warm the atmosphere.

In the case of production forests, trees are removed for human use. In terms of carbon budgets relating to land-use change, it makes no difference whether trees decay in the forest (the natural situation) or outside the forest (the managed situation). (The fate of wood products is discussed separately, below.) Likewise, it makes no difference whether trees decay or are removed as individual stems, or groups of stems ("stands").

In New Zealand, the term 'stand' usually refers to an even-aged block of trees with uniform management practices. The word 'forest' implies a larger scale unit, consisting of stands of various ages, so that at any one time some stands are being felled and replanted while others are growing. For example, New Zealand's largest man-made forest (Kaingaroa) is a mature forest and is continuously being logged but contains substantially more carbon than the landscape of sparse scrub that it supplanted in the 1920s. It is not true, therefore, that forests store carbon for only one rotation of trees, and then the carbon is released again. The reality is that all forests act as sinks, that is net absorbers of CO_2, during their establishment phase, and then act as reservoirs during their maintenance phase. A reservoir contains carbon, but is not necessarily a sink or a source. There is no reason why this maintenance phase should not continue indefinitely. Afforestation is a temporary expedient, not because trees are felled or decay naturally, but because there is a finite area of low carbon-density land that can be converted to a high carbon-density land use.

---

3. These words are defined in the Framework Convention on Climate Change. 'Sink' is defined as: "any process, activity of mechanism which removes a greenhouse gas, an aerosol or a precursor of a greenhouse gas from the atmosphere." The FCCC defines a 'reservoir' as "a component or components of the climate system where a greenhouse gas or a precursor of a greenhouse gas is stored." Only if a reservoir increases in size does it constitute a net sink, and then only for the period during which it is expanding. The term 'stocks' refers to the quantity of carbon present in a reservoir, whereas 'flows' (or flux) refers to inputs and outputs to that reservoir.

1.2 THE IMPORTANCE OF WOOD PRODUCTS

When a tree is removed from a forest and used to make a product, this is equivalent, in carbon terms, to an artificial extension of the tree's life (or part of the tree's life). In accounting for the carbon in the wood product, no new concepts are required.

Let us consider a "wood product pool" for each type of product (e.g., paper, buildings, and furniture). Again, these product pools are sinks only if there is a net inflow of carbon. They certainly contain carbon, but is that quantity of carbon increasing? Although new products are being manufactured daily, products are also routinely being destroyed. The change in size of the "wood products pool" in New Zealand is so small that it can be disregarded.[4] This is the thinking in the IPCC Guidelines Reporting Instructions.[5] The Guidelines assume that "emissions from the combustion or decay of wood and wood products are assumed to take place in the country in which the wood was harvested and within a year of harvesting." Canada, however, disputes that wood products are unimportant and has calculated[6] that Canadian forest products are being added to the global carbon reservoir at the rate of 23.5 Mt/a.

Wood products substitute for non-wood materials that use considerable fossil fuel and emit carbon dioxide as a chemical byproduct during their manufacture. Care should be taken, however, not to exaggerate this effect. Buchanan[7] has calculated that New Zealand's emissions would only be reduced by 1.5% if almost half its buildings were constructed with timber, rather than concrete, aluminum or steel.

Wood can also be substituted for fossil fuels to meet energy needs, with some 7% of primary energy in New Zealand in 1990 coming from combustible renewables and waste. Emissions from such sources are not included in Framework Convention target-setting, because they merely recycle atmospheric carbon rather than providing net additions.

The overwhelming importance of the carbon stored in standing trees, relative to other factors, has been summarised by Ford-Robertson.[8] More work is currently underway to assess methodologies of accounting for harvested wood products.

---

4. J.P Maclaren, *New Zealand's Planted Forests as Carbon Sinks*, 75(1) COMMONWEALTH FORESTRY REV. (1996), at 100-103.
5. IPCC, *IPCC Guidelines For National Greenhouse Gas Inventories*, Ch. 5, land-use change and forestry (1996).
6. J.A Greenough, M.J Apps & W.A. Kurz, *Influence Of Methodology And Assumptions On Reported National Carbon Flux Inventories: An Illustration From The Canadian Forest Sector*, Paper presented at the IEA Greenhouse Gas R&D Programme Conference: Technologies for Activities Implemented Jointly, 26-29 May, 1997.
7. A.H. Buchanan, *Energy, Carbon Dioxide And Timber Engineering*, in D. HAMMOND, FORESTRY HANDBOOK, NZ Institute of Forestry (1995), at 38-39.
8. J.B. Ford-Robertson, *Carbon Balance Calculations For Forest Industries: A Review*, NZ FORESTRY, May, 1997, at 32-36.

1.3 FORESTRY AND THE EFFECT ON GREENHOUSE GASES OTHER THAN CO_2

Methane and nitrous oxide emissions are largely a product of pastoral farming, and forestry is attributed with being one of the main causes of livestock reductions in recent years. In the 1996 agricultural survey,[9] it is noted that there has been a decline between 1995 and 1996 of 6.4% in beef cattle numbers and 2.9% in sheep numbers. This is somewhat offset by an increase of 1.1% in the deer herd, and 2.1% in dairy cattle. Sheep numbers on 30 June 1996 were estimated to be 47.39 million, down from the high of 70 million in 1981, and are expected to decline further. The Ministry for the Environment[10] projects substantial declines in both methane and nitrous oxide emissions.

Any encouragement of afforestation would tend to result in the displacement of livestock and the reduction of methane emissions, and probably nitrous oxide. In this context, it is also worth noting that there is some evidence that forest soils are net sinks for methane.[11]

1.4 THE AMOUNT OF SEQUESTED CARBON IN NEW ZEALAND'S FORESTS

As a result of deforestation over the last thousand years, the country has a lower carbon-density than is theoretically achievable and sustainable. New Zealand has the potential to increase the reservoir capacity (the "stocks") of carbon retained by living and dead vegetation on and below the surface of the land. There are three broad categories of land use that are of interest in the context of carbon sequestration. These are indigenous forests (tall, primary forests that in New Zealand are largely unavailable for exploitation); abandoned farmland; and commercial, or plantation, forests. In New Zealand, this consists almost exclusively of exotic species. Each of these categories must be considered separately, as their behaviour with respect to carbon sequestration differs.

Indigenous forests. It is commonly stated that New Zealand's indigenous forests are in decline, possibly as a result of damage from the introduced Australian possum. There is insufficient evidence at present to support or refute this view, although the matter is under active investigation. A deterioration in tree health does not necessarily indicate a reduction in standing carbon, as dead wood can retain carbon for many decades. Furthermore, it is not clear whether any observed tree decline is a temporary perturbation or a permanent reduction in high forest cover. Thus, there are no incontrovertible data for carbon sequestration or emissions attributable to indigenous forests.

---

9. *National Beef Herd Decreases 6.4%*, CHRISTCHURCH PRESS, 5 June 1997, at 22.
10. Ministry for the Environment, *Second National Communication On Policies And Measures Relating To New Zealand's Greenhouse Gas Emissions* (1997).
11. Sources cited in J.P. Maclaren, *Environmental Effects Of Planted Forests In New Zealand*, 198 FOREST RESEARCH INSTITUTE BULLETIN (1996), at 94.

Abandoned farmland. Removal of subsidies and declining terms of trade for agricultural products has resulted in the abandonment of considerable areas of marginal farmland. This land is reverting to woody shrubs, both indigenous and exotic, and—if uncontrolled— will probably climax in tall forest cover. Without doubt, it constitutes a carbon sink, the magnitude of which cannot yet be determined, but it is also under investigation.

Plantation forests. The area of land under commercial plantation forestry accounts for only 6% of New Zealand's total land area, but is undergoing rapid changes in carbon density and therefore has a disproportionate influence on the carbon budget of land use change.

For plantation forests, the most current and best estimates are provided by Wakelin and Te Morenga.[12] They state that a 5-year rolling average of the sequestration for the year ending April 1, 1990 was 5.8 MtC/a, and that this was estimated to decline to 5.5 MtC/a by 2000. This reduction is likely to be transient. Assuming a "medium" rate of new land planting, the 1990 rate of sequestration will be surpassed in 2002, and remain at a high level (5.8-9.3) well into the next century. Assuming a "high" rate of new land planting, the same result follows, except that annual sequestration may attain 14 MtC/a. To put these figures in perspective, annual sequestration by commercial forests is already a high proportion of the approximately 7.0 MtC that was emitted in 1990 from combustion of fossil fuels.

2. New Zealand's Position Regarding Carbon In Soil

Tate *et al.*[13] estimate that there is almost twice the quantity of carbon in New Zealand's soils (to 1 m depth) as there is in all vegetation, which indicates the importance of this carbon reservoir. It is the change in soil carbon over time, however, that is of interest in determining the effect on atmospheric carbon concentrations. Although some data have been obtained for individual situations, it is difficult to generalise results to provide a meaningful estimate of soil carbon changes for all of New Zealand. This issue is currently being researched.

A reasonable assumption is that the conversion of farmland to commercial forestry would result in a loss of soil carbon if the original soil was high in organic matter, e.g., peatland. In contrast, a skeletal soil (sand-dune, bare pumice, or eroded hillside) would result in the accumulation of soil carbon. In New Zealand, planting often occurs on the latter, and conversion from organic soils is quite rare. Most calculations of New Zealand's carbon sequestration have omitted any changes that may be taking place in soil carbon as a result of afforestation, and in doing so it is generally considered that these calculations are

12. S.J. Wakelin & L.Te Morenga, *Carbon Sequestration By Plantation Forests In New Zealand.* [calculations revised in December 1996], Contract report for the Ministry of Forestry and the Ministry for the Environment (1996).
13. K.Tate et al., *Estimation Of Carbon Stored In New Zealand's Terrestrial Ecosystems*, Manaaki Whenua - Landcare Research (1993).

conservative, i.e., underestimating the magnitude of the sink.

3. Carbon Sequestration—The Debate

3.1 IF FORESTRY IS NOT A PERMANENT SOLUTION, WHY BOTHER WITH IT ?

Afforestation is not a permanent solution to greenhouse gas emissions, but it may enable New Zealand to "buy time" until new technologies become available, including those that increase fuel efficiency or that substitute for fossil fuels. It is a low-cost option, and indeed, it can be argued that it is a zero-cost option. New-land planting has taken place at the average rate of almost 80 000 ha/yr. since 1993 for reasons unrelated to global warming. There is debate about the extent of pasture that is suitable and available for conversion to forestry, but some estimates place it as high as 5.5 million hectares.[14] Depending on the rate of planting, this could provide a substantial sink for up to 100 years.

3.2 THE RISK OF ADOPTING A "NET" APPROACH

Disease (or fire, wind, etc.) do not constitute a major threat to the New Zealand government's "net" approach, (whereby greenhouse emission reduction targets are based on the difference between fossil-fuel emissions and sequestration by forests), in that the carbon stocks in standing forests are unlikely to decline significantly under any plausible scenario. This is reviewed more fully in a contract report by Maclaren and Wakelin.[15] On the other hand, the risk of a deliberate, anthropogenic decrease in carbon stocks is not negligible.

The behaviour of future forest owners cannot be modelled, and there is currently no mechanism for influencing their behaviour. The potential of carbon sequestration is directly related to the rate of new-land planting, and this is beyond the control of New Zealand government agencies. Similarly, the carbon-density of a forest estate is very sensitive to the average harvesting age. A reduction from a felling age of 28 to 25, a not unlikely possibility, will have major repercussions on targets based on a net approach. Conversely, an increase of felling age from 28 to 35 would provide a major windfall gain in sequestration.

Wakelin and Te Morenga[16] estimated the consequences of an increase, or a reduction, in average rotation age. An increase of rotation age from 28 to 35 years will ultimately add approximately 20 tC/ha to the carbon-density of the estate. A decrease to 25 years will result in a loss of about 8 tC/ha. When multiplied by the

---

14. J.P.Maclaren, *Environmental Effects Of Planted Forests In New Zealand*, 198 FOREST RESEARCH INSTITUTE BULL. (1996), at 45.

15. J.P. Maclaren & S.J. Wakelin, *Carbon Offset Forestry - The Risk Factor*, Contract report for the Ministry for the Environment (1992).

16. Maclaren, *supra* note 11.

1.7 m ha of the existing estate, (let alone a possibly far larger future estate), these represent major implications for the national carbon budget.

Reliance on sequestration has been criticised because it depends on a continued planting programme on pasture. In the absence of a state-owned forestry agency, or government subsidies, such planting is largely at the whim of small growers and their perception of current and future markets. Furthermore, there is minimal legal impediment to prevent deforestation and a return to pasture. Thus, there is a risk that any agreement that grants New Zealand a credit for new land planting will also penalise New Zealand if the forest estate shrinks in area. This may be an unlikely scenario, but should be noted as a possibility. Far more likely is the risk that future forest owners will decide to reduce their average rotation age, thus decreasing the average carbon-density stand. This could have dire consequences for a government policy that is based on a net approach, but provides no incentive to forest growers to support that approach.

3.3 DECLINING QUALITY OF STATISTICS

Although it is relatively simple to justify New Zealand's methodology and calculations for carbon sequestration to the satisfaction of any authority with competence in forest mensuration, it is more difficult to furnish explanations that are comprehensible to those without the necessary forestry expertise. The calculations of sequestration are dependent on reliable forestry statistics, which—in past decades—have been provided by a few major forest owners. The privatisation of the NZ Forest Service, and the recent proliferation of small forest growers, has complicated data collection and it is not yet clear if the quality of statistics can be maintained, even with new techniques such as satellite imagery.

3.4 MISSION OF INDIGENOUS FORESTS

The current "net approach" advocated by the NZ Government omits indigenous forests from consideration, in accordance with the IPCC Guidelines. If a future international agreement stipulates that New Zealand can adopt a net approach, but only if it includes all forested land, then this creates some difficulties. First, there is no easy means of quantifying changes in these estates, at least with the degree of precision that may be necessary. Indigenous forests, soils and scrublands are being addressed under current research workplans, but results are uncertain and may be many years away. Second, it is possible that overall carbon loss may be occurring in these indigenous estates, and this may counteract the sequestration from commercial forests. On the other hand, reversion of marginal land—if included—could have the opposite effect.

3.5 YEAR-TO-YEAR VARIATION IN SEQUESTRATION

Estate modelling has shown that there is considerable year-to-year variation in the carbon contained by New Zealand's standing forests. Short-term changes in market conditions may be reflected in the volume of wood harvested, and the finance available for new-land planting. Despite this, forest managers strive to manipulate their estates so as to provide a smooth flow of products in order to supply processing plants, to service long-term contracts, and to minimise disruptive changes in labour or machinery requirements. They do this by allowing the *growing stock*, or standing volume, of wood in the entire estate to fluctuate. The felling age and the rate of new-land planting is allowed to vary somewhat to meet these constraints.

The large year-by-year variation in the standing stock of carbon in the national plantation forest estate suggests that it is misleading to use comparisons of carbon stocks based on a single year. A rolling average based on a longer period provides an indication of trends. The authors of this paper have opted (arbitrarily) for a 5-year smoothing. In other words, the figure provided for 1990 is the mean of the figure for that year, the previous two years and the subsequent two years. Shorter term smoothing (such as 3 years) has been avoided, because this only partly compensates for year-by-year fluctuations. Longer term smoothing (such as 21 years) is also rejected because this would require good estimates of sequestration for the previous 10 years and for the subsequent 10 years. In the case of the 1990 baseline, the quality of forestry statistics in 1980 is questionable. In the case of the 2000 reporting date, projections based ten years out (i.e., 2010) can be challenged on the justifiable grounds that the future is uncertain.

Whereas year-by-year accounting may be appropriate for emissions based on fossil fuel combustion, which can be increased or reduced at the turn of a switch, a longer time-frame may be more suitable for forestry. Trees planted in the ground do not sequester much carbon in their first year, but a process is set into motion that will eventually, when averaged over a long time period, remove a considerable quantity of carbon. If this is seen as desirable, a system must be formulated that rewards those who initiate this process, even if the full benefit to global society is not realised for decades.

Given that global warming is a long-term threat, it would not be unreasonable to evaluate policies based on likely responses over a term of 100 years. This is the central figure provided by the IPCC when the Global Warming Potential of various greenhouse gases is being calculated. The other figures provided are for 20 years and 500 years. The former is too short to account for lag effects, such as those that are implicated in forestry, and the latter is too long a planning horizon for almost any society. If the "100 year timespan" is accepted by other Parties, then the long-term average carbon density of forests that are established today must be considered. This is not to say that timing should be ignored as a consideration. Carbon sequestration that occurs today is obviously more valuable than the same quantity of sequestration that is delayed by several decades. The traditional approach to this difficulty in economics is to use Discounted Cash Flow analysis, and the same technique could perhaps be used in this situation.

3.6 CHOICE OF BASELINE AND TARGET DATES

New Zealand appears to be disadvantaged by the Framework Convention on Climate Change 1990 baseline, as sequestration was abnormally high for that year and neighbouring years. Similarly, New Zealand is disadvantaged by the year 2000 for goal-setting. Sequestration is likely to be abnormally low for that year, or adjacent years. In any case, a decade is far too short of a period to allow policies based on land-use change to have any impact. The Kyoto Protocol appears to have resolved that 2008 should be the baseline and 2012 should be the first target date, but this is not obvious from the text.

4. Some Options For Reporting On Forest Sequestration

Even if there is full agreement on a base set of forestry data, and on calculations derived from that data, there is plenty of opportunity for disagreement as to how the results should be expressed in national reports. For example, a recent Canadian paper[17] examined seven methodologies, five of which were consistent with the IPCC Guidelines, with a wide disparity in "bottom line" figures for carbon sequestration by forests. Below, four major methodologies are considered, plus alternatives to each. These are labelled Options 1a, 1b, 2a, 2b, 3a, 3b, 4a and 4b.

In New Zealand, concentrating on the commercial/plantation forests, the lead of the IPCC guidelines can be followed with three levels of reporting:

- IPCC default annual growth per hectare, less harvested biomass (option 1a);
- NZ default annual growth per hectare, less harvested biomass (option 1b);
- More sophisticated approaches using forest inventory data (options 2 to 4).

The following calculations were based upon work undertaken for New Zealand's Greenhouse Policy Coalition in September 1997. They were designed to clarify the options available to negotiators at the Kyoto Conference in December 1997. The Kyoto Protocol endorsed an approval similar to option 4b.

17. Greenough, *supra* note 6.

4.1 OPTION 1A. IPCC DEFAULT

For plantations, the IPCC default guidelines provide an average gross increment in biomass (i.e., growth averaged over the full rotation length), from which any harvested biomass must be deducted. Default figures for Douglas-fir and loblolly pine are given as 6.0 and 4.0 tonnes dry matter per hectare per year, of total biomass.[18] The default value for biomass-to-carbon conversion is given as 0.5. In other words, temperate plantations can be assumed to grow at the rate of 2.0-3.0 tonnes carbon per hectare per year. The following calculation uses the mean of these two figures.This average growth rate can be multiplied by the area in plantations (1,245,944 ha for 1990, and 1,862,533 ha for 2000 with a middle-of-the-road scenario of forest expansion) to give the gross carbon sequestration for the whole plantation resource for those years. From this, the carbon extracted through harvest is deducted. This latter quantity encompasses not only the actual stemwood that is removed, but the stumps, roots, branches, etc., that are associated with the harvest and are assumed to decay immediately. Tables 1 and 2 give the results of this calculation:

TABLE 1. Sequestration for 1990 and 2000, using the IPCC default figures.

Gross sequestration (MtC/yr)		Harvest removal (MtC/yr)		Net sequestration (MtC/yr)	
1990	2000	1990	2000	1990	2000
3.1	4.7	0.4	0.5	2.7	4.2

Figures are given for actual dates, with no rolling averages used.
Column 5 is the difference between columns 1 and 3, and column 6 is the difference between columns 2 and 4.

TABLE 2. Change in net emissions over the period 1990-2000, using the IPCC default figures

Carbon sequestration (MtC/yr)		Emissions (MtC/yr)		Increase in net emissions
1990	2000	1990	2000	MtC
2.7	4.2	7.0	8.5	0.0

Column 5 is the difference between column 1 minus column 3 and column 2 minus column 4.

The IPCC calculation is 295 tonnes dry matter per hectare.[19] Converting this to carbon gives 132.5-177.7 tC/ha, or a mean of 155. Multiplying this by the change in plantation area between 1990 and 2000 indicates an increase in carbon stocks of 95.6 Mt. Note, however, that the default biomass stocks were probably intended to be representative of natural forests, rather than plantations. The latter have a lower

18. *See* Table 5 of the IPCC Workbook.
19. *See* Table 5.6 in the IPCC Reference Manual.

carbon density, commensurate with their lower average age-class.

4.2 OPTION 1B. NZ DEFAULT

The IPCC "strongly encourages" national experts to substitute country-specific default values, wherever possible, for the IPCC default values. In New Zealand, an area-weighted average of all the plantation stands in New Zealand has a carbon density of 219 tC/ha at age 28, the typical rotation age.[20] In the simplistic IPCC default methodology, this would imply an annual growth of 7.8 tC/ha/yr. In actual fact, the growth of plantations is non-linear, both in terms of stem volume and in the biomass/stemwood ratio. For example, if the plantation estate consisted mainly of stands aged 20-28 years, the gross carbon increment would be 9.8 tC/ha/yr. In contrast, stands aged 0-5 years would grow at the rate of only 1.2 tC/ha/yr.

Using the 7.8 figure as a country-specific default value, Tables 3 & 4 show the changes in net sequestration between 1990 and 2000:

TABLE 3. Sequestration for 1990 and 2000, using the NZ default figure

Gross sequestration (MtC/yr)		Harvest removal (MtC/yr)		Net sequestration (MtC/yr)	
1990	2000	1990	2000	1990	2000
9.7	14.5	0.4	0.5	9.3	14.0

Figures are given for actual dates, with no rolling averages used

Column 5 is the difference between columns 1 and 3, and column 6 is the difference between columns 2 and 4.

TABLE 4. Change in net emissions over the period 1990-2000, using the NZ default figure

Net sequestration (MtC/yr)		Emissions (MtC/yr)		Increase in net emissions
1990	2000	1990	2000	MtC
9.3	14.0	7.0	8.5	-3.2

Column 5 is the difference between column 1 minus column 3 and column 2 minus column 4

Using a New Zealand default value for average carbon stocks of 98 tC/ha,[21] the change in carbon stocks over the period 1990-2000 can be derived as 60.4 MtC.

---

20. NZ Forest Research Institute, *Data Collection Strategy For Improving Estimates Of Carbon Sequestered By Planted Forests*, A Contract Report For The Ministry Of Forestry (1996).

21. Wakelin, unpublished data.

4.3 MORE SOPHISTICATED APPROACHES (OPTIONS 2-4)

The IPCC Guidelines states that it is "highly desirable' for countries that have access to much more detailed data on changes in forest stocks, to reformat that data and present it in a form comparable to others working with less detail. Since 1990, New Zealand has been using a sophisticated "estate modelling" approach, which takes into account the actual areas and biomass of each type of plantation forestry crop, broken down by age-classes. The methodology is detailed in the Appendix to this chapter.

Results can be expressed in various ways. One approach is to exclude all forests extant in 1990, and to consider only land afforested or deforested subsequent to that date (Options 2a and 2b). This avoids any fortuitous credit or penalty for events that occurred before global warming became an international issue. In New Zealand, the young average age-class of the plantation estate in 1990 resulted in a high level of net sequestration for that year, which creates a "hurdle" that may have to be exceeded in subsequent years.

Comparisons of sequestration between two dates (say, 1990 and 2000) can be made in several ways. One method is to compare the annual net emissions for the year 2000 with those of the year 1990 (Options 2a, 3a, 4a). Events occurring in intermediate years are deemed irrelevant. This method could be called the *flux method of comparison*, because it compares the flows at two dates. The alternative is to accumulate the annual emissions and sequestration for 1990-2000 and to assess the total at the end of the period (Options 2b, 3b, 4b). This method does not ignore events occurring in intermediate years. It could be called the *stocks method of comparison*, because it assesses cumulative stocks of carbon either emitted or sequestered. Its main disadvantage is that it does not identify trends or enable goals to be set.

Lastly, it is possible to calculate emissions above a certain *allowable* level, namely the emissions that occurred in 1990, and to compare them with sequestration from forest planting (Options 4a and 4b). This approach is based on the understanding that the current aim of the FCCC is to stabilise emissions at 1990 levels, not to reduce them to zero. If an "allowable emissions approach" is taken, there is no doubt that sequestration would exceed additional emissions, i.e., New Zealand would be able to show a "carbon dioxide credit" for all years post-1990.

The "more sophisticated approaches" considered below are by no means an exhaustive list of the possibilities.

4.4 OPTION 2A. POST-1990 FLUX

This option considers only forests that have been created or destroyed after the baseline date of 1990. (Here it is essential to distinguish 'deforestation' and 'afforestation,' which are permanent changes of forest cover, from 'harvesting' and 'replanting' which are temporary and are merely part of a crop rotation.)

In order to model growth of future plantings, it is assumed that new stands will grow the same way as the weighted-average of past stands. This is a conservative assumption, because the genetic quality of the new planting stock is superior to that historically planted, and—equally important—because new land planting is taking place on farmland, with far higher productivity than traditional forest sites.

Multiplying the estimated and projected new plantings for the period 1990-2000 (years ending March 31 1991 to 2001) by their carbon content, as derived from the weighted-average national carbon yield table, the following rates of sequestration have been obtained (see Table 5):

TABLE 5. Carbon Sequestration by post-1990 new land plantings

New planting		**Cumulative sequestration since 1990**	**Annual sequestration**
Year	**ha**	**MtC**	**MtC**
1991	15,400	0.0	0.0
1992	50,200	0.0	0.0
1993	61,600	0.0	0.0
1994	98,200	0.1	0.0
1995	70,900	0.2	0.1
1996	77,800	0.4	0.2
1997	70,000	0.9	0.5
1998	70,000	2.4	1.6
1999	70,000	5.7	3.2
2000	55,000	9.7	4.0

It is not necessary to include carbon removed in harvest, as for Tables 1 and 3, because no harvesting will take place in these young stands. In this approach, only new stands are being considered and the oldest will be aged 10 in the year 2000.

The results for this option are shown below:

TABLE 6. Change in net emissions over the period 1990-2000, considering only new forests, and using estate modelling with the "flux" method of expression.

Carbon sequestration (MtC/yr)		Emissions (MtC/yr)		Increase in net emissions
1990	2000	1990	2000	MtC
0.0	4.0	7.0	8.5	-2.5

Figures are given for actual dates, with no rolling averages used.

Column 5 is the difference between column 1 minus column 3 and column 2 minus column 4.

4.5 OPTION 2B. POST-1990 STOCK

This option is the same as 2a, but the *cumulative* emissions and sequestration are assessed, and expressed in annual terms, rather than direct comparisons of annual flows at the start and the end of the period. The results for this option are also derived from Table 5 above, with an increase in carbon stock of 9.7 MtC or an average annual figure of 1.0 MtC. Cumulative emissions since 1990 are 77.8 MtC or an average annual figure of 7.8 MtC.

TABLE 7. Change in net emissions over the period 1990-2000, considering only new forests, and using estate modelling with the "stocks" method of expression.

Carbon sequestration (MtC/yr)		Emissions (MtC/yr)		Increase in net emissions
1990	2000	1990	2000	MtC
0.0	1.0	7.0	7.8	-0.2

Column 5 is the difference between column 1 minus column 3 and column 2 minus column 4

4.6 OPTION 3A. TOTAL ESTATE FLUX

Since 1990, New Zealand has reported on carbon changes using an approach that takes account of all stands in the plantation estate, whether existent prior to 1990 or not. The actual age-class structure of the estate is an important consideration. The calculation methodology is summarised in the Appendix at the end of this chapter. Using this method, the most recent results by Wakelin and Te Morenga[22] for the carbon sequestered in 1990 and 2000, are 5.8 and 5.5 Mt/a, respectively (5 year rolling averages). The decline over this period is the result

22. Wakelin & Te Morenga, *supra* note 12.

of changes in age-class structure, an effect which overshadows the increasing size of the estate. Putting these figures into calendar years and taking into account the emissions from indigenous forest harvesting, scrub clearance and wildfires, the Ministry for the Environment calculates the sequestration rates to be 5.6 and 5.2, respectively.

Estimated annual sequestration[23] using the estate modelling approach, as applied by the NZ Ministry for the Environment, is illustrated (solid line) in Figure 1, and compared to estimated annual gross emissions from all sources except land use change and forestry (dashed line). The lower line (dotted) represents the annual emissions above the 1990 baseline (see Option 4a explanation below).

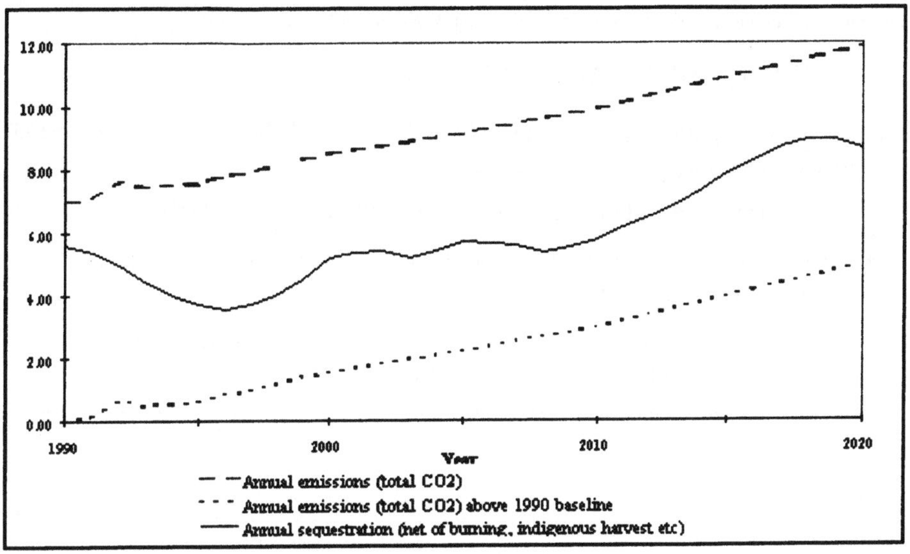

Figure 1. Annual sequestration and emissions (estate modelling approach)

Net emissions are calculated by the difference between the top two lines. The lines are close together in 1990 due to a high sequestration rate resulting from historic planting patterns. Indeed, the sequestration in 1990 is higher—in this middle-of-the-road scenario—than at any time in the following decade. This creates a high "hurdle" so that, while net emissions fluctuate over the period, they always remain well above the 1990 baseline level. In other words, relative to 1990, the two top lines diverge.

As far as New Zealand's policy of including forest sinks is concerned, therefore, 1990 is an unfortunate choice of baseline year. A date in the mid-1990s would have been associated with lower rates of sequestration (resulting in 4 MtC higher net emissions), and would have been an easier target to exceed.

23. Medium new land planting scenario. *See* Appendix at the end of this book.

The results from this option are given below:

TABLE 8. Change in net emissions over the period 1990-2000, considering all plantation forests, and using estate modelling with the "flux" method of expression.

Carbon sequestration (MtC/yr)		Emissions (MtC/yr)		Increase in net emission
1990	2000	1990	2000	MtC
5.6	5.2	7.0	8.5	2.0

Sequestration figures are 5-year rolling averages.

Column 5 is the difference between column 1 minus column 3 and column 2 minus column 4.

4.7 OPTION 3B. TOTAL ESTATE STOCK

Figure 2 shows results from the same scenario described above, this time reported as "stocks". The lines on the graph are as follows:

- accumulated carbon emissions since 1990 (top line, dashed);
- accumulated carbon sequestered by the plantation estate since 1990 (central line, solid);
- accumulated carbon emissions over and above "allowable emissions" (bottom line, dotted) (see 4b).

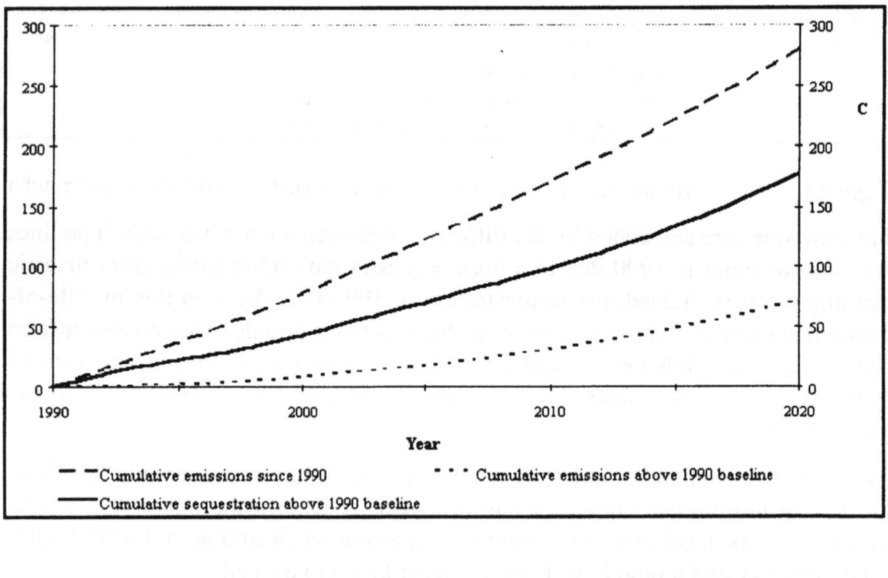

Figure 2. Accumulated emissions and sequestration since 1990

Under this option there is an increase in carbon stock since 1990 of 43.3 MtC, or an average annual figure of 4.3 MtC. As in Option 2b, cumulative emissions since 1990 are 77.8 MtC, or an average annual figure of 7.8 MtC. Table 9 gives the net result:

TABLE 9. Change in net emissions over the period 1990-2000, considering all plantation forests, and using estate modelling with the "stocks" method of expression.

Carbon sequestration (MtC/yr)		Emissions (MtC/yr)		Increase in net emissions
1990	2000	1990	2000	MtC
0.0	4.3	0.0	7.8	3.5

Column 5 is the difference between column 1 minus column 3 and column 2 minus column 4.

4.8 OPTION 4A. TOTAL ESTATE FLUX WITH ALLOWABLE EMISSIONS

This approach assumes the intent is to stabilise emissions of carbon at 1990 levels, i.e., the 7.0 MtC emitted in 1990 is the "allowable emissions" each year. The method therefore accounts for only those emissions above the 1990 baseline, i.e., in each year the emissions level in 1990 (7.0 MtC) is subtracted from the total annual emissions. The adjusted emissions are illustrated in the bottom line (dotted) in Figure 1. The *net* emission is then calculated from the difference between the middle and the bottom line. Results are provided below in Table 10.

TABLE 10. Change in net emissions over the period 1990-2000, for emissions above the 1990 baseline, with the "flux" method of expression.

Carbon sequestration (MtC/yr)		Emissions (MtC/yr)		Increase in net emissions
1990	2000	1990	2000	MtC
5.6	5.2	0.0	1.5	2.0

Column 5 is the difference between column 1 minus column 3 and column 2 minus column 4

4.9 OPTION 4B. TOTAL ESTATE STOCK WITH ALLOWABLE EMISSIONS

A comparison of the middle and bottom lines (dotted) in Figure 2 shows that the cumulative "allowable emissions" are less than cumulative sequestration by the plantation forest estate. The net sequestration will therefore be positive, i.e., in year 2000 the plantation forest estate can be seen to sequester all the increase in emissions, plus an additional 35 MtC.

The result, in this case, is an increase in carbon stock since 1990 of 43.3 MtC, or an average annual figure of 4.3 MtC (the same as in Option 3b). Cumulative emissions since 1990 are in this option only 8.3 MtC, or an average annual figure of 0.8 MtC.

TABLE 11. Change in net emissions over the period 1990-2000, for emissions above the 1990 baseline, with the "stocks' method of expression.

Carbon sequestration (MtC/yr)		Emissions (MtC/yr)		Increase in net emissions
1990	2000	1990	2000	MtC
0.0	4.3	0.0	0.8	-3.5

Column 5 is the difference between column 1 minus column 3 and column 2 minus column 4

5. Conclusion

This paper has described the key features associated with the measurement of carbon sequestration in planted forests in New Zealand. As illustrated below, the choice of option for expressing the measurement of net sequestration can lead to significant variation in results. This issue requires consideration in policy deliberations where estimates of carbon sequestration form a significant component in the determination of appropriate policies and policy measure concerning climate change.

TABLE 12. Summary of results under each option

	Net increase in forest stock (MtC)	Carbon seqjuestration (MtC/yr).		Emissions (MtC/yr)		Increase in net emissions (MtC/yr)
Option	1990-2000	1990	2000	1990	2000	
1a	95.6	2.7	4.2	7.0	8.5	0.0
1b	60.4	9.3	14.0	7.0	8.5	-3.2
2a	9.7	0.0	4.0	7.0	8.5	-2.5
2b	9.7	0.0	1.0	7.0	7.8	-0.2
3a	43.3	5.6	5.2	7.0	8.5	2.0
3b	43.3	0.0	4.3	0.0	7.8	3.5
4a	43.3	5.6	5.2	0.0	1.5	2.0
4b	43.3	0.0	4.3	0.0	0.8	-3.5

It must be noted that the results of options 2b, 3b and 4b are reported in the above table as annualised fluxes derived from the change in carbon stocks.

Subsequent to these calculations, the Kyoto Protocol endorsed a stocks approach, similar to the option 4b above. In other words, the Protocol is based on an increase in the stocks over a commitment period, except that the agreed period is from the beginning of 1998 to the end of 2012, rather than from 1990 to 2000.

The authors of this paper, as representatives of a scientific organisation, attempt to distance themselves from policy debate, and therefore are unwilling to express an opinion on the policy, as opposed to the scientific merits, of the various accounting options of national carbon budgets. While lobby groups may attempt to use one particular approach to support their respective viewpoints, the choice of approach should be distinguished from the merits of the calculations on which it is based.

11. JOINT IMPLEMENTATION: A SURVEY OF PRINCIPLES AND PRACTICAL ISSUES

PETER ALSOP
Policy Analyst, Ministry of Commerce
PO Box 1473, Wellington, NZ

1. Introduction

Joint Implementation (JI) is the process whereby a Government or private entity from one country invests in an emission-reducing project in another country and claims credit for the emission reductions. This chapter will focus on the 'pure' form of JI in which a developed country invests in a project located in a developing country.

JI has received increasing international attention under the United Nations Framework Convention on Climate Change (FCCC). This is not surprising, as the concept of JI is based on the simple proposition that cooperation between countries in projects to reduce or sequester greenhouse gas emissions will lower the total global cost of emission abatement to below that which would be incurred by countries acting alone. However, despite theoretical advantages and support for JI, neither political barriers nor international institutional impediments to implementation should be underestimated. The JI debate is extremely controversial and sensitive. It cannot easily be separated from a much broader diplomatic agenda in which North-South discord is endemic.

Many developed countries and a wide array of industries vigorously support JI. They view it as an opportunity to attain emission reductions at a cheaper cost, reducing possible negative effects on economic growth while conferring benefits on developing countries. Overall, they view JI as a 'win/win' proposition from an economic and ecological standpoint. While acknowledging that benefits can result from JI projects, most developing countries view JI as a side-step by developed countries of international equity and their obligations to lead the way in climate protection.[1]

*The views represented in this paper are those of the author, not the New Zealand Ministry of Commerce or the New Zealand Government.

1. The developed/developing division does not depict two mutually exclusive viewpoints. Rather, it provides a convenient bisection from which to analyse the debate. The terms 'developed' and 'developing' are used predominantly throughout this chapter. Northern/Southern (or north/south), first world/third world and industrialised/emerging are also common terms used in the literature to describe the development 'status' of countries.

A. Gillespie and W. C. G. Burns (eds.), Climate Change in the South Pacific: Impacts and Responses in Australia, New Zealand, and Small Island States, 209–231.

This chapter undertakes an analysis of JI from developed and developing country viewpoints and outlines the key issues that have been identified in the literature. It then moves on to a discussion of JI in the context of the South Pacific and the positions of Australia, New Zealand and AOSIS. A definitive conclusion relating to the future of JI is beyond the scope of this chapter; however, some concluding remarks identify key issues that still need to be addressed.

2. What is Joint Implementation?

JI is defined as:

> A process by which an investor may provide financial and/or technical assistance to achieve cost-effective greenhouse gas emission reductions or sequestration in a host country in exchange for credit for the reductions achieved by the project . . . [which can be credited] toward meeting its national (international) obligations for greenhouse gas emission reduction.[2]

JI belongs to a family of market-based instruments whose distinct theoretical advantage lies in the attainment of a targeted level of environmental quality at a lower cost than that which would be incurred through use of traditional 'command-and-control' environmental regulation.[3]

In simple terms, JI is bilateral trade.[4] Like standardised international trade, JI is built on the theory of comparative advantage.[5] Countries will purchase the requisite amount of mitigation wherever it is cheapest. JI has received attention in the literature since the early 1990s. Interest has subsequently increased with the FCCC, which provides for JI as a mechanism for the mitigation of greenhouse gas emissions.[6]

The motivation for JI is that developing countries have predominantly less infrastructure than developed countries and, within the former, industrial processes, power generation and residential and commercial energy use tend to be substantially less efficient. Energy sectors of developing countries (and some former cen-

---

2. A.A. Niederberger, *AIJ: Review of Issues for the Pilot Phase,* Swiss Federal Office of Environment, Switzerland (1996), at 10.

3. Market-based instruments alter the quantities or prices of goods or services through the use of property allocations, taxes or subsidies, instead of relying on centralised mandates of specific technologies or production processes, as do traditional command and control regulatory policies. *See* T.C. Heller, *Environmental Realpolitik: JI and Climate Change,* 3(2) INDIANA J. LEGAL STUDIES, (1996), at 295-340. R. Hahn & R. Stavins, *Economic Incentives for Environmental Protection*, 82 AM. ECON. REV. (1992), at 464-468; T. Tietenberg, *Economic Instruments for Environmental Regulation*, 6(1) OXFORD REV. ECON. POL'Y (1990), at 17-33; J.B. Opshoor & H.B. Vos, *Economic Instruments for Environmental Protection,* OECD (1989).

4. Three or more countries can work together on a JI project, however, this will most likely increase economic, political and legal complications. Projects normally involve two parties. Nevertheless, technical, financial or geographical circumstances may require a multi-party project.

trally planned countries) are expected to expand rapidly over the next half century to match the level of industrialisation of developed nations.[7] This growth may result in large increases in greenhouse gas emissions if nothing is done to dramatically increase energy efficiency and the use of renewable energy sources.

JI is one way of helping manage the growth of developing country emissions. The ability of developed countries to invest, contribute technologies and jointly implement energy efficiency and renewable energy projects is viewed as an effective way of limiting developing country emissions during the industrialisation process. As developing countries build infrastructure and prepare to meet future energy demands, there is an opportunity to adopt new, highly efficient technologies.[8] Conversely, future capital investments in industrialised nations will go largely toward the gradual replacement of existing facilities. This would likely be a somewhat slower process through which to introduce new technologies. This process is also believed to be more expensive than reducing an equivalent quantity of emissions in developing countries.[9] A corollary of this result is that a given expenditure on reducing emissions should generate greater emission reductions in developing countries.

Proponents of JI believe that the cost differential between emission reductions in developed and developing countries provides the incentive to undertake JI projects and provide a mechanism for lowering the total global cost of achieving a given environmental objective. The rationale is that greenhouse gas emissions have the same environmental consequences regardless of where in the world they occur. In other words, emission reductions in developing countries have the same environ-

---

5. Given its 'trade characteristics,' JI has been promoted as a step toward establishing an international tradeable permit system. "[JI] initiatives could help build confidence in future emission trading." *See* S. Upton, (Minister for the Environment, New Zealand), *Climate Change: The Perplexing Environmental Issue for NZ*, THE SOUTHLAND TIMES, 3 March 1997. Thus, some discussions of JI are entrenched in a wider discussion of international trading. *See* R. Sapsford, *Joint Initiatives and Action: Issues and Opportunities for NZ.* Contract report to the Ministry for the Environment (1997); K. Hanslow, et al., *Climate Change - Trade and Welfare Effects*, 1(3) AUSTRALIAN COMMODITIES (1994), at 344-354. Much of the technical, institutional and financial expertise needed for JI is also needed to construct a trading system. If maximum efficiency gains were extracted from JI, the marginal cost of abatement would be equalised between industrialised and developing countries - a theoretically identical result to that achieved by an international trading regime (with full global participation). JI is also referred to as a "project" system since emission reductions occur on a project by project basis. Additionally, because JI projects are voluntary, JI has characteristics of a Voluntary Agreement.
6. The Kyoto Protocol to the FCCC provides for JI projects (see Section 3) between developed countries (Article 6) and between developed and developing countries (Article 12). The term 'JI' has historically been associated with projects between developed and developing countries. Keeping with tradition, this form of JI will be the focus of this paper. Additionally, JI between developed countries could be accommodated in an international emissions trading system under Article 17 of the Protocol. That is, Article 6, which provides for projects between developed countries could essentially become redundant. The project would simply represent a contractual detail of the trade - hence the name 'project level trading,' which is sometimes used to refer to JI.
7. For a discussion of emissions growth in developing countries, *see* Intergovernmental Panel on Climate Change, *Climate Change: The IPCC Scientific Assessment* (1995).

mental benefit as reductions in developed countries; for example, one tonne of CO_2 emitted in Australia has the same global warming effect as one tonne of CO_2 emitted in Poland or China.[10]

3. JI and International Law

Recent attention on JI has led many to view it as a new and innovative tool. However, it is not only within the ambit of climate change that the tool has been utilised. The clearest precedent is the 1987 Montreal Protocol to the 1985 Vienna Ozone Convention, which set a timetable for the gradual elimination of consumption and production of chlorofluorocarbons (CFCs).[11] Although Parties had specific individual obligations, provision was also made for joint action.

The 1993 Rhine 'Salts' Treaty to the 1976 Convention on the Protection of the Rhine Against Pollution by Chlorides is another example of JI.[12] The Treaty enables riparian provinces in the Netherlands who share the Rhine to engage in joint funding of downstream measures to divert chlorides to the North Sea rather than requiring upstream polluters to remove them at the source. JI is also permitted under the 1994 Oslo (Second Sulphur) Protocol of the 1979

---

8. The International Energy Agency (IEA) estimates that 30% of primary energy could be saved by use of state of the art steel production equipment in developing countries. Further, available efficiency gains could reduce emissions dramatically (4% more efficiency for 10% of all coal fired coal plants reduces CO_2 emissions by 50 million tonnes per annum). The IEA estimates the market for JI projects over the next 15 years at US$5-20 billion. *See* I. Puhl & L. Solsbery, *IEA's work on Methodological Issues for AIJ*, Presentation given to the FCCC Roundtable, Geneva (1996); J. Goldemberg, *Energy Needs in Developing Countries*, 269 SCI. (1994), at 1058-1059. Goldemberg reports that developing country emissions are projected to exceed those of developed countries by 2020. An OECD analysis came to a similar conclusion, projecting developing country emissions to increase from the present share of 30 percent of global emissions to just over 50 percent by the year 2020. The aggregate electricity generating capacity of Brazil, China, India, Indonesia, and Russia could quadruple by 2020, http://www.oecd.org/subject/climatechange/kyoto.htm. Goldemberg & Reid point out that, "although developing country world emissions will equal those of industrialised nations by 2020, it will take considerable time before their cumulative releases - a better index of overall contributions to global warming - match those of industrialised countries." J. Goldemberg, & W.V. Reid, *Developing countries are combating climate change*, 26(3) ENERGY POL'Y (1998), at 233-237.
9. D. Harvey & E.J. Bush, *Joint Implementation: An Effective Strategy for Combating Global Warming?*, 39(8) ENVIRONMENT (1997), at 14-20, 36-44. That emission reductions are cheaper in developing counties is the widely held view. However, some research counters this claim. *See* T. Jackson, *JI and Cost-Effectiveness under the FCCC*, 23(2) ENERGY POL'Y (1995), at 117-138.
10. .J. Yellen, *Testimony Before the House Commerce Committee*, March 4, 1998), at 9. .
11. Article 2.5 of the Protocol, allows all CFC producers to, within certain limits, transfer/receive production quotas to/from other countries. Article 2.8(a) permits Parties that are members of a regional economic grouping, such as the European Community (EC), to jointly fulfil their obligations. *See* D. PEARCE, BLUEPRINT 4: CAPTURING GLOBAL ENVIRONMENTAL VALUE, Earthscan (1995); O. KUIK, et al., JOINT IMPLEMENTATION TO CURB CLIMATE CHANGE: LEGAL & ECONOMIC ASPECTS (1994).
12. Pearce, *supra* note 11.

Convention on Long Range Transboundary Air Pollution.[13]

Under the 1992 Framework Convention on Climate Change, Parties are able to jointly fulfil their climate protection responsibilities. Article 4(2)(a) of the FCCC states: "[P]arties may take such policies and measures [to mitigate climate change] jointly with other Parties and may assist other Parties in contributing to the achievement of the objective of the [FCCC]." Additionally, Article 3(3) requires Parties to take "into account that policies and programs to deal with climate change should be cost-effective to ensure global benefits at the lowest possible cost." It also notes, "efforts to address climate change may be carried out cooperatively by interested Parties." Clearly, Parties to the FCCC may work together for the purpose of cost-effective climate protection.

JI was introduced into the climate negotiation process by Norway at the Intergovernmental Negotiating Committee (INC) meeting in 1992. Subsequent discussions at INC meetings prior to the First Conference of the Parties (COP1) failed to reach consensus on the nature and role of JI within the FCCC. COP1, held in Berlin in April 1995, was no exception. Agreement on a protocol legitimating JI was not achieved. This was primarily due to disagreements between developed and developing countries on the nature of commitments to limit greenhouse gas emissions after 2000.[14]

Developed countries largely supported JI, arguing that deeper targets would be ineffective without universal commitments, as the bulk of emissions growth in the future would take place in developing countries, especially in Asia. Developing countries, including the G-77 led by China, stymied efforts to establish a JI system. They largely insisted that developed Parties deepen their own emission reductions in the next century, contending that they were historically responsible for climate change.

As a compromise, JI was permitted on a four-year trial basis (1996-1999 inclusive), known as the 'Pilot Phase.' To distinguish from the full fledged concept of JI, projects during the pilot phase are referred to as 'Activities Implemented Jointly' or 'AIJ.'[15] The pilot phase was established to explore the concept of JI, gain experience with projects, better define methodological and implementation issues and to consider the progression of JI beyond the pilot phase in the future. During the pilot phase, however, developed countries could not credit any JI emission reductions toward their FCCC commitments.

---

13. *Id.* Article 2.7 states "[T]he Parties to the Protocol may .... jointly implement the obligations set out in Annex II." Annex II sets targets for emissions of sulphur dioxide (SO_2).

14. P. Lefale, *AIJ: Opportunities and Expectations of Pacific Island Countries*, paper delivered to Closing the Communication Gap Workshop (1997).

15. Decision 5/CP.1 of FCCC/CP/1995/7/Add.1. AIJ and JI are used interchangeably in the literature and are functionally equivalent. The key distinction is that credits are unavailable for AIJ in the pilot phase. For the purpose of this chapter, the more generic term 'Joint Implementation' (JI) is generally used to include both JI and AIJ projects.

Eligibility criteria for AIJ projects, which also identify key issues of the JI debate, were established. These highlight four distinct concerns about JI projects.[16] The first pertains to environmental and local priorities. That is, AIJ projects should be compatible with and supportive of national environment and strategic development priorities of the 'host' country and contribute to cost-effectiveness in achieving global benefits (see section 5). The second concern is that projects must have host and home country acceptance. Accordingly, AIJ projects require prior acceptance, approval and endorsement by Governments of participating Parties. Reduction additionality is the third concern. The rationale is that AIJ projects should bring about real, measurable and long-term environmental benefits related to the mitigation of climate change that would not have occurred in the absence of such activities (see section 6). Finally, financial additionality needs to be addressed. That is, the financing of AIJ projects shall be additional to the financial obligations of Parties to the Convention (through the Global Environment Facility) as well as to current official development assistance.

The Third Conference of the Parties (COP3) was held in Kyoto, Japan, in December 1997. After intense formal and informal negotiations, Parties to the FCCC adopted the Kyoto Protocol.[17] The Protocol includes emission limitation and reduction commitments (i.e., targets) for the first commitment period (2008-2012) for developed countries who are listed in 'Annex B' of the Protocol.[18] Subject to ratification of the Protocol, Annex B countries will be legally bound to reduce emissions to their target levels.[19]

Throughout COP3, both developed and developing countries called for technology transfer and the achievement of sustainable development in developing countries. Some developing countries, particularly China, continued to express reservations about JI as a cost-effective means of achieving these objectives.

JI between developed and developing countries can be found in the Protocol in

---

16. N.J. Cutright, *Joint Implementation: Biodiversity and Greenhouse Gas Offsets*, 20(6) ENVT'L MGT., (1996), at 913-918. *See also* Niederberger, *supra* note 2, at 4.

17. For a summary of the COP3 meeting and the Kyoto Protocol *see* P. Chasek, *Report of the Third Conference of the Parties to the UNFCCC*: 1-11 December 1997; International Institute for Sustainable Development, 12(76) EARTH NEGOTIATIONS BULL., 13 December 1997; M. Jefferson, *Climate Change 1997 - COP3 to the UNFCCC and the Kyoto Protocol.* World Energy Council (1997); *A Brief Analysis of the Kyoto Protocol*, 9(24) GLOBAL ENVT'L CHANGE REP. (1997), at 1-8.

18. It should be noted that not all 'developed' countries are Annex B countries. For example, Singapore is richer than New Zealand and has emissions that are twice as high, yet it is not classified as an Annex B state.

19. However, there is provision in the Protocol for countries to adjust their targets for (a) transfers "for acquisitions of portions of other Annex B Parties' targets (i.e., international trading under Article 17) and; (b) emission reducing (JI) projects between developed and developing countries under the Clean Development Mechanism (Article 12); (c) emission reducing (JI) projects between developed countries (Article 6); and, changes in carbon stocks. This adjusted target is referred to in the Protocol as a Party's 'assigned amount.' Pursuant to Article 3.1, to be in compliance, a Party's emissions for the commitment period must be no more than its assigned amount.'

Article 12 which establishes a "Clean Development Mechanism" (CDM). The purpose of the CDM is to assist developing countries "in achieving sustainable development and to assist [developed countries] in achieving their [targets]."[20] Simply stated, this provision is JI by another name. The CDM will supersede the current AIJ provision of the FCCC in 2000.

The CDM encourages voluntary participation by developed and developing countries (including entities within those countries)[21] in JI projects that bring about real, measurable and long-term benefits related to the mitigation of climate change. Under the CDM, certified JI emission reduction credits will be able to be used by developed countries to achieve their targets. This contrasts with the AIJ pilot phase where any credits resulting from JI projects cannot be accrued toward FCCC commitments. Article 3(12) of the Protocol states: "Any certified emission reduction units which a Party acquires from another Party in accordance with the provisions of Article 12 [CDM projects] shall be added to the assigned amount for the acquiring party." This includes CDM projects undertaken between 2000 and the start of the first commitment period (2008) which can be credited toward a Parties target during the first commitment period (2008-2012).[22]

Another notable difference between the FCCC and the Kyoto Protocol is that (subject to development) the CDM will require a share of JI project proceeds to be used to cover administrative expenses and assist developing counties to meet the costs of adaptation to the adverse effects of climate change.

The CDM lays the foundation for a positive future for JI. The Fourth Conference of the Parties (COP4), scheduled to be held in Buenos Aires in November 1998, will seek to resolve outstanding methodological issues and in doing so, take the CDM a step closer to implementation and further clarify the future of JI.[23]

---

20. Article 12(2). As noted earlier, Article 6 of the Kyoto Protocol also establishes JI between developed countries. Notwithstanding that this is generically referred to as 'JI,' it is essentially project-level trading and could be included within an international trading regime (subject to design and rules which permitted project-based trades). For a discussion of the potential rules for an international trading system see a recent submission (June 1998) to the UNFCCC process by Canada (on behalf of Australia, Iceland, Japan, New Zealand, Norway, the Russian Federation, Ukraine, and the United States) Available on the Internet, http:www.unfccc.de/fccc/docs/1998/sb/miscØ1a1r1. *See* document FCCC/SB/1998/Misc.1/Add.1/Rev.1.

21. Article 12(9): "[P]articipation under the [CDM] . . . may involve private and/or public entities, and is to be subject to whatever guidance may be provided by the executive board of the [CDM]." The executive board and operational entities, which will oversee, authorise and verify CDM projects, are yet to be established.

22. Article 12(10) of the Protocol states that "[C]ertified emission reductions obtained during the period 2000 up to the beginning of the first commitment period [2008] can be used to assist in achieving compliance in the first commitment period [2008-2012]."

23. Article 12(7) of the Protocol states that "the . . . meeting of the Parties to this Protocol shall, at its first session [Nov. 1998], elaborate modalities and procedures with the objective of ensuring transparency, efficiency and accountability through independent auditing and verification of project activities."

4. Investor (Developed) Views

An "investor" in a JI project may be a developed country government or private entities, such as an industrial sector or individual firm. Individual emitters in developed countries may seek emission reduction credits to meet domestic obligations under a national regulatory framework, such as a Voluntary Agreement or a domestic emissions trading system, subject to their government recognising such credits. They are called "investors," as they invest human and financial resources in projects. Developed country (investor) views largely focus on the cost-effectiveness of JI and the benefits for developing countries of technology transfer.

Developed countries have argued that it makes sense for emission reductions to take place in developing countries where they are effectuated most cheaply. The justification for this argument is that the same environmental benefit is achieved through emission reductions in developing countries as could be achieved in developed countries, but at a lower cost. Exploiting opportunities in developing countries could result in substantial economic savings. Results from economic modelling has indicate that the cost of emission reduction in developed countries could be reduced by 85% in 2000 and 70% in 2010 by JI.[24]

Proponents of JI also believe that it could enable developing countries to bypass inefficient, carbon intensive energy technologies. The adoption of cleaner, more advanced, alternatives could be achieved through technology transfers and building of institutional capacity, hastening progress toward sustainable development. Such transfers could be used to retire, or prevent installation of, 'dirty' technologies, such as coal-fired power stations, that commit nations to high emissions trajectories due to the slow rate of capital turnover for large plants

The progress towards sustainable development may also be promoted by some of the purported indirect benefits of JI. These may eventuate as the results of projects that focus on providing energy to satisfy basic human needs such as food, shelter, sanitation, education and health care and, as a consequence, can help improve the standard of living and decrease fertility rates in developing countries.[25]

---

24. J.C. Bollen et al., *A Framework for the Assessment of the Global Potential of JI*, Report Nr. 481507011. National Institute of Public Health and the Environment (1995). This refers to OECD countries only. An Australian study came to similar conclusions. ABARE (Australian Bureau of Agricultural and Resource Economics and Department of Foreign Affairs and Trade), *Global Climate Change: Economic Dimensions of a Cooperative International Policy Response Beyond 2000*. ABARE (1995). ABARE estimates that the global cost saving from using JI to help attain stabilisation of emissions at 1990 levels by 2010 is 60%; or US$12 billion in 2010 and US$47 billion in 2020. Bohm also concludes that the use of JI amongst four Nordic nations could reduce the total cost by around 50%. P. Bohm, *Joint Implementation as Emission Quota Trade: An Experiment Among Four Nordic Countries*, Nord Energy (1997). The potential economic merits of JI have been widely discussed. *See* S. Barret, *The Strategy of Joint Implementation in the FCCC*, United Nations (1994); R. Loske & S. Oberthur, *Joint Implementation under the Climate Change Convention*, 6(1) INT'L ENVT'L AFF. (1994), at 45-58; THE FEASIBILITY OF JOINT IMPLEMENTATION (C.J. Jepma ed., 1995).

5. Recipient (Developing Country) Views

A developing country that permits a JI project to take place in its jurisdiction is referred to as the 'recipient' or 'host' country. Since under the FCCC and the Kyoto Protocol developing countries are not required to achieve legally binding emission limitation or reduction commitments, a host country will normally participate in a JI project to receive money and/or other local benefits (such as employment and pollution reduction) that can stem from the project.[26]

The JI equity debate involves two issues. The first issue pertains to the question of responsibility or demonstration of leadership by developed countries as the FCCC requires. The second consideration is whether JI exploits the vulnerability of developing countries or perpetuates neo-colonial relationships, thereby leading to long-term harm.[27] The equity debate is also embedded in a wider context in which developed and developing countries argue over historic responsibility for causation of the climate change problem, responsibility for mitigating any adverse effects, and the most appropriate basis for this mitigation.

The G-77 bloc has argued that JI is unacceptable because it is not consistent with the equities of global environmental politics.[28] They contend that JI would constitute an evasion of historical responsibilities by developed countries, that emissions must be eliminated within their own jurisdiction and that no limits may be imposed

---

25. Harvey & Bush, *supra* note 9, at 20. Benefits reported to have stemmed from the Wisconsin Electricity Rio Bravo Project include: greenhouse gas emission reductions (CO_2 and other gases); improved water and air quality; soil conservation; sustainable economic development; regional employment; community services; environmental education; improved human health; land productivity; and preservation of cultural heritage and biodiversity. *See* Cutright, *supra* note 16. For a discussion of a "human needs" approach to energy development *see* T. Goldemberg et al., *Energy for Development*. World Resources Institute (1987).

26. For a detailed discussion of recipient views, *see* P. Ghosh et al., *Joint Implementation of Climate Change Commitments: Opportunities and Apprehensions*, Tata Energy Research Institute (P. Ghosh & J. Puri eds., 1994), at 13-34. Due to differences in abatement and sequestration costs and other factors (such as economic and political structure), the potential for JI projects will vary from country to country. The largest recipients are expected to be China and economies in transition. *See* Niederberger *supra* note 2, at 33.

27. L. Yaker, *Joint Implementation from a Southern Perspective*, in THE FEASIBILITY OF JOINT IMPLEMENTATION (C.J.Jepma ed., 1995); Harvey & Bush, *supra* note 9, at 39.

28. China has led the political opposition to JI in the G77 bloc. This initiative seems rooted in a wider international agenda in which China emphasises its position as a major sovereign power, resilient to 'western' intrusions. JI would undercut this stance. On the other hand, China is aware of capital needs for energy and transportation infrastructure and is increasingly concerned about domestic pollution resulting from fossil fuel use. Japan and Korea, in particular, have associated interests in energy sector JI in China to reduce their vulnerability to acid rain. *See* Heller, *supra* note 3, at 336. "China's inefficient burning of coal and other fossil fuels damages not only the planet's atmosphere but also the health of Chinese living in industrial cities choked by smog. . . . A recent World Bank study estimates that if China doesn't improve its environment, its urban residents face 600,000 premature deaths and 5.5 million cases of chronic bronchitis a year by 2020 because of exposure to fine particles in the air." *See* P. Landers, *Heated Debate: A New US-China Battleground - The Environment*, FAR EASTERN ECON. REV., 18 December 1997, at 17 & 20.

on compensatory economic growth in developing countries. As such, they view JI as a way for developed countries to shirk their climate protection responsibilities since it affords them the opportunity to purchase emission offsets abroad - essentially buying their way out of having to make emission reductions "at home."[29]

Some developing countries suggest that JI violates the spirit of the FCCC, which calls upon developed countries to take the lead in reducing emissions.[30] JI critics have maintained that developed countries are making unfair attempts to link financial and technological transfers with their commitments under the FCCC. Additionally, some developing countries have argued for unconditional financial reparation for past climate exploitation by developed countries. JI projects, which are conditional on investments in environmental quality improvement, would not meet this requirement.[31]

Scepticism of JI by some developing countries is furthered by a mistrust of market forces and from experience with industry and international organisation investment projects.[32] As such, they may feel vulnerable to the implicit objectives of JI 'investors' due to a perceived inability to defend their own interest in complex deals. Their financial situation may also weaken their bargaining power in negotiating the direction and details of JI investments.[33]

Sceptics also maintain that the cheapest and easiest ("low hanging fruit") mitigation opportunities would occur first, leaving more complex and expensive reductions to be made in the future.[34] This could increase the cost for developing countries in the future should they ever be required to take on emission reduction targets. Developing countries also fear that some projects, such as reforestation efforts or forest protection projects, could create a situation where 'investor' countries determine forestry policies instead of local and indigenous peoples. Some also fear that JI could encourage developing countries to shift development priorities toward areas targeted by JI investors.[35] As such, the potential to attract capital and the appeal of short-term benefits from technology transfer and foreign investment may skew decisions and produce negative long-term consequences.

---

29. Heller, *supra* note 3, at 306; Harvey & Bush, *supra* note 9, at 39. Viewed from a different angle, developed countries, during the process of industrialisation, have used more than their fair share of the global emissions budget, denying developing countries an equitable share of the global atmospheric 'commons.'
30. "The Parties should protect the climate system . . . in accordance with their common but differentiated responsibilities" (FCCC: Article 3(1)). "The developed country Parties . . . shall take all practicable steps to promote, facilitate and finance . . . environmental sound technologies and know-how to . . . particularly developing country Parties" (FCCC: Article 4(3) & 4(5)).
31. Heller, *supra* note 3, at 330. Article 12(8) of the Kyoto Protocol requires " . . . a share of the proceeds from certified project activities [to be] used to cover administrative expense as well as to assist developing country Parties that are particularly vulnerable to the adverse effects of climate change to meet the costs of adaptation". This could be seen as a political compromise to meet ongoing calls by developing countries for unconditional financial reparation.
32. Niederberger, *supra* note 2, at 34.

Finally, the potential impact of JI on the rate of technological development is a mutual concern of developed and developing countries. The reduction of global emissions largely depends on how quickly developed countries can develop advanced energy technologies and deploy them to developing countries. If JI became widespread, the pressure on developed countries to make major emissions reductions at home would be reduced. This could encourage developed countries to postpone immediate and necessary investments in energy research and development. It has been argued that this has the potential to delay the rate of technological development and increase the long-term costs of future emission reductions.[36]

33. Harvey & Bush, *supra* note 9, at 39. Notwithstanding this, a JI 'deal' is voluntary and thus represents an agreement between two willing parties. Therefore, suggestions that JI will undermine the true desires of developing countries could be factored into a 'deal' in which a developing country acts in a risk averse manner to protect its local, social and environmental objectives. Much of the community development literature emphasises community ownership, control and management as essential for successful development projects. These elements facilitate self reliance, empowerment and community autonomy. *See* S. Bygrave, *Utility Management Of Village Scale Solar Photovoltaic Projects in Developing Countries: The Case of the Solar Energy Company and Notoue Village, North Tarawa, Kiribati,* Solar '97 Conference, Australian National University (1997), at 1-3. D.C. Korten, COMMUNITY-BASED RESOURCE MANAGEMENT, in Community Management: Asian Experience and Perspectives (D.D. Korten ed., 1986); R. CHAMBERS, RURAL DEVELOPMENT: PUTTING THE FIRST LAST (1983). The negotiation and direction of project details could also extend to sharing the actual emission reduction credits created by CDM projects. If an international trading regime became operational, developing countries (and/or entities within these countries) could potentially become involved by selling verified CDM credits in the trading market. This CDM-trading financial 'link' could encourage developing countries to become more involved in projects and negotiate project details to their benefit.
34. Heller, *supra* note 3.
35. Harvey & Bush, *supra* note 9, at 40. For example, host governments may abandon local environmental controls, maintain subsidies or trade controls and relax pricing reforms in order to market JI projects. A nation like China may be tempted to move away from its stated goal of reducing coal subsidies, providing a disincentive to switch away from coal- based electricity generation. JI projects involving clean coal technologies (with lower emissions) would only involve "true" reductions if the 'baseline' (discussed in section 6) against which improvements are measured did not assume the elimination of coal subsidies. JI projects may therefore be better qualified when located in nations which enter into agreements committing them to environmental standards for other policies related to greenhouse gas emissions, such as energy pricing reforms. *See* Heller, *supra* note 3, at 333. A transparent project negotiation process involving the local community, non-government organisations (NGOs) and other interested parties may also help in this regard. *See* Niederberger, *supra* note 2, at 34.
36. P. Zollinger & R.C. Dower, *Private Financing for Global Environmental Initiatives: Can the Climate Convention's "Joint Implementation" Pave the Way?,* World Resources Institute (1996); Niederberger, *supra* note 2, at 21; Harvey & Bush *supra* note 9, at 36. This concern could be addressed by placing a restriction ('ceiling') on the quantity of emission reductions that can be achieved using JI. "The EU has restated its demand that the use of flexibility mechanisms be capped under the Kyoto Protocol to ensure that industrialised countries achieve their targets for limiting greenhouse gas emissions mainly through domestic action." *The EU wants 'Concrete Ceiling' on Emissions Trading,* 10(6) GLOBAL ENVTL. CHANGE REPORT (1998), at 1-3. Any such restriction would, as in the case of international trading, be politically contentious (other countries do not want any restrictions); would limit the extent to which Parties can exploit least cost opportunities in developing countries and, accordingly, could increase the overall global cost of achieving the Protocol's environmental objective.

6. JI in Operation - Practice and Pitfalls

JI programmes have been established in a number of countries including the United States, Australia, Germany, Japan, Norway, Poland, Sweden and Switzerland.[37] The World Bank also administers a programme.[38] Due to the trial nature of these programmes, they have typically been referred to as 'pilot programmes.' To stimulate interest in these programmes, governments have tended to offer incentives for firms to participate.

37. For a discussion of these programs, *see* Niederberger, *supra* note 2, at 39-48. *See also* the FCCC's home page on the Internet: http://www/unfccc.de/fccc/ccinfo/aijproj.htm. An outline of the Australian programme is contained in the 'JI in the South Pacific' section. The United States has promoted JI in regional and international forums under the 1993 United States Climate Action Plan. The Plan calls for voluntary climate change mitigation measures by various sectors. The "United States Initiative for Joint Implementation" (USIJI) is one part of the Plan. It was the first national JI pilot programme to have a formal set of acceptance criteria, guidelines and a project evaluation process and is currently the largest programme in operation. By December 1997, the United States alone had approved 28 AIJ projects in 12 countries (US Delegation to COP3, 1997); 40% of these projects involve carbon storage in forests; 28% supply side and end use energy efficiency improvements; 16% renewable energy technologies; 8% fuel switching; and, 8% measures to capture methane. These percentages are based on the first 25 projects. United States Government, *JI: Second Report to the Secretariat of the UNFCCC: Accomplishments and Descriptions of Projects Accepted Under the USIJI.* USIJI Secretariat (1997), at 3.

38. The World Bank programme was initiated in collaboration with the Norwegian Government in April 1996. The six objectives of this programme are to: maximise learning about AIJ; promote the long term objective of the FCCC, contribute to client (host) country development; explore solutions to methodological issues of AIJ; promote partnership and expand private sector participation; and, implement a variety of AIJ projects within a given time frame. World Bank, *The AIJ Program at the World Bank*, Global Climate Change Unit: Global Environment Division (1997).

JI projects usually are backed by incentives.[39] For example, government may provide funding in the form of financial and fiscal incentives to complement private sector investment to help overcome project transaction and administration costs.[40] Direct, non-financial incentives are the second type of incentives. These usually accord companies the right to apply JI emissions reductions to domestic commitments. They also include match-making services provided by governments, international organisations or non-government organisations (NGOs), such as information provision and technical assistance (e.g., workshops, guidance documents, issue papers and advice hotlines). Finally, "soft" or indirect incentives are also considered important. Such considerations refer to public recognition of climate protection measures; the right to use programmes/logos for advertising and public relations and opportunities to contribute to the design criteria for a 'fully fledged" JI regime in the future.[41]

To date, private sector investors in JI have largely been electric utilities and some energy intensive production industries. Projects have largely focused on carbon dioxide (CO_2) because of its importance as a greenhouse gas and because CO_2 emissions are easier to quantify than other greenhouse gas emissions. More specifically, the projects have focused on CO_2 absorption through creation of forest (carbon) sinks.[42] Some examples of JI to date are:

Project Type	**Project Title**	**Partners**	**Lifetime (yrs.)**	**Estimated Total CO_2 "Saved" (tonnes)[a]**
Afforestation	CARFIX - Sustainable forest management	Costa Rica & USA	25	21,776,749
Energy Efficiency	Emission reduction at power plants in Romania	Netherlands & Romania	5	1,092,000

39. Niederberger, *supra* note 2, at 28-29.

40. Transaction costs, such as finding a willing project partner and negotiating an agreement, are an impediment to the implementation of JI projects. Hourcade and Baron estimate that transaction costs could fall in the range of 10-30% of investment costs. Transaction costs can be reduced, for example, by rules or institutions that foster information provision and provide clear guidelines for project development. J. Hourcade & R. Baron, *International Economic Instruments and Climate Change,* OECD (1993). For a discussion of transaction costs as they specifically relate to JI, *see* D.J. Dudek & J.B. Weiner, *Joint implementation Transaction Costs under the Climate Change Convention,* OECD (1996).

41. For a detailed discussion of incentives for private sector investment, *see* ECON Centre for Economic Analysis, *Domestic Climate Regimes and Incentives for Private Sector Involvement in JI,* Norway: Commissioned by the World Bank (1997).

42. Zollinger & Dower, *supra* note 36, at 5; Harvey & Bush, *supra* note 9, at 16.

Forest Preservation	Rio Bravo Carbon Sequestration Pilot Project	Belize & USA	40	4,801,478
Reforestation	Reforestation in Vologda	Russian Federation & USA	60	858,000
Fuel Switching	District Heating in Decin	Czech Republic & USA	27	607,150
Fugitive Gas Captwure	RUSAGAS Fugitive Gas Capture Project	Russian Federation & USA	25	30,000,666
Renewable Energy	Boiler Conversion	Latvia & Sweden	10	254,000

a. To put these figures in context, consider that New Zealand's gross CO_2 emissions in 1995 were around 27,000,000 tonnes. The UNFCCC homepage contains a comprehensive listing of all officially sanctioned JI projects. See Internet, http://www/unfccc.de/fccc/ccinfo/aijproj.htm.

JI projects are expected to grow out of a wide range of industrial, agricultural, residential and, importantly, energy and transportation opportunities.[43] Development in the latter two areas has been slow because the energy and transportation sectors in many developing countries are in the process of transition. As such, uncertainties about subsidies and privatisation make financial returns in these key sectors excessively risky.[44]

43. Heller, *supra* note 3, at 298-299. JI projects in energy sectors may involve supply side activities that reduce emissions by changing the inputs to productive processes or by increasing their efficiency. Examples are switching fuels from coal to natural gas and improving the capacity of transmission lines that carry electricity from generators to users. Alternatively, JI may operate on the demand side through programs such as financing consumer use of compact fluorescent lighting that lowers the consumption of emission-intensive products or services.

44. Heller, *supra* note 3, at 299. Although uncertainty over subsidies has provided a disincentive for JI projects within these sectors, subsidies have generally been reduced in recent years, which is positive from a climate perspective. Between 1991 and 1996, total fossil fuel subsidies in 14 developing countries (accounting for 25% of global emissions) declined from US$60 billion to US$33 billion (45% reduction). *See* World Bank, *Expanding the Measure of Wealth: Indicators of Environmentally Sustainable Development,* World Bank (1997). Reduced subsidies lead to higher fuel prices and reduced rates of consumption and emissions. Although China's emissions grew by 228 million tonnes of carbon (MtC) between 1980 and 1990, energy pricing reforms (including subsidy reductions) prevented an additional 155 MtC from being emitted over that period. None of these carbon savings specifically resulted from dedicated domestic climate change policies, but rather were a side benefit of other policy changes to meet social and public health needs. However, this has also been the case in many developed countries. For example, in the United Kingdom, "two thirds of the nation's emission reductions expected by 2000 will result from the replacement of coal-fired power plants with combined cycle gas turbines - a result of energy privatisation over the past decade." *See* Goldemberg, *supra* note 8, at 236.

Aside from Costa Rica, which is playing a very active role in JI, developing countries have been slow to promote programmes or the implementation of JI projects.[45] The reasons include prioritisation and scepticism about JI's practical benefits. Such concerns often stem from a lack of knowledge of JI. In the case of Costa Rica, the President has played an important role by granting JI high priority. Some other recipient countries have moved to establish a secretariat for JI activities to examine its potential and provide a contact for investors searching for JI projects (e.g., India and Poland).[46]

An important impetus for developing country participation in JI is an agreement between the governments of Norway and China to upgrade a coal-fired power plant, signed in January 1998. Norway will contribute (US) $5 million to the project and hopes to get CO_2 reduction credits in the year 2000 when the CDM clause of the Kyoto Protocol becomes operational. If developed according to current plans, the cogeneration project will increase the energy efficiency of the coal fired power plant by 40%.[47]

Despite the potential for JI, implementation at the international level has been slow.[48] There have been three primary reasons for this. The first pertains to transaction and administration costs, which can be substantial. Complementary services (e.g., legal and monitoring services and information provision) to manage these costs have not yet been developed and firms are reluctant to explore the market accordingly.[49] Moreover, JI financing does not easily lend itself to 'software' transfers (human skills and resources) as emission reductions resulting from the use of software are not easily quantified.[50] Software is required to implement, maintain and operate new technology and adapt it to local conditions. Finally, national policies to create incentives to invest (e.g., a domestic carbon tax or emissions trading regime) in JI projects have yet to be implemented in developed

45. The first review of the pilot phase in 1996, released at the Second Conference of the Parties (COP2) in Geneva, listed 17 host countries where projects had been proposed or launched. Costa Rica hosts the largest number of official JI projects and is endeavouring to establish its own Carbon Fund as part of the Certified Transferable Offsets (CTO) programme. This fund would create an investment portfolio of forestry and renewable energy projects. Investors could purchase certified emissions reductions (offsets) from the fund without becoming directly involved in the projects. The advantage of this programme is that JI investors do not have to carry out a feasibility analysis to establish the viability of a project. Instead, the investor would simply buy an offset "credit," lowering transaction costs and investor risk. Zollinger & Dower, *supra* note 36, at 4.

46. From a broader climate perspective, many developing countries are actively promoting energy efficiency and renewable energy. Mexico, India and Brazil have all launched specific energy efficiency and renewable energy programmes. China's Ministry of Energy now requires all large industrial boilers to be converted to (more environmentally friendly) cogeneration facilities. *See* Goldemberg, *supra* note 8.

47. GECR, *supra* note 36. The project will take place in the Henan province south of Beijing.

48. For example, as of October 1996, only 4 of 15 of the official USIJI (United States Initiative on Joint Implementation) projects had been fully financed and only 9 of the 44 proposals submitted to the International Utility Efficiency Partnerships Programme (IUEP) were being considered for further development. *See* Zollinger & Dower, *supra* note 36, at 5.

nations.[51] Additionally, firms have been unsure whether the 'reduction additionality' of projects will be verified (discussed below) and whether projects will be recognised by governments at the national level to count toward domestic obligations under a national regulatory framework.

'Reduction additionality' is a formal criterion of the AIJ pilot phase adopted at COP1.[52] The COP1 decision stated that "AIJ [projects] should bring about real, measurable and long term environmental benefits related to the mitigation of climate change which would not have occurred in the absence of such activities."[53] To determine the extent of reduction additionality, investors and recipients are jointly required to establish an emissions 'baseline' by which the JI project can be assessed. The baseline is the 'business as usual' level of emissions prior to a JI project being implemented. To compute the emission reduction 'credit' resulting from a JI project, it is necessary to know both the original baseline and the emission level after the project is completed. Calculating these levels is often a difficult and controversial task.[54]

The major concern over additionality is that, although some AIJ programs have

---

49. Heller, *supra* note 3, 298. "The biggest impediments for industry are a lack of information about potential markets; lack of finance to explore market opportunities and difficulty in establishing credibility with potential buyers" *See also* C.L. Sonneborn, S. Bygrave, & A. Pears, *Activities Implemented Jointly, Sustainable Energy Industries Association of Australia,* Internet, http:// www.spirit.com.au/~gregoryw/ papers/aij.html (1997). S. Bygrave, *Potential for the diffusion of Australian renewable energy technologies to countries in the Asia-Pacific region,* 4(2) AUSTRALIAN J. ENVT'L MGT. (1997), at 130-143. Bygrave surveyed 17 Australian renewable energy companies on what they perceive as impediments to overseas projects. Respondents believed the most common impediments were lack of information about overseas markets and government policy in recipient countries. Other impediments cited included Australian (domestic) government policy and competition with conventional energy sources and overseas aid organisations.

50. A more appropriate software transmission tool is considered to be general development aid. *See* Harvey & Bush, *supra* note 9, at 38. Successful technology transfers involves the deployment of appropriate hardware (materials) and software. "The concept of technology, and what it is comprised of, needs to be reconsidered so that critical elements of technology are transferred in the technology transfer process . . . [I]nformation exchange, institutional support, training and education are central tenets of the technology transfer process. . . . Such an approach will ensure that technologies are directed to basic needs, capable of local operation and maintenance, equitable and consistent with the social and cultural context into which they are introduced." S. Bygrave, *Revisiting Appropriate Technology And Community Development As Key Concepts To Assist Renewable Energy Projects In Developing Countries.* Solar '97 Conference, Australian National University (1997), at 1,2,5. Technology is not simply hardware or objects, but incorporates knowledge, information, methodology, processes and practices. *See* J.D. Eveland, *Diffusion, Technology Transfer and Implications: Thinking and Talking About Change,* 8(2) KNOWLEDGE (1986), at 303-322. D. Mansell, *Appropriate Technology: An Overview,* in ISLAND TECHNOLOGY FOR DEVELOPMENT IN THE SOUTH PACIFIC (T. Marjoram ed., 1993).

51. Heller, *supra* note 3, at 298. With domestic policy measures in place, JI projects would be favoured whenever the cost per unit of avoided emissions is less than the cost they would incur in reducing their own emissions domestically.

52. For a detailed discussion of additionality, *see* L. Carter, *Modalities for the Operationalisation of Additionality.* Prepared for presentation at the UNEP International Workshop on AIJ, Germany, 5-6 Mar. 1997.

national guidelines,[55] there are no internationally agreed methods for the determination of additionality. Moreover, it is widely accepted that determining whether a particular project would have occurred anyway, even if methodologies were available, is a difficult and sensitive task.[56]

Closely related to additionality is the concept of 'incremental cost.' This is the additional cost of implementing an emission reduction project over and above the cost that would have been incurred in the absence of the project. Article 4(3) of the FCCC states that a JI investor is required "to meet the full incremental cost of implementing measures" (e.g., the additional cost required to build a lower emitting gas electricity plant as opposed to a coal-fired one. The cost that the recipient country would have incurred in the absence of the project is known as the "base-

53. United Nations, Decision 5/CP.1 of FCCC/CP/1995/7/Add.1, *Report of the Conference of the Parties on its First Session.* Held at Berlin 28 March - 7 April, 1995. United Nations. A project is deemed "additional" when "it would not have been implemented in the absence of support given through JI or any other international support specifically intended to promote emission reductions (such as the Global Environmental Facility (GEF). [For example, a lower emitting utility plant would not have been built and a higher emitting utility plant would have been built in the absence of JI.] If a project is not additional, its emission reductions cannot be attributed to JI, and should not be counted as such." For JI projects between developed countries with emission reduction targets (under Article 6 of the Protocol), reduction additionality, although required under Article 6(1)(b), is not strictly necessary. An emission reducing project under Article 6 results in a reduction in the target of the 'host' country (since emissions have been reduced there) and a corresponding (equal) increase in the target of the 'investor' country (i.e., an offsetting plus and minus). So long as both countries agree to the project, the level and whether the reductions are additional to what would have occurred in the absence of the project of reduction (the offsetting plus and minus) and their targets are adjusted accordingly, the achievement of the environmental objective is not compromised (i.e., the exact project emission reduction level is essentially irrelevant since the offsetting adjustment to the targets results in a 'zero-sum gain.')

54. Harvey & Bush, *supra* note 9, at 19. For example, baseline projections are sometimes required for ventures that have not been implemented and will be modified in design by the JI project in question. For example, a developing country's plan to construct a coal-fired electricity plant which is subsequently modified by a JI proposal to switch the fuel from coal to gas and reduce emissions accordingly. There is also potentially a "moral hazard" problem associated with the projection of the baseline. It is in the interest of both the investor and recipient to exaggerate the level of emission abatement - making the marketable 'credit' more valuable for the recipient country and the emission reduction larger for the investor country.

55. For example, the additionality criterion of the United States Initiative on Joint Implementation (USIJI) states in section V: "the Evaluation Panel must find a project submission . . . A(4) Will reduce or sequester greenhouse gas emissions beyond those referred to in [the absence of JI]," United States Government, *Activities Implemented Jointly: Second Report To The FCCC*, 26 June 1997.

56. In practice, additionality criteria have been difficult to define and apply. *See* Carter, *supra* note 53, at 2. Instead of being a useful screening concept of questionable projects, the lack of a single international criteria has resulted in a non-transparent decision-making axiom, lacking consistency and accountability. The difficulty in determining additionality is largely derived from the fact that the "effect of low-cost abatement [JI] projects are particularly uncertain, since such projects may be close to being profitable and, hence, may be carried out by the market itself in the near future [in the absence of JI]." IPCC, *Economic and Social Dimensions of Climate Change, Contribution of Working Group III to the Second Assessment Report* (1996).

line cost."[57] While conceptually simple, incremental cost is also difficult to evaluate in practice. Determining how much a given JI project will cost is far from straight forward. Behind every cost analysis lies a host of value-laden decisions about which costs to include, what time frame and discount rate to use and which methodology to adopt.[58]

Finally, concern has arisen over the absence of adequate monitoring and enforcement mechanisms. In a domestic context, a regulatory agency can enforce the law, however, in an international context, states rarely enforce legal obligations against each other. There are no existing international organisations with equivalent legal competence to regulate JI projects and undertake extensive monitoring, verification and enforcement. Certain non-governmental organisations may have a role to play in some of these areas.[59]

7. JI in the South Pacific

7.1 AUSTRALIA

The Australian AIJ pilot initiative, known as "International Greenhouse Partnerships"[60] (IGP), was developed jointly by industry and government and is now largely industry-run. There was a strong incentive for Australia to explore the possibility of lower cost options overseas, given the relatively high domestic costs of emissions abatement. IGP was announced at the Second Conference of the Parties (COP2) in July 1996 by the Australian Minister for the Environment, Senator Hill.

---

57. Theoretically, the recipient would still cover the baseline cost. However, in the negotiation of project specifics, a recipient country may incorporate a premium to help cover this cost. The extent of this premium would depend on the relative bargaining power of the two parties and the differences in their marginal cost of abatement.

58. Harvey & Bush, *supra* note 9, at 37.

59. *See* K. Yamin, *The Use of Joint Implementation to increase compliance with the Climate Change Convention: International Legal and Institutional Questions*, 2(4) RECIEL (1993), at 348-353; R. Schmalensee, *Greenhouse Policy Architecture and Institutions*. Paper prepared for National Bureau of Economic Research Conference - Economics and Policy Issues in Global Warming: An Assessment of the IPCC Report, Colorado 23-24 (1997). For example, NGOs could provide information, build confidence in JI projects and ensure that proposed projects do not generate negative social and environmental impacts. NGOs could also serve as partners in project implementation and as inspectors to monitor project performance and verify the extent of emission reductions. Some recipient country NGOs have become more pragmatic and are prepared to bargain and use JI as a mechanism to achieve their own development and social goals. They demand project data must be transparent and that NGOs must be involved in monitoring and auditing the performance of projects. Involvement of competent NGOs (such as the Environmental Defence Fund, Greenpeace or the World Resources Institute who have already brokered JI projects for the private sector) in all projects would help ensure the accuracy, reliability and credibility of claims over JI emission reductions. *See* Niederberger, *supra* note 2, at 34. Notwithstanding this potential, this may be a difficult task given the donation, volunteer-dependent nature of many NGOs.

60. For more information on IGP, *see* DFAT (Department of Foreign Affairs and Trade), *Australia and Climate Change Negotiations: An Issues Paper*, (1997), at 103-118; Department of Primary Industries and Energy home page, http://www.dpi.e.gov.au/resources.energy/environment/greenhouse/aij/index.html.

He stated:

> It is clear that actions by developed countries alone will not be sufficient to mitigate climate change. Over the longer term, all countries will need to contribute to limiting global emissions. This will require effective international partnerships and cooperation. In this context I take pleasure in announcing an Australian pilot initiative on Activities Implemented Jointly (AIJ), which will be developed jointly with Australian industry. Its primary focus will be in the Asia-Pacific region. The pilot initiative reflects our view that AIJ is a win-win situation for all countries. At an appropriate time there needs to be proper international recognition of these mutual benefits.[61]

IGP was designed to facilitate voluntary, cost effective emissions abatement. Additional objectives are to enhance trade and investment links in environmental technology and services areas,[62] the facilitation of cooperation with developing countries to address climate change,[63] encouraging investment in capital, technologies and know-how in developing countries and the advancement of a more comprehensive form of JI in the future.

Participants in IGP can credit emission reductions toward their 'Cooperative [Voluntary] Agreement' under the Greenhouse Challenge.[64] The Australian government supports companies participating in the IGP in a number of ways, including providing information, ensuring projects meet the guidelines and monitoring requirements, negotiating and facilitating recipient government endorsement of projects and minimising compliance costs where possible.

---

61. Senator Hill, Address to COP2, 17 July 1996, *United Nations FCCC: Conference of the Parties Second Session: Statement by Senator Hill,* http://www.environment.gov.au/portfolio/minister/env/96/mr17jul-unfccc.html.

62. Bygrave, *supra* note 50, at 132, reports that a range of technologies could be exported from Australia to the Asia-Pacific region, including solar heating and electricity, hydro, wind, geothermal, biomass (for both burning and alcohol fuels) and energy storage. The market for renewable energy technology in the Asia-Pacific region is estimated at between AU$15-40 billion. *See* Commonwealth of Australia, *Annual Report 1994-1995: Department of Industry, Science and Technology.* Australia Government Publishing Service (1995); W. Parer, *Coalition Policy on Renewable Energy,* 17(2) SOLAR PROGRESS (1996), at 16-17.

63. "We are playing a key role in assisting developing nations to reduce greenhouse gas emissions. Australia is developing an AIJ programme which will allow developing nations to benefit from our technology and expertise." Senator Hill, *Greenhouse - Developing a New Strategy,* Internet, http://www.environment.gov.au/portfolio/minister/env/97/mr28feb97.html, 28 February 1997. "The [AIJ] programme would provide, ... most importantly, a win for the global environment." Senator Hill (The Hon. Robert), *New Australian Move on Greenhouse,* http://www.environment.gov.au/portfolio/minister/env/96/mr18jul2.html, 18 July 1997.

64. "The AIJ initiative is, in many respects, the international dimension of the Greenhouse Challenge programme and will add further to the contribution Australian industry can make to reducing global greenhouse gas emissions." Senator Hill, *Historic Agreement Between Australia and Indonesia on Climate Change,* Internet, http://www.environment.gov.au/portfolio/minister/env/96/mr24oct.html, 24 Oct. 1996

In early 1997, two AIJ agreements involving Australian firms investing in the South Pacific were finalised. The first involves the installation of a 10 kilowatt solar power supply system in Fiji. The second involves improving air conditioning efficiency in the Solomon Islands. Both were expected to be completed by July 1998.[65]

In addition to these two projects, a number of other AIJ initiatives have also been undertaken.[66] These include a statement of intent signed between Australia and Indonesia[67] and the preliminary development of two AIJ projects in Indonesia. Finally, workshops in Melbourne, Sydney, New Delhi and Jakarta have been staged or sponsored to explore commercial opportunities for establishing AIJ projects.

In November 1997, in the lead up to Kyoto, Prime Minister Howard announced "Australia's Response to Climate Change." The announcement was a package of new and existing measures to address greenhouse gas emissions in Australia. The Government committed to provide (AU)$180 million over five years for these measures. AIJ projects and developing country involvement were emphasised by Prime Minister Howard, indicating a clear commitment to the future of JI in Australia. He stated:

> We have persistently stressed the need to involve developing countries as their participation is crucial to any lasting solution to the global warming problem. [As part of the specific measures of Australia's response to climate change, there will be] additional funding for the AIJ office [of (AU) $6 million] to facilitate commercial projects in developing countries and help meet the additional transaction costs incurred by business in undertaking an AIJ project. This will provide Australian industry with a greater incentive to undertake such projects.[68]

---

65. B. Barr, *Personal Communication*, AIJ Australia Office, 15 December 1997 .
66. DFAT, *supra* note 63, at 117.
67. "The signing of the Statement of Intent demonstrates that Australia is following through with the commitments it made [at COP2]. [It] will enable both countries to work together with their respective industrial sectors in a practical, business-like way on a wide range of projects to stabilise greenhouse gas concentrations." Hill, *supra* note 67.
68. J. Howard, *Safeguarding the Future: Australia's Response to Climate Change*, Internet, http://www.pm.gov.au/media/pressrel/speech/1997/GREEN.html, 20 November 1997

7.2 NEW ZEALAND

New Zealand has advocated a least-cost approach to the mitigation of greenhouse gas emissions climate-supporting mechanisms such as international emissions trading and JI. Despite the merits of JI, New Zealand has no formal JI initiative in place. JI would likely be of value to New Zealand given a potentially high marginal cost of emissions abatement.[69] In recognition of this fact, an allocation of funds from the Government's 1996/1997 Vote Environment 'Green Package' was utilised to investigate, identify and explore issues related to the development of JI policy for New Zealand.

The report concluded and recommended, among other things, that:[70]

- Interest in JI from New Zealand industry has arisen both as a result of the threat of domestic regulation and in response to a perceived market opportunity;[71]
- JI is viewed by some major emitters primarily as a business opportunity rather than a way of reducing mitigation costs;
- A more extensive study of the business opportunities associated with JI would be appropriate, possibly funded by a trade-related organisation such as TRADENZ; and
- New Zealand should investigate 'piggy-backing' on another pilot programme, such as the Australian IGP programme, for cost-effectiveness.

Although the report was a preliminary investigation, the study indicated an intent by the Government to explore JI. It also illustrated a keen interest by business to become involved in JI projects.

New Zealand has also consistently articulated that reductions by developed countries alone will be ineffective unless commitments by developing countries are forthcoming in the future. Minister for the Environment, the Honourable Simon Upton, has recognised that JI can help play an important bridging role in this regard:

---

69. *See* W. McKibbin & D. Pearce, *Impacts on the New Zealand Economy of commitments for abatement of CO_2 Emissions*, Centre for International Economics, Contract report to the Ministry of Commerce (1997), at viii. "[T]he tax [marginal cost of abatement for NZ] required in 2010 is almost four times that required in the United States and one and a half times that required in Australia [for stabilisation at 1990 levels by 2010 for CO_2 only without international trading]. This is predominantly due to the large hydro base of New Zealand's electricity generation.
70. Sapsford, *supra* note 5, at 13-14.
71. Discussions were held with representatives of Fletcher Challenge Ltd, Environmental Intermediaries and Trading Group, The Treasury, NZ Sustainable Forests Ltd, ECNZ and Forest and Bird. In addition, brief written or verbal replies were received from Carter Holt Harvey, Comalco, NZ Dairy Group, Cement and Concrete Association of NZ and Greenpeace NZ. *See* Sapsford, *supra* note 5, at 42.

> I believe Kyoto should seek to establish a timetable for the evolution of commitments beyond developed countries. .... It is here that JI strategies could prove particularly valuable. And the reality is, that if credit were given for the reductions achieved through [JI] projects, the flow of capital and technology to countries like China and India could be sufficient to enable them to completely leapfrog the energy-intensive development phase that [developed countries] have had to pass through.[72] That is why there is growing interest by developed countries in assisting the transfer of leading edge technologies to those economies. Such [JI] initiatives could build confidence in future emissions trading. ... If this sounds ambitious and complex, it is. But only a flexible, genuinely multinational effort has a chance of making progress.[73]

7.3 AOSIS

Vulnerability to the effects of climate change has led countries within the Alliance of Small Island States (AOSIS) to take an active interest in climate negotiations. The AOSIS position on many aspects of the negotiations is aligned with that of the G-77 and China group.

The formal G77 and China JI position has typically been to argue against JI. Some countries have strayed from this 'parent' position and indicated strong support for JI. AOSIS is one such example. Indonesia is at the forefront within the Asia-Pacific region, having recently agreed to the establishment of AIJ projects, including two by Australia.[74]

Pacific Island countries (PICs) have also acknowledged that JI can help address national priorities if projects are compatible with, and supportive of social, economic and sustainable development. They maintain that projects should also ensure the interest of local people and be complementary to existing programmes such as the South Pacific Regional Environment Programme (SPREP).

JI opportunities are expected to emerge from the energy and land-use sectors of PICs. Energy projects, which will assist in reducing petroleum imports necessary to meet present energy demand, are of high priority. Projects resulting in improved land-use practices, forest preservation, reforestation and sustainable forest management are also deemed beneficial.

If JI projects are to flourish in PICs, a number of deficient administrative capabilities will need to be addressed. These pertain to domestic weaknesses and limitations, such as financial shortages, ineffective management of public resources,

---

72. S. Upton, *Climate Change Speech to the Energy Federation of NZ*, 31 Oct. 97.
73. Upton, *supra* note 5.
74. DFAT, *supra* note 59, at 117. Information on the AOSIS position in this section sourced from Lefale, *supra* note 14, at 1-10.

political instability and shortage of technical know-how.

8. Conclusion

Clearly, JI is far from straight forward in its application and is no panacea for climate change. Although the foundations are in place, the future of JI is by no means certain. There are a myriad of methodological, institutional, financial and political barriers that need to be overcome if the full potential of JI is to be realised.

To ensure the credibility, durability and political acceptability of JI projects, the full spectrum of social, economic and environmental costs and benefits should be considered. Further, developing a common analytical framework that countries can agree upon is critical.

Striking a balance may well be the key. On one hand, placing too many restrictions or requirements on JI projects will risk reducing the incentives for investors. On the other hand, to maintain the integrity of a JI system, comprehensive and effective monitoring and enforcement mechanisms appear to be essential.

JI has the potential, however, to be a viable and cost-effective instrument for climate protection. The establishment of the 'Clean Development Mechanism' under the Kyoto Protocol, which officially recognises JI emission reduction credits from projects between developed and developing countries, is a notable and important step forward.

Within the South Pacific, JI has received worthy attention. Both Australia and New Zealand have strongly advocated least cost flexible mechanisms, including JI, throughout the international negotiations. In practice, Australia has led the way with the establishment of 'International Greenhouse Partnerships.' A report by New Zealand suggests that it consider JI as one possible policy response to meet its FCCC obligations. AOSIS nations, who have diverged from the negative view of JI held by most G77, have recognised the benefits of JI and are gaining valuable experience through participation in projects.

Outside the South Pacific, China, who has typically led the G77 opposition to JI, has shown signs of weakening in its opposition with the agreement in early 1998 to a project with Norway. Participation by important countries such as China can only bolster the future prospects for JI as a successful strategy for climate protection. Confidence must continue to be built among developing countries. If carefully managed, JI could enable developing countries to prosper from JI-related benefits and, importantly, make a valuable contribution to the global response to climate change.

12. THE IMPACT OF CLIMATE CHANGE ON PACIFIC ISLAND DEVELOPING COUNTRIES IN THE 21ST CENTURY

WILLIAM C.G. BURNS
Editor-in-Chief, Journal of International Wildlife Law & Policy
PMB 805, 2124 Kittredge St., Berkeley, California 94704 USA

There, on the Islands of the Blessed, by the edge of the turbulent Oceans, with carefree hearts live the heroes favored by fortune, for whom the fertile earth thrice yearly brings forth sweet and abundant harvests. — Hesiod, *Works and Days*

Pacific Island Countries are microcontributors of greenhouse gases yet they are likely to be affected most. Some large nations have blatantly shown their lack of concern on these issues and they seem to believe that their economy and well-being are much more important than the survival of the people from PICs. This sounds inhumane to us. How can we make our voice heard? Are we over-reacting? One thing is for sure: we are vulnerable. — Janita Pahalad, *Fiji Meteorological Service*

1. Introduction

The projected increase in global mean surface temperatures of 1.3° -4.0° C by the year 2100[1] will constitute "a change, although gradual, unparalleled in recent millennia."[2] While most nations may suffer deleterious consequences from climate change,[3] small island states[4] may face the most dire and immediate consequences. This chapter will focus on the possible impacts of climate change on Pacific Island Developing Countries (PIDCs)[5] and the prospects for averting these impacts under the United Nations Framework Convention on Climate Change.

1. Tom M.L. Wigley, *The Science of Climate Change*, Report of the Pew Center on Climate Change (1999), at 21. Wigley's projections are a shift upwards from the IPCC's predictions in its Second Assessment of 08-3.5° and reflect projected SO_2 emissions in scenarios being developed for the IPCC's upcoming Special Report on Emissions Scenarios. *Id.*. The current "best estimate" of the IPCC is a 3° C increase by 2100. Hadley Centre for Climate Prediction and Research, *Climate Change & Its Impacts*, http://www.meto.govt.uk/sec5/CR_div/Brochure98/science.html (1998); Global Network of Environment & Technology News, *Global Warming Forecast Raised One Degree*, http://www.gnet.org, December 19, 1998.
2. *The Calamitous Cost of a Hotter World*, WLD. PRESS REV., July, 1995, at 9 (reprinted from DER SPIEGEL).
3. Margaret S. Torn, Evans Mills & Jeremy Fried, *Will Climate Change Spark More Wildfire Damage?*, LBNL Rep. No. 42592, November, 1998, SCAR Global Change Programme, *A Summary of Global Change in the Antarctic*, http://www.antcrc.utas.edu.au/scar/newsletter2/2summary.html (1998); Kenneth Blackman, *Global Warming Worries Indigenous People*, INTER PRESS SERVICE, August 13, 1998; William C. Burns, *The Second Session of the Conference of the Parties to the United Nations Framework Convention on Climate Change: More Heat than Light?*, 1996 COLO. J. INT'L ENVTL. L. & POL'Y Y.B. 153, 153-54.

A. Gillespie and W. C. G. Burns (eds.), Climate Change in the South Pacific: Impacts and Responses in Australia, New Zealand, and Small Island States, 233–250.

2. Impact of Climate Change on Small Island States: Coastal Areas and Ecosystems

At the outset, it must be emphasized that regional assessments of climate change remain fraught with uncertainty:

> [General circulation models have difficulty in reproducing regional climate patterns, and large discrepancies are found among models. In many regions of the world, the distribution of significant surface variables, such as temperature and rainfall, are often influenced by the local effects of topography and other thermal contrasts, and the coarse spatial resolution of the GCMs can not resolve these effects.[6]

While researchers have developed several strategies to improve regional predictions, including the use of nested model and statistical downscaling,[7] most efforts to project climate change in the Pacific region are still conducted with atmosphere-ocean GCMs with insufficient horizontal resolution to simulate island climate.[8] However, in recent years several models have yielded reasonable correlations between regional climate projections and observed climatic indicia, justifying preliminary assessments of the impacts of climate change on PIDCs over the next century.[9]

2.1 SEA-LEVEL RISE

Based on varying assumptions of future population and economic growth, land-use pat-

---

4. The United Nations defines "small island states" as islands with less than 10,000 square kilometers in land mass and with less than 500,000 inhabitants. France Bequette, *Small Islands: Dreams and Realities*, UNESCO COURIER, March, 1994, at 23. Among the 22 island states in the South Pacific, of which 15 are politically independent, four have land area of less than 100 square kilometers, and 11 have land area between 100 and 1000 square kilometers. Paul F. Holthus, *Coastal and Marine Environments of Pacific Islands: Ecosystem Classification, Ecological Assessment, and Traditional Knowledge for Coastal Management*, in SMALL ISLAND STATES: MARINE SCIENCE & SUSTAINABLE DEVELOPMENT 342 (George Maul ed., 1996); J. R. Campbell, *Contextualizing the Effects of Climate Change in Pacific Island Countries*, *in* CLIMATE CHANGE: DEVELOPING SOUTHERN HEMISPHERE PERSPECTIVES 354 (Thomas W. Giambelluca & Ann Henderson-Sellers eds., 1996).
5. For a comprehensive list of Pacific island states, *see* John C. Pernetta, *Impacts of Climate Change and Sea-Level Rise on Small Island States*, GLOBAL ENVTL. CHANGE, March, 1992, at 19, 20.
6. Silvina A. Solman & Mario N. Nunez, *Local Estimates of Global Climate Change: A Statistical Downscaling Approach*, 19 INT'L J. CLIMATOLOGY 835, 835-36 (1999).
7. Stephen P. Charles, et al., *Validation of Downscaling Models for Changed Climate Conditions: Case Study of Southwestern Australia*, 12 CLIMATE RES. 1, 1-2 (1999); K. YA. KONDRATYEV & A.P. CRACKNELL, OBSERVING GLOBAL CLIMATE CHANGE 381 (1998); Solman & Nunez, *supra* note 6, at 836.
8. Robert T. Watson, et al., *The Regional Impacts of Climate Change*, Special Report of IPCC Working Group II (1998), at 340.
9. *See* Kondratyev & Cracknell, *supra* note 7, at 383-84; Gerald A. Meehl, *Pacific Region Climate Change*, 37(1) J. OCEAN & COASTAL MGMT. 137, 145 (1997).

terns, and energy policies, the Intergovernmental Panel on Climate Change (IPCC) projects an increase in sea levels of between 15 and 95 centimeters by the year 2100, with a best estimate of 49 centimeters.[10] Moreover, levels could continue to rise for several centuries after greenhouse gas emissions are stabilized.[11]

In the South Pacific, recent research indicates that sea levels have been increasing by as much as 25 millimeters per year, more than ten times the global trend this century.[12] While these dramatic increases may be attributable to ENSO in the short-term,[13] recent research indicating that regional climate change in the future may closely track El Nino effects leads to the inference that sea levels may increase substantially over the next century in the Pacific.[14] The IPCC projects that thermal expansion alone will raise sea levels in the southwest Pacific by 28-32cm at the time of CO_2 doubling,[15] although regional projections remain highly speculative because dynamic ocean effects have yet to be effectively modelled.[16]

In terms of land loss, low-lying atoll nations in the South Pacific, including the Marshall Islands, Tuvalu, Nauru, Kiribati and Tokelau, are extremely vulnerable to even small rises in sea level. A one-meter rise could result in the loss of 80 per cent of the Majuro atoll in the Marshall Islands, home to half of the nation's population,[17] and 12.5 per cent of the landmass in Kiribati.[18] Erosion will further reduce land area on low-lying PIDCs, as well as increase the swampiness and salinity of land that remains above sea level.[19] The most vulnerable areas will be those previously reclaimed from

10. Intergovernmental Panel on Climate Change, *Contribution of Working Group I to the IPCC Second Assessment Report*, IPCC-XI/Doc.3 (1995), at SPM.35. As the IPCC notes, its projections are premised on the assumption of minimal melting of the Greenland and Antarctic ice sheet. If this assumption proves incorrect, which the IPCC admits is a possibility, ocean levels could be elevated much more than originally predicted. *Id.* One researcher at Victoria University in New Zealand recently warned that the Western Antarctic Ice Sheet may be on the point of melting, which could result in a *six-meter* rise in sea-levels in less than a century. *Ice Sheet 'On Point of Melting,'* THE PRESS (New Zealand), January 28, 1999, http://www.press.co.nz/04/99012833.htm. *See also* Grover Foley, *The Threat of Rising Seas*, 29(2) THE ECOLOGIST 76, 76 (1999); Leonie Haimson, *Concerns Grow Over West Antarctic Ice Sheet*, GLOBAL CHANGE, Winter, 1999, at 5.

11. IPPC, *The IPPC Assessment of Knowledge Relevant to the Interpretation of Article 2 of the UN Framework Convention on Climate Change: A System* (1995), at sec. 3.16.

12. *South Pacific Climate Change*, 26 TIEMPO (1998), http://www.cru.uea.ac.uk/tiempo/floor0/archive/issue26/t26art2.htm; Small Islands Developing States Network, www.sidsnet.org.

13. Chairpersons of the Third SPREP Meeting on Climate Change and Sea Level Rise, *South Pacific Climate Change*, 26 TIEMPO (1997), <http://www.cru.uea.ac.uk/tiempo/floor0/archive/issues26/t26art2.htm>.

14. Scott Curtis & Stefan Hastenrath, *Long-Term Trends and Forcing Mechanisms of Circulation and Climate in the Equatorial Pacific*, 12 J. CLIMATE 1134, 1144 (1999); Meehl, *supra* note 9, at 145-46.

15. IPCC, *The Regional Impacts of Climate Change*, Special Report of IPCC Working Group II (1998), at 341.

16. Richard J.T. Klein, *Assessment of Coastal Vulnerability to Climate Change*, 28(2) AMBIO 182, 182 (1999); Stefan Rahmstorf, *Shifting Seas in the Greenhouse?*, 399 NATURE 523, 523-24 (June 10, 1999).

17. Paul Watson, *Rising Sea Levels Threaten Islands Global Warming a Major Issue for Islanders Living on Pacific Atolls*, TORONTO STAR, December 4, 1997, at A2.

18. IPCC Special Report, *The Regional Impacts of Climate Change: An Assessment of Vulnerability*, <http://www.ipcc.ch> (1997). Two uninhabited islands of Kiribati, Tebua Tarawa and Abanuea, have already disappeared under rising sea levels. Alex Kirby, *Islands Disappear under Rising Seas*, BBC Online, <http://news.bbc.co.uk/low/english/sci/tech/newsid_368000/368892.stm>.

the sea, including parts of Kiribati and the Marshall Islands.[20]

Land losses in coastal regions of PIDCs could be particularly serious. Most PIDCs are characterized by high population density, especially in coastal regions,[21] and PIDC populations could double in the next 25 years.[22] Roads and urban centers, generally concentrated in coastal areas in PIDCs, will be threatened,[23] and economically strained PIDCs may be unable to afford the cost of rebuilding expensive infrastructure.[24] Moreover, the economic well-being of most PIDCs is primarily derived from activities carried out in low-lying coastal zones.[25]

Construction of sea walls could ameliorate these impacts for some time; however, this will prove cost-prohibitive for most PIDCs.[26] For example, in the Marshall Islands, effective sea walls would cost more than $100 million; yet, the nation's annual GDP is only $80 million.[27] In addition, the use of such defenses would interfere with the landward migration of coastal ecosystems, such as wetlands. This migration could be a critical method of accommodation by such ecosystems to cope with rising sea levels.[28]

Mangrove forests are a critical ecosystem in many PIDCs, supporting commercially valuable fish and crustacean species, as well as rare fauna, providing a source of construction and fire wood, tannin and herbal medicines, and helping to ameliorate the impacts of storms.[29] Mangrove communities can often cope with sea-level rise where

---

19. John Connell & John Lea, *Global Warming: Meeting the Planning Challenge in Island States*, in SECOND SPREP MEETING ON CLIMATE CHANGE AND SEA RISE IN THE SOUTH PACIFIC REGION 151 (John Hay & Chalapan Kaluwin eds., 1993).
20. *Id.*
21. IPCC, *supra* note 8, at 338. For example, in Tuvalu, Kiribati, and Vanuatu, 90 per cent of the inhabitants live along the coasts. World Meteorological Organization, *A Change of Climate in Paradise Small Island States Under Threat*, http://www.wmo.ch/web-en/pardsen.html.
22. *Migration to Australia No Answer to Pacific Island Woes*, JAPAN ECON. NEWSWIRE, February 28, 1995; Charles Wallace, *Paradox of Poverty at the Heart of Pacific*, THE GUARDIAN, March 21, 1995, at 12 (LEXIS, World Library).
23. Michael J. Edwards, *Climate Change, Worst-Case Analysis and Ecocolonialism in the Southwest Pacific*, 8 PACIFICA REV. 63, 64 (1996).
24. Christopher S. Lobban & Maria Schefter, Tropical Pacific Island Environments 333 (1997).
25. For example, foreign exchange earnings from tourism provide more than 50 per cent of total revenues for many small island states. Rising sea levels could adversely affect tourism through the loss of beaches from erosion or inundation, as well as salinization of freshwater resources. IPCC, *supra* note 8.
26. IPCC, *Summary For Policymakers, Scientific/Technical Analysis of Impacts, Adaptations, and Mitigation of Climate Change*, http://www.unep.ch./ipcc/sumwg2.html (1995).
27. Sierra Club, *Rising Sea Levels are Flooding Small Islands Out of Existence*, SC-ACTION #70, July 15, 1998. Tuvalu's efforts to reinforce its shoreline with concrete blocks has proved to be unavailing, with strong waves breaking up the concrete. *Pacific Island Seeks to Avoid Fate of Atlantis*, CHI. TRIBUNE, December 20, 1994, at 15.
28. PETER D. MOORE, GLOBAL ENVIRONMENTAL CHANGE 105 (1996).
29. Joanna C. Ellison & David R. Stoddart, *Mangrove Ecosystem Collapse During Predicted Sea-Level Rise: Holocene Analogues and Implications*, 7 J. COASTAL RES. 159, 159 (1991); P.B. TOMLINSON, THE BOTANY OF MANGROVES 231-32 (1986). Mangroves also help to detoxify contaminants in PIDC waters and serve as sediment filters. Vance P. Vincente, *Littoral Ecological Stability and Economic Development in Small Island States: the Need for an Equilibrium*, in MAUL, *supra* note 4, at 274.

sedimentation rates are commensurate with or exceed local sea-level rise.[30] However, most small island states are characterized by microtidal, sediment-poor environments. Thus, mangroves are expected to suffer reductions in geographical distribution from projected sea-level rise over the next century.[31] Sago stands, an important source of food and building materials in many coastal areas of Papua New Guinea, but particularly around the gulf and western region and the Sepik area, may also be severely affected by a rise in sea level.[32]

An equally serious threat is posed to the coral reefs that ring many PIDCs. Coral reefs are rocklike ridges, composed of calcium carbonate, formed from the harder outer skeletons, or polyps, of coral animals.[33] Most polyps sub-divide as they grow and their skeletons fuse together, creating complex coral colonies.[34] Reefs have been termed the "tropical rainforests of the ocean," serving as the home for up to one-quarter of all marine species.[35] Additionally, reefs protect coastal areas from erosion and storms,[36] and serve as a "sink" or absorber of carbon dioxide, helping to reduce the level of this potent greenhouse gas in the atmosphere.[37]

In the context of small island nations, coral reefs are "an extensive and vital"[38] component of the ecosystem. Coral reefs serve as a buffer against coastline erosion,[39] a function which will become even more critical in the future if climate change intensifies storm surges[40] and sea-level rise threatens coastal regions of PIDCs.[41] Reefs are the home for fish species that provide up to 90 per cent of the protein needs of PIDC inhabitants,[42] as well as supporting the livelihood of small-scale fishers in the region.[43] Moreover, coral reefs are the primary source of carbonate sand that constitutes the

30. M.D. Hendry & G. Digerfelt, *Palaeogeography and Palaeoenvironments of a Tropical Coastal Wetland and Adjacent Shelf During Holocene Submergence, Jamaica*, 73 PALAEOGEOGRAPHY, PALAEOCLIMATOLOGY, PALAEOECOLOGY 1 (1989).

31. Dexter Hinckley, *Assessing the Condition of Tropical Island Ecosystems and Their Responses to Climate Change*, unpublished manuscript supplied to the author, at 10; Joanna Ellison, *How South Pacific Mangroves May Respond to Predicted Climate Change and Sea-Level Rise*, this volume. Mangroves may also be adversely affected by increases in temperature, water salinity and ambient CO_2 associated with climate change in the region. *Id.*

32. Peter Hulm, *A Climate of Crisis, Global Warming and the Island South Pacific* (1989).

33. SUE WELLS & NICK HANNA, THE GREENPEACE BOOK OF CORAL REEFS 14 (1992).

34. Roger Highfield, *Why Coral Reefs Matter*, DAILY TELEGRAPH, March 1, 1995, at 16. Other constituent elements of reefs include shells, foams, and calcareous algae. John W. McManus, *Coral Growth and Sea-Level Rise*, United Nations University Electronic Seminar on Global Warming, November 20, 1995 (Internet document available from author).

35. Peter K. Weber, *Saving the Coral Reefs*, FUTURIST 28, 28 (July-August 1993). A single reef may contain as many as 3,000 different species of marine life. *Fact Sheet: The Coral Reef Initiative*, DEPT. STATE DISPATCH, December 26, 1994. Overall, coral reefs support 1-9 million species and a far greater number of phyla than rainforests. Peter F. Sale, *Recruitment in Space and Time*, 397 NATURE 25, 26 (1999).

36. LYNNE T. EDGERTON, THE RISING TIDE 32 (1991).

37. *Coral Reefs Act As Sponge For Carbon Dioxide*, NIKKEI WKLY., August 7, 1995 (LEXIS, World Library). Coral reefs absorb approximately 1.1 billion metric tons of carbon dioxide per year, equal to approximately 2 per cent of annual discharges. *Id.*

38. Hinckley, *supra* note 31, at 7.

39. Laura Tangley, *Will Coral Reefs Be the First Victims of Global Warming?*, EARTHWATCH, April 1991, at 27.

majority of beach deposits on PIDCs.[44]

Researchers believe that reefs' maximum sustained vertical accretion rate of 10 millimeters per year will be adequate to keep up with sea-level rise of one meter or less,[45] and that even slowly accreting reef flats should be able to cope.[46] Historical evidence supports this conclusion, with most reefs keeping up with sea-level rises of 20 centimeters per decade between 14,000 and 6,000 year ago.[47] Indeed, some researchers argue that rising sea levels may be beneficial for coral reef ecosystems by inducing vertical growth.[48]

However, under the stress of rising sea levels, reefs are likely to develop at deeper average depths in the future, exposing coastlines to greater wave and current effects in the future.[49] Moreover, the synergism between rising sea levels and anthropogenic damage to reefs, such as destructive fishing practices[50] and pollution,[51] may impede reef recovery.[52]

---

40. "Storm surge is the elevation of water generated by strong wind-stress forcing and a drop in atmospheric pressure." Pierre Daniel, *A Real Time System for Forecasting Hurricane Storm Surges Over the French Antilles*, in MAUL, *supra* note 4, at 146. *See also*, Graeme D. Hubbert & Kathleen L. McInnes, *A Storm Surge Inundation Model for Coastal Planning and Impact Studies*, 15 J. COASTAL RES. 168, 168 (1999).

41. *Coasts at Risk*, GLOBAL CHANGE, Oct. 1998, at 10; Hubbert & McInnes, *supra* note 40, at 184; Steering Committee of the Climate Change Study, *Climate Change Science: Current Understanding And Uncertainties* 46 (1995).

42. Weber, *supra* note 35, at 28; David N. Zurick, *Preserving Paradise: Environmental Degradation in South Pacific Island Countries*, 85(2) GEO. REV. 157 (1995). Reefs surrounding Palau support more than 2000 species of fish. Donald Hinrichsen, *Requiem for Reefs?*, INT'L WILDLIFE, March 13, 1997, at 12.

43. Michael E. Huber, *An Assessment of the Status of the Coral Reefs of Papua New Guinea*, 29(1-3) MARINE POLLUTION BULLETIN 65, 65 (1994); Sarah Lonsdale, *Hopes Rise for Coral Rainforests of the Sea*, THE OBSERVER, April 25, 1993 (LEXIS, World Library). The fishing industry is one of the three primary economic sectors in most PIDCs. Canadian International Development Agency, *Environmental Change, Vulnerability and Security in the Pacific*, 1 AVISO, Jan. 1999, http://www.gechs.org/aviso/January1999.html. In Micronesia, fish products as a percentage of export revenues increased from 43 per cent to 86 per cent during the period of 1988-1993, and in the Marshall Islands this figure increased from 20-80 per cent. Anjali Acharya, *Small Islands: Awash in a Sea of Troubles*, 8(6) WORLD WATCH 24 (1995). Stocks are already declining in many PIDCs as a consequence of overharvesting. Zurick, *supra* note 42, at 160; Simon Haydon, *For Pacific Islands, Global Warming Could Mean Extinction*, REUTERS, June 6, 1992 (LEXIS, World Library).

44. SPREP, *ICRI Pacific Regional Workshop* 92 (1995).

45. McManus, *supra* note 34; Prepared Statement of Fred T. Mackenzie, *Global Climate Change and the Pacific Islands*, Hearing Before the Senate Committee on Energy and Natural Resources, S. Hrg. 102-664 (1992), at 16.

46. A.J. Edwards, *Impact of Climate Change on Coral Reefs, Mangroves and Tropical Seagrass Ecosystems*, *in* CLIMATE CHANGE: IMPACT ON COASTAL HABITATION (D. Eisma ed., 1994), at 209.

47. Clive R. Wilkinson, *Global Change and Coral Reefs: Impacts on Reefs, Economies and Human Cultures*, 2 GLOBAL CHANGE BIO. 547, 553 (1996).

48. IPCC, *supra* note 10, at 342; C.M. Roberts, *Coral Reefs: Health, Hazards and History*, 8 TRENDS IN ECOLOGY & EVOLUTION 425 (1993).

49. Robert W. Buddemeier, *Coral Reef Responses to Climate Change: Issues for Pacific Island Nations*, in HAY & KALUWIN, *supra* note 19, at 98.

50. SPREP, *supra* note 37, at 82 (dynamite fishing); Marla Cone, *A Toxic Solution: The Growing Use of Cyanide to Stun and Catch Tropical Fish is Killing Off Coral Reefs, Researchers Say*, L.A. TIMES, Orange County Edition, November 13, 1995, at Metro B:2.

The impacts of sea-level rise could be exacerbated by an increase in the frequency and intensity of storm surges as ocean temperatures rise.[53] Storm surges could further degrade coral reefs[54] and flood land surfaces, rendering them uninhabitable.[55] Also, heightened carbon dioxide levels in the atmosphere increase the acidity of seawater, reducing the availability of calcium carbonate levels, thus making it more difficult for coral-forming organisms to create their skeletons.[56]

2.2 TEMPERATURE RISE

The IPCC's most recent assessment predicts "moderate warming" in PIDCs over the next century.[57] However, a recent run of a global coupled ocean-atmosphere GCM projected increases of sea-surface temperatures in the Pacific region (under conditions of CO_2 doubling) of 3.49° C in the eastern Pacific and 2.21° C in the western Pacific.[58] These results "resemble not only the climate anomalies associated with present day El Nino-related events in many areas, but also the decadal timescale climate anomalies observed during the 1980s.[59] Coral reefs have extremely narrow temperature tolerances of between 25-29° C, with some species in many PIDCs currently living near their threshold of tolerance.[60] Water temperature increases of 1-2° C over an extended period can result in coral "bleaching," whereby dinoflagellates, endosymbiotic algae species that live in coral fish, will be expelled or reduced.[61] Because coral derive most of their food from dinoflagellates,[62] they can quickly die in their absence.[63] Moreover, bleaching could reduce the capability of reefs to adjust to rising sea levels.[64] While reefs can recover from the effects of episodic bouts of warming, a persistent elevation

51. United States Geological Service, *Hurricane Effects on Wildlife and Ecosystems*, http://biology.usgs.gov/pr/newsrelease/1998/12-8.html; SPREP, *International Coral Reef Initiative Pacific Region Strategy*; March 1996, at 3; Ruth Flanagan, *Corals Under Siege*, EARTH, May, 1993, at 28.
52. Phillip Dustan, *Coral Reefs: Harbingers of Global Change?*, in CORAL REEFS: CHALLENGES & OPPORTUNITIES FOR SUSTAINABLE MANAGEMENT 140 (Marea E. Hatziolos, Anthony J. Hooten & Martin Fodors eds., 1998); Wilkinson, *supra* note 41, at 554.
53. *Id.* at 553.
54. Weber, *supra* note 35, at 32.
55. Geoffrey Lean, *'Paradise' Islands United Against Sea Level Threat: Alarm over Global Warming*, THE INDEPENDENT, March 13, 1994; John Madeley, *Island Paradise is Often in Troubled Water*, THE GAZETTE, April 23, 1994.
56. John Roach, *Coral Reefs Threatened by Increasing Carbon Dioxide*, nationalgeographic.com, http://www.ngnews.com/news/1999/05/051799/coral_3210.asp; *Global Warming a Threat to Barrier Reef*, DEUTSCHE PRESS-AGENTUR, April 2, 1999 (LEXIS, World Library).
57. Robert T. Watson, et al., *The Regional Impacts of Climate Change*, Special Report of IPCC Working Group II (1998), at 340.
58. Meehl, *supra* note 9, at 142-43.
59. *Id.* at 145.
60. IPCC, *supra* note 9, at 342; J. Legget, *The AOSIS Summary of Scientific and Policy Issues*, February, 1991.
61. Clive Wilkinson, et al., *Ecological and Socioeconomic Impacts of 1998 Coral Mortality in the Indian Ocean: An ENSO Impact and a Warning of Future Change?*, 28(2) AMBIO 188, 191 (1999); Sarah Carpin, *Global Warming Killing the Spectacular Seychelles Reefs*, SAN ANTONIA NEWS-EXPRESS, January 12, 1999, http://www.mcbi.org/maritimes/news05.htm; Anette E. Chadwick-Furman, *Reef Coral Diversity and Global Change*, 2 GLOBAL CHANGE BIO. 559, 566 (1996).

of ocean temperatures can result in permanent destruction of reef building capacity.[65]

In 1998, with global mean surface temperature reaching its highest level in recorded history, and sea-surface temperatures increasing 2-3° C in many tropical areas where reefs are found, as much as 4-6° C in some areas,[66] the world's reefs "appear to have suffered the most extensive and severe bleaching and subsequently mortality in modern record."[67] Approximately 95 per cent of shallow water corals in the Maldives died, as well as 75 per cent of corals in the Seychelles Marine park system.[68] Massive bleaching also occurred on the coasts and islands of India, Kenya, Tanzania and Australia's Great Barrier Reef.[69]

It is difficult to definitively link reef bleaching and mortality to temperature increases because other stressors, such as pollution, low salinity and El Nino can also contribute to the phenomenon.[70] However, a U.S. Department of State study of the 1998 bleaching events concluded that only warming could have caused such extensive bleaching throughout virtually all of the disparate reef regions of the world.[71] Moreover, the authors of a multi-factorial analysis of coral bleaching events in the 1980s concluded that "of all the stresses which could potentially cause widespread mass bleaching, only excessively high temperature was present in all cases."[72]

Reefs can recover from bleaching if conditions improve, with coral larvae re-settling on the reef structure to renew the building process. However, if elevated sea surface temperatures and bleaching persists, new building will not occur and the reef frame can gradually erode, resulting in habitat destruction and mortality.[73] For example, one year

---

62. Dinoflagellates, such as zooxanthellae, live within coral fishes and engage in a symbiotic relationship. Coral polyps provide shelter for the zooaxanthellae and their waste provides a source of nutrients; in turn zooxanthellae supply the coral polyps with carbohydrates for food and construction of the limestone skeletons in which they live. LOUISE B. YOUNG, ISLANDS: PORTRAITS OF MINIATURE WORLDS 202 (1999); WELLS & HANNA, *supra* note 33, at 14.

63. Hinckley, *supra* note 31, at 8. *See also* Robert W. Buddemeier & Daphne G. Fautin, *Coral Bleaching as an Adaptive Mechanism: a Testable Hypothesis*, 43 BIOSCI. 320, 322 (1993).

64. Joby Warrick, *Hot Year Was Killer for Coral*, WASH. POST, March 5, 1999, at A03; McManus, *supra* note 34. Warming trends may also result in increased cloud cover, robbing zooxanthellae of sunlight critical for conversion of carbon dioxide. WELLS & HANNA, *supra* note 33, at 61.

65. Anthony D. Socci, *Coral Reef Bleaching: Ecological and Economic Implications*, Internet document, February 7, 1996, ENVIRONMENT-L, available from author; IPCC, *supra* note 9, at 342.

66. Robert W. Buddemeier & Daphne G. Fautin, *Coral Bleaching as an Adaptive Mechanism: a Testable Hypothesis*, 43 BIOSCI. 320, 322 (1993).

67. Rafe Pomerance, *Coral Bleaching, Coral Mortality, and Global Climate Change*, Report to the U.S. Coral Reef Task Force by the Bureau of Oceans and International Environmental and Scientific Affairs, March 5, 1999, http://www.state.gov/www/global/global_issues/coral_reefs/990305_coralreef_rpt.html.

68. Wilkinson, *supra* note 61, at 191.

69. *Id.* at

70. G. Parker-Muller & C.F. D'Elia, *Interactions Between Corals and their Symbiotic Algae*, in LIFE AND DEATH OF CORAL REEFS 96-113 (C. Birkeland ed., 1997); M. Warner & W.K. Fitt, *Mechanisms of*

71. Pomerance, *supra* note 67.

72. The Global Coral Reef Alliance, *Coral Reef Bleaching and Sea Surface Temperature*, <http://www.fas.harvard.edu/~goreau/bleaching.intro.html>.

after the 1998 mass-bleaching incident, 80-90 per cent of the bleached corals in the most severely affected areas of the Indian Ocean had died, including previously resistant species.[74]

Temperature rises in the future in the region may also threaten fish and other species in other ways. While projected temperature increases are not anticipated to have a substantial impact on small island fisheries, they may have a negative effect on productivity in some areas, such as shallow lagoons, by increasing hypersaline conditions.[75] Moreover, clam and sea-turtle fisheries could be sensitive to temperature changes.[76] Warmer water in bays and lagoons might also stimulate the growth of toxic algae, threatening humans that consume fish and bivalves from the region, as well as predator species, such as cormorants and pelicans.[77]

On the other hand, some fisheries might benefit from warming. Temperature-induced changes in current patterns might increase nutrient upwellings, providing more food for some species.[78]

2.3 FRESHWATER RESOURCES

The threat of sea-level rise to the water resources of PIDCs is of great concern. Freshwater supplies are often extremely limited on islands because very few of them have lakes or permanent watercourses.[79] For many PIDCs, the primary source of water is Gyben-Hertzberg "lenses" of fresh water floating on salt water in the porous sands under the islands.[80]

Projected rises in sea levels may result in the intrusion of saltwater into the freshwater lens, as well as shrinking the lens as the high tide line encroaches on the land.[81] More-

---

73. U.S. Global Change Research Program, *Coral Reef Bleaching: Ecological and Economic Implications*, Environment-L, Internet, February 7, 1996.

74. Wilkinson, *supra* note 61, at 191. Moreover, the synergism between rising sea levels and anthropogenic damage to reefs, such as destructive fishing practices and pollution, may impede reef recovery. United States Geological Service, *Hurricane Effects on Wildlife and Ecosystems*, <http://biological.usgs.gov/pr/newsrelease/1998/12-8.html>; Nancy MacKinnon, *Destructive Fishing Practices in the Asia-Pacific Region*, *in* Hatziolos, *supra* note 51, at 32; Phillip Dustan, *Coral Reefs: Harbingers of Global Change?*, *in id.*, at 554.

75. A.E. Alm, *Climatic Changes and Socio-economic Impacts*, in CLIMATIC CHANGE IN THE INTRA-AMERICAS SEA (George A. Maul ed., 1993), at 333.

76. E.D. Gomez & C.A. Belda, *Growth of Giant Climates in Bolinao, Philippines*, *in* GIANT CLAMS IN ASIA AND THE PACIFIC, Australian Centre for International Agricultural Research Monograph No. 9, (J.W. Copland & J.S. Lucas eds., 1988), at 178; J.S. Lucas et al., *Selecting Optimum Conditions for Ocean-Nursery Culture of Tidacna Gigas*, in COPLAND & LUCAS, *id.*, at 129.

77. Hinckley, *supra* note 31, at 3-4.

78. A. Bakun, *Global Greenhouse Effects, Multi-Decadal Wind Trends, and Potential Impacts on coastal Pelagic Fish Populations*, 195 ICES MARINE SCI. SYMP. 316 (1993).

79. Bequette, *supra* note 4; IPCC, *Scientific-Technical Analyses of Impacts, Adaptations, and Mitigation of Climate Change*, Report of Working Group II (1995), at 8.

80. Steering Committee of the Climate Change Study, *supra* note 34, at 46; Orman E. Granger, *Geography of Small Tropical Islands: Implications for Sustainable Development in a Changing World*, in MAUL, *supra* note 61, at 166.

over, there is the danger that water tables will rise close to or above the land surface, resulting in full potential evapotranspiration that can ultimately destroy the resource.[82]

Sea-level rise could result in diminution of freshwater volume in some PIDCs by as much as 50 per cent.[83] This would severely affect the quality and quantity of potable water available in many PIDCs,[84] perhaps rendering some countries uninhabitable well before the specter of submergence by seawater.[85]

Sea-level rise is not the only threat posed to water resources by climate change. Storm surges precipitated by ocean warming[86] could damage freshwater supplies through salinization.[87] In addition, "most low-lying atoll countries rely almost entirely on rainwater."[88] Thus, changes in rainfall patterns in these nations might have an adverse impact on water supplies.[89]

2.4 TERRESTRIAL RESOURCES

Forest and agricultural resources are two of the most important economic resources in PIDCs.[90] Forests are also important habitats for many island species, help reduce soil erosion, and exert a significant influence on local and regional climates.[91]

While non-tropical forests probably will be most adversely affected by climate change,[92] PIDC forests may also face threats in the future. In addition to possible forest losses from rising sea levels,[93] temperature rises and changes in soil water availability could have adverse implications for PIDC forests.[94] Moreover, increased

---

81. LOBBAN & SCHEFTER, *supra* note 24, at 332.
82. Richard V. Cant, *Water Supply and Sewerage in a Small Island Environment: The Bahamian Experience*, in MAUL, *supra* note 61, at 338.
83. JOHN HOUGHTON, GLOBAL WARMING 95 (1994); Mackenzie, *supra* note 39, at 15.
84. Peter Roy & John Connell, *Climatic Change and the Future of Atoll States*, 7 J. COASTAL RES. 1055, 1064 (1991).
85. Yvette Collymore, *Negotiating the Future of Island-States*, INTER PRESS SERVICE, March 6, 1992 (LEXIS, World Library). Diminution of potable water supplies rendered the Phoenix Islands of Kiribati uninhabitable in the 1960s. Connell & Lea, *supra* note 13, at 151.
86. *See infra* note 47 & accompanying text.
87. Wilkinson, *supra* note 37, at 554.
88. Small Islands Developing States Network, *supra* note 8.
89. IPCC, *supra* note 9, at 344.
90. *See* Canadian International Development Agency, *supra* note 37.
91. IPCC, *supra* note 9, at 343.
92. Roger Sedjo & Brent Sohngen, *Impact of Climate Change on Forests*, Resources for the Future Climate Issue Brief # 9 (2d Ed., 1998), at 1420. Existing boreal forest cover could decline by as much as 50-90 per cent, Kevin Jardine, *Finger on the Carbon Pulse*, 24 THE ECOLOGIST 220-23 (1994). *See also* Mary E. Clark, *Consequences of Global Change for Earth's Biosphere*, in GLOBAL WARMING AND THE CHALLENGE OF INTERNATIONAL COOPERATION: AN INTERDISCIPLINARY ASSESSMENT 51 (Gary C. Byner ed., 1992) (shift of deciduous forests in U.S. to Canada); Simon Retallack, *Wildlife in Danger*, 29(2) THE ECOLOGIST 102, 102 (1999); Jerry M. Melillo, *Warm, Warm on the Range*, 283 SCI. 183, 184 (1999).
93. Connell & Lea, *supra* note 13, at 151.
94. IPCC, *supra* note 9, at 343.

temperatures may result in more frequent outbreaks of pathogens and pests deleterious to forests, as well as the frequency and intensity of fires.[95] However, increased amounts of carbon dioxide may enable some forest species to use water and nutrients more efficiently, ameliorating or wholly offsetting some of the negative impacts of climate change.[96]

In recent years, rapid population growth has placed serious strains on the food resources of many PIDCS.[97] Moreover, devastating droughts have become a regular occurrence in many PIDCs, including Fiji, Papua New Guinea, the Marshall Islands and the Federated States of Micronesia.[98] Climate change could exacerbate these trends. Seawater intrusion precipitate could reduce the fertility of coastal soils, resulting in a reduction of crop yields.[99] Production of taro, a crop grown in depressions and pits excavated down to the freshwater lens, would be severely affected in nations such as Tuvalu and Kiribati.[100] Coconut and breadfruit yields, as well as those of other trees with roots that reach down into the water lens, would likely be adversely affected as well.[101] Shoreline erosion will also disrupt crop production in coastal areas.[102]

Projected changes in temperature and precipitation are also likely to have a negative impact on food production as a consequence of reduced solar radiation, shorter growing duration and increased sterility of some cultivars from temperature increases, as well as changes in water availability.[103] As a result, maize yields could decline by 30-50 per cent, sugarcane by 10-35 per cent and taro 35-75 per cent.[104] An increase in violent weather events[105] could also threaten freshwater supplies and destroy crops,[106] as

---

95. IPCC, *Climate Change Impacts on Forests,* Working Group II (1998), http://www.usgcrp.gov/ipcc/html/chap01.html. Heightened temperatures in Siberia, Canada and Alaska have resulted in increased infestation by pests of spruce and pine trees. Clive Cookson, *UN Conference on Climate Change: Case for Action Strengthens as Signs Point to Global Warming,* FIN. TIMES, March 28, 1995, at 8.

96. IPCC, *supra* note 9, at 343. However, recent research indicates that carbon dioxide's stimulatory effects may decrease as trees age. *Report: High Carbon Dioxide Boosts duke Forest Growth By 25 Percent,* <http://www.eurekalert.org/releases/dumc-rhc051099.html> (1999).

97. *Id.* at 344.

98. SPREP, *Research Shows Major Change in Pacific Climate,* Press Release, August 6, 1998, SEA-SPAN-L Internet discussion list; *Globalization, erosion of trade preferences undermine small island states, Commission Told,* M2 PRESSWIRE, April 26, 1999 (LEXIS, World Library); *Government Appeals for Drought Assistance,* BBC SUMMARY OF WORLD BROADCASTS, March 6, 1999 (LEXIS, World Library).

99. IPCC, *Climate Change 1995: Impacts, Adaptations and Mitigation of Climate Change: Scientific-Technical Analyses. Contribution of Working Group II to the Second Assessment Report of the Intergovernmental Panel on Climate Change* (1996), at secs. 13.6.4, 13.7.

100. Michael Field, *Small Pacific States Threatened by Industrialized World's Carbon Emissions,* AGENCE FRANCE PRESSE, August 27, 1998 (LEXIS, World Library); IPCC, *supra* note 9, at 345; Steering Committee of the Climate Change Study, *supra* note 35, at 46.

101. LOBBAN & SCHEFTER, *supra* note 24, at 332.

102. Steering Committee of the Climate Change Study, *supra* note 35, at 46.

103. IPCC, *supra* note 9, at 345; Pernetta, *supra* note 5, at 23.

104. U. Singh, *Potential Climate Change Impacts on the Agricultural Systems of the Small Island Nations of the Pacific,* Draft paper, IFDC-IRRI (1994); Hadley Centre for Climate Change, *Impacts of Climate Change on Food Supply,* http://www.meto.govt.uk/sec5/CR_div/Brochure98/food.html (1999).

105. *See* sec. IID, *infra.*

could a surge in pathogens and pests.[107]

As is true with forests, increased carbon dioxide fertilization might have a beneficial effect on some crops in PIDCs.[108] However, this would most likely be offset by enhanced activity of soil organisms, growth in aggressive weeds, and the negative impact of moisture stress.[109]

Many PIDCs are already highly dependent on foodstuff imports, with some nations spending as much as 40 per cent of their GDP on food imports.[110] Future declines in food production would necessitate even greater expenditures on imports, further straining PIDC economies and undermining their independence.[111]

Endemic species in many PIDCs, often exceptionally vulnerable to extinction or extirpation because of small populations and minimal exposure to competition, predators, or disease,[112] may face diminution from increased temperatures, as well as altered species distribution and composition.[113]

2.5 INFRASTRUCTURE AND VIOLENT WEATHER EVENTS

Projected buildups in greenhouse gas emissions will likely raise ocean temperatures and ocean surface water to above 26° C in the next century.[114] This could result in a greater exchange of energy and add momentum to the vertical exchange processes critical to the development of tropical typhoons and cyclones.[115] As a consequence, some

106. Bill Aalbersberg, *Climate Change, Agriculture and Land Use in the South Pacific*, in SPREP, *supra* note 13, at 116; LOBBAN & SCHEFTER, *supra* note 24, at 125.

107. Peter Bunyard, *A Hungrier World*, 29(2) THE ECOLOGIST 86, 90 (1999).

108. IPCC, *supra* note 9, at 344; Robert Wilkinson & Teresa Rounds, *Climate Change and Variability in California*, National Center for Ecological Analysis and Synthesis (1998), at 46. C3 plants (named for the first product of photosynthesis, which is an organic compound with three carbon atoms), which include temperate grasses and cereals such as wheat and rice, as well as root crops important in PIDCs, such as taro, yams and cassava, might photosynthesize more efficiently under heightened levels of carbon dioxide. Aalbersberg, *supra* note 91, at 115-16. However, this would be less true for C4 species, such as maize, sorghum and sugar cane. Bunyard, *supra* note 93, at 90.

109. *Id.;* United Kingdom Meteorological Office, *Impacts of Climate change on Food Supply*, http://www.meto.govt.uk/sec5/CR_div/Brochure98/food.html; IPCC, *supra* note 9, at 345. Moreover, "there is some evidence that nutritional value of plants may decrease as a result of CO_2 fertilization. Wilkinson & Rounds, *supra* note 94, at 46.

110. Richard Mieremet, *Vulnerability Assessment to Accelerated Sea Level Rise Case Study of Majuro Atoll*, Hearing Before the Senate Committee on Energy & Natural Resources, *supra* note 38, at 133.

111. PIDCs will also likely have to pay more for food imports in the future, as projected declines in global food production linked to climate change may increase prices by 17 per cent above the level they otherwise would have been. United Kingdom Meteorological Office, *supra* note 95.

112. Lobban & Schefter, *supra* note 24, a 257 (1997).

113. E.D. Gomez & C.A. Belda, *Growth of Giant Clams in Bolinao, Philippines*, *in* GIANT CLAMS IN ASIA AND THE PACIFIC, Australian Centre for International Agricultural Research Monograph No. 9 (J.W. Copland & J.S. Lucas eds., 1988), at 178.

114. NASA Goddard Institute for Space Studies, *How Will the Frequency of Hurricanes Be Affected By Climate Change?*, http://www.giss.nasa.gov/research/intro/druyan.02/ (1999); Thomas R. Karl, Neville Nicholls & Jonathan Gregory, *The Coming Climate*, SCI. AM., http://www.sciam.com/0597issue/0597karl.html (1997).

researchers estimate that the occurrence of tropical typhoons and cyclones could increase by as much as 50-60 per cent,[116] and their intensity by 10-20 per cent.[117]

However, there is by no means universal agreement that climate change will visit an increase in violent weather events on PIDCs. Some researchers believe that the purported linkage between increased ocean temperatures and violent weather events is overly simplistic, citing other factors that influence storm development, including atmospheric buoyancy, instabilities in the wind flow, and vertical wind shear.[118] Moreover, some climate scientists argue that ocean circulation changes associated with climate change may counter the effects of added warmth.[119] In its most recent regional assessment, the IPCC concluded that "model projections suggest no clear trend, so it is not possible to state whether the frequency, intensity, or distribution of tropical storms and cyclones will change."[120]

If climate change does result in an increase in violent weather events, it could prove disastrous for many PIDCs. Violent weather events take more than 15,000-23,000 lives annually, with a disproportionate number occurring in developing countries.[121] From an economic perspective, violent weather events are perhaps the most important of the natural hazards that coastal regions in the tropics and subtropics must contend with.[122] As indicated earlier, much of the critical infrastructure and socioeconomic activities of PIDCs are concentrated in coastal areas.[123] Moreover, violent weather events can devastate the agricultural sector of island nations, including critical cash crops such as coconuts and breadfruit, and accelerate the accretion of coastal areas.[124] Overall, in recent years, the economic damage caused by violent weather events has reached as

115. *Insurers Refuse to Cover Global Warming Risks*, THE INDEPENDENT, May 8, 1992, at 11.

116. NASA, *supra* note 98; M.E. Schlesinger, *Model Projections of CO_2-Induced Equilibrium Climate Change*, in CLIMATE & SEA LEVEL CHANGE: OBSERVATIONS, PROJECTIONS & IMPLICATIONS 186 (R.A. Warrick, E.M. Barrow & T.M. Wigley eds., 1993); R.J. Haarsman, *Tropical Disturbances in a GCM*, 8 CLIMATE DYNAMICS 247 (1993).

117. IPCC, *supra* note 9, at 341; Thomas R. Knutson, Robert E. Tuleya & Yoshio Kurihara, *Simulated Increase of Hurricane Intensities in a CO_2-Warmed Climate*, 279 SCI. 1018, 1018, February 13, 1998.

118. Karl, *supra* note 97; G.J. Holland, *The maximum intensity of tropical cyclones*, 54 J. ATMOSPHERIC SCI. 2519-2541 (1995).

119. Bette Hileman, *Climate Observations Substantiate Global Warming Models*, CHEM. & ENG. N., November 27, 1995, http://pubs.acs.org/hotartcl/cenear/951127/pgl.html.

120. IPCC, *supra* note 9, at 341. *See also* J.F. Royer et al., *A GM Study of the Impact of Greenhouse Gas Increase on the Frequency of Occurrence of Tropical Cyclones*, 38 CLIMATIC CHANGE 307, 322 (1998) (Study using high resolution atmospheric model found substantial *reduction* in number of tropical storms, especially in the Southern Hemisphere, in doubled carbon dioxide simulation); David Schneider, *The Rising Seas*, SCI. AM. 112, 117, March, 1997. *But see* Knutson, *supra* note 102, at 118.

121. Royer, *supra* note 103, at 308; James Lewis, *Sea-Level rise: Some Implications for Tuvalu*, 18(8) AMBIO 458, 459 (1989).

122. S.C.B. Raper, *Observational Data on the Relationships Between Climatic Change and the Frequency and Magnitude of Severe Tropical Storms*, in Warrick, *supra* note 98, at 192; Kahori Sakane, *Island States Seek Speedy Measures*, DAILY YOMIURI, December 4, 1997, at 3 (LEXIS, World Library).

123. *See infra* notes 17-19 and accompanying text.

124. LOBBAN & SCHEFTER, *supra* note 18, at 125; Haydon, *supra* note 36; Steering Committee of the Climate Change Study, *supra* note 35, at 46.

high as 47 per cent of the GNP of affected small island states.[125] PIDCs are particularly vulnerable to violent weather events because they usually lack early warning systems that would help them ameliorate the impact of storms and typhoons,[126] as well as adequate provisions for disaster relief.[127] The economic impact of these disasters has also been exacerbated in recent years by the withdrawal of coverage by insurance companies, including for critical infrastructure, such as electrical facilities.[128]

2.6 INCREASES IN THE INCIDENCE OF DISEASE

Disease outbreaks can exact a terrible toll on societies:

> Beyond the illness and death, there are losses in work productivity, land lost to use due to the presence of new disease vectors, threats to tourism (by algal blooms and [vector-borne diseases] in new areas), and there can be losses of food exports and other commercial activities. Diseases occurring across taxa ultimately threaten the development of societies.[129]

The World Health Organization has identified climate change as one of the primary threats to public health in the next century.[130] In the Pacific, climate change could exacerbate the incidence of several serious infectious diseases that beset PIDCs, including malaria, dengue fever and cholera. Malaria is endemic to many PIDCs and there have been many serious epidemics over the past two decades.[131] Rising temperatures and increased precipitation associated with climate change could increase the longevity of vectors carrying the disease, as well as enhance conditions for their incubation and reproduction.[132] As a consequence, the number of cases worldwide could increase by 50-80 million annually.[133] While it is anticipated that most of the

---

125. *Small Islands and Sustainable Development*, 10 TIEMPO 5, 5 (1993).

126. *South Pacific Climate Change*, *supra* note 8; *Global Conference is Told of NGO Draft on Problems of Small Island Developing States*, FED. NEWS SERVICE, May 9, 1994 (LEXIS, World Library); A.M. MANNION, GLOBAL ENVIRONMENTAL CHANGE 177 (1991).

127. *Small Islands and Sustainable Development*, *supra* note 109.

128. *Id.*; Nick Nuttall, *Islanders Ready for Pacific Evacuation*, THE TIMES, November 29, 1997 (LEXIS, World Library); Denis Staunton, *Islands Appeal for Emissions Cut to Lessen Danger From Rising Seas*, THE IRISH TIMES, March 30, 1995, at 11 (LEXIS, World Library).

129. Paul R. Epstein, *Climate Ecology and Public Health* (1995), at 8; *see also Deadly Global Warming*, BIRMINGHAM POST, November 6, 1998, at 8 (LEXIS, World Library).

130. Timothy E. Wirth, *Our Global Future: Climate Change*, Dept. of State Dispatch, October, 1997, at 16.

131. Charles P. Wallace, *The 'Pacific Paradox': Islands of Despair*, L.A. TIMES, March 16, 1995, at A1; Martin Parry, *Guinea*, AGENCE FRANCE PRESSE, February 19, 1998 (LEXIS, World Library); Phyllida Brown, *Infant Infection May Stop Malaria Running Wild*, NEW SCI., September 26, 1996, 21.

132. Rita R. Colwell & Jonathan A. Patz, *Climate, Infectious Disease and Health*, American Academy of Microbiology (1998), at 6-8.

133. IPCC, *Summary for Policymakers: Impacts, Adaptation and Mitigation Options*, Contribution of Working Group II to the IPCC Second Assessment Report, IPCC-XI/Doc.4 (1995), at SPM-2. *See also* Anthony J. McMichael & Andrew Haines, *Global Climate Change: The Potential Effects on Health*, 315 BRIT. MED. J. 805 (1997) (Percentage of world's population living in areas of potential malaria transmission may increase from 45 per cent to 60 per cent as consequence of projected increases in temperature).

projected increases will occur in temperate-zone countries, PIDCs might also see an increase in incidence, further straining their health care systems.[134]

In recent years, dengue fever, an incurable disease spread by the *Aedes aegypti* mosquito, has become widespread in PIDCs.[135] While its victims usually recover in a month, it can develop into a severe and potentially fatal form, dengue hemorrhagic fever.[136] The explosion of cases of dengue fever over the past few years, associated with El Nino,[137] may be a portent of the possible impacts of higher temperatures and rainfall anomalies associated with climate change.[138]

Temperature increases and a higher incidence of flooding and water shortages could also result in an increase of the incidence of waterborne and food-borne infectious diseases in PIDCS.[139] Cholera is a debilitating and sometimes fatal disease caused by the bacterium *Vibrio cholerae.* It is spread by bathing in or drinking contaminated water, or ingestion of contaminated food.[140] Warmer sea-surface temperatures may result in plankton blooms, which are a food source for the copepods upon which the cholera bacterium attach themselves.[141] Ocean warming could also increase the abundance of biotoxins such as ciguatera, resulting in higher incidence of seafood contamination and biotoxin poisoning.[142] This could further undermine the nutritional status of some PIDCs.[143]

3. The United Nations Framework Convention on Climate Change and Small Island States

The primary legal instrument to confront the possible ramifications of climate change is the United Nations Framework Convention on Climate Change (FCCC),[144] which entered into force in 1994.[145] However, while the parties pledged to "achieve . . . sta-

134. IPCC, *supra* note 9, at 348; Edwards, *supra* note 17, at 65.

135. *Dengue Fever Grips the South Pacific*, DEUTSCHE PRESS-AGENTUR, December 31, 1997 (LEXIS, World Library); S. Hales, P. Weinstein & A. Woodward, *Dengue Epidemics in the South Pacific: Driven by El Nino Southern Oscillation*, 348 LANCET 1664-5 (1996).

136. *Ministry Warns Travelers of Dengue Fever Epidemics*, THE DAILY YOMIURI, April 29, 1998; M.J. Cardosa, *Dengue Vaccine Design: Issues and Challenges*, 54(2) BRIT. MEDICAL BULLETIN 395, 395 (1998).

137. *Dengue Fever on the Rise, Warn Thai Health Chiefs*, AGENCE FRANCE PRESSE, April 24, 1998; *Smart Money's Betting the Farm on El Nino*, TORONTO STAR, February 28, 1998.

138. IPCC, *supra* note 9, at 349; *El Nino Helps Scientists Collect Data on Warming*, CHICAGO TRIBUNE, August 13, 1998, 7. A simulation model has recently projected that a one degree centigrade rise in temperatures could increase potential transmission risks for dengue fever by 31-47 per cent, placing up to 195 million additional people at risk by the middle of the next century. Anthony J. McMichael, Jonathan Patz & R. Sari Kovats, *Impacts of Global Environmental Change on Future Health and Health Care in Tropical Countries*, 54(2) BRIT. MEDICAL BULLETIN 475, 482 (1998).

139. IPCC, *supra* note 9, at 349.

140. Colwell & Patz, *supra* note 116, at 10.

141. *Id.* at 11.

142. IPCC, *supra* note 9, at 349.

143. *Id.*

144. 31 ILM. 849 (1992).

145. The UNFCCC had been ratified by 179 nations as of June 1999, http://www.unfccc.de.

bilization of greenhouse gas concentrations in the atmosphere at a level that would prevent dangerous anthropogenic interference with the climate system,"[146] their record to this point has been disheartening.

Initially, the major greenhouse gas emitting states agreed to "aim" to reduce their greenhouse gas emissions to 1990 levels by 2000. Yet, all industrialized nations flouted this pledge, leading the OECD to conclude that emissions from industrialized nations would actually rise by 11-24 per cent over the next fifteen years.[147]

At the Third Conference of the Parties of the FCCC, the parties adopted the Kyoto Protocol,[148] under which industrialized nations agree to reduce their collective emissions of six greenhouse gases by at least 5 percent below 1990 levels by 2008-2012.[149] However, hostility to the Protocol by powerful sectors in the United States, including organized labor, fossil fuel producers, and influential members of the Senate,[150] may thwart its adoption.[151] This would severely undercut the treaty's effectiveness, as the United States is responsible for approximately one-quarter of greenhouse gas emissions.[152] Moreover, U.S. refusal to adopt the Protocol might result in European nations and Japan also balking.[153] This could doom the agreement, as it requires ratification by Annex I nations producing at least 55 per cent of greenhouse gas emissions before it

---

146. UNFCCC, *supra* note 128, at art 2.

147. *US Greenhouse Gas Emissions Continue to Climb*, GLOBAL ENVTL. CHANGE REP., April 26, 1996, at 2; *see also Canada Report Exposes Limits of Voluntary Efforts*, GLOBAL CHANGE, Winter, 1999, at 13; Ute Collier, *Climate Change Policies in the European Union*, 15 WGES NEWSL., Spring, 1996, at 6.

148. *Kyoto Protocol to the United Nations Framework Convention on Climate Change*, FCCC/CP/1997/L.7/Add. 1, Dec. 10, 1997.

149. *Id.* at art. 3 (1). For a detailed analysis of the Third Conference of the Parties, *see* Peter G.G. Davies, *Global Warming and the Kyoto Protocol*, 47 INT'L & COMP. L. Q. 446 (1998).

150. *Mine Workers Gore's Treaty Will Cost 1.7 Million U.S. Jobs; Union Bosses Should Quiz Veep on 'Economic Armageddon,'* PR NEWSWIRE, February 19, 1999 (LEXIS, World Library); Rebecca Sultana, *Campaign Against Global Warming: The Differing Positions*, THE INDEPENDENT, February 13, 1999 (LEXIS, World Library); *A New Disinformation Campaign*, RACHEL'S ENVIRONMENT & HEALTH WKLY., April 30, 1998; Gretchen Vogel & Andrew Lawler, *Hot Year, But Cool Response in Congress*, 280 SCI., June 12, 1998, at 1684.

151. It is anticipated that the Protocol will not even be submitted to the Senate for consideration until after the 2000 presidential election. *International Energy Agency: There is No Time to Lose*, PETROLEUM ECONOMIST, May 12, 1999 (LEXIS, World Library). Recently, the Clinton administration and industry have been emphasizing voluntary programs and tax incentives to reduce emissions. *GCC Industry Voluntary Actions Highlighted; Administration Focus, Legislative Efforts in Congress Emphasize Importance*, PR NEWSWIRE, April 22, 1999 (LEXIS, World Library); *Clinton Calls for Clean-Air Fund to Tackle Greenhouse Gas Emissions*, POWER ECON., March 31, 1999; Danielle Knight, *Industry and Greens Debate Climate Change Bill*, INTER PRESS SERVICE, February 11, 1999 (LEXIS, World Library). Congress has recently evinced its contempt for Kyoto by slashing FY 2000 appropriations for renewable energy and energy efficiency programs by 10 per cent from 1999 levels while substantially increasing funding of fossil fuel programs. *Renewable Energy Funding Takes a Hit*, May 28, 1999, http://www.enn.com/news/enn-stories/1999/05/052899/energy_3460.asp.

152. Alex Barnum, *Can World Unite, Halt Climate Threat?*, S.F. CHRONICLE, November 28, 1997, at A21.

153. Robert A. Reinstein, *In the News*, FED. NEWS SERVICE, April 29, 1999; *Risky Business*, GLOBAL CHANGE, Oct. 1998, at 2. Moreover, recent evidence indicates that the EU has yet to formulate climate policy measures to effectuate the commitments it made at Kyoto. Environment News Service, *Europe Faces Climate Policy Credibility Gap*, May 22, 1999, http://ens.lycos.com/ens/may99/1999L-05-19-06.html.

will come into effect.[154]

Moreover, because many greenhouse gases persist in the atmosphere for decades "their radiative forcing – their tendency to warm Earth – persists for periods that are long compared with human life spans."[155]

As a consequence of this, and the exclusion of developing countries from reduction commitments,[156] full implementation of the Kyoto Protocol will reduce projected warming by a mere one-twentieth of one degree C, and have a minimal effect on elevation of sea levels.[157] Finally, the FCCC currently only binds developed countries and economies in transition to the reduction of greenhouse gas emissions.[158] However, given the tremendous projected increases in greenhouse gas emissions in developing countries over the next century,[159] the future effectiveness of the FCCC is contingent on engaging these nations in the regime's mission.[160] Yet, there is great trepidation among developing countries about possible economic impacts associated with reducing emissions, as well as a sense of unfairness given the tremendous disparity in per capita emissions between industrialized and developing nations.[161] Developing countries repulsed an effort at the Third Conference of the Parties in Kyoto to establish emission limitation objectives for wealthier developing states.[162] At the Fourth Conference of the Parties in Buenos Aires, Kazakhstan announced its intention to join Annex I and thus assume commit-

154. Kyoto Protocol, *supra* note 131, at art. 24. The Protocol will enter into force 90 days after "not less than 55 parties to the [UNFCCC], incorporating Parties included in Annex 1 which accounted in total for at least 55 per cent of the total carbon dioxide emissions for 1990 of the Parties included in Annex 1" have ratified it. As of August, 1999, 84 nations have signed the Protocol, but only 14, all developing nations, have signed it. UNFCCC Secretariat, Kyto Protocol, Status of Ratification, <http://www.unfccc.de/resource/kpstats.pdf> (1999).

155. Bette Hileman, *Climate Observations Substantiate Global Warming Models*, CHEMICAL & ENGINEERING N., November 27, 1995, http://pubs.acs.org/hotartcl/cenear/951127/pgl.html *See also*, Thomas R. Karl, et al., *The Coming Climate*, SCI. AM., May, 1997, at 224 ("as much as 40 percent of [carbon dioxide] tends to remain in the atmosphere for centuries"); Silvio Kusidio, *Climatic Changes are no Longer Preventable, Warn Experts*, DEUTSCHE PRESSE-AGENTUR, March 22, 1995 (LEXIS, News file) ("Even a worldwide stabilization of the emissions would not a prevent a rise in the greenhouse gases . . . for the next 200 years . . .").

156. *See infra* notes 56-58 and accompanying text.

157. Martin Parry, et al., *Buenos Aires and Kyoto Targets Do Little to Reduce Climate Change Impacts*, 8 GLOBAL ENVTL. CHANGE 285, 285 (1998).

158. UNCED, Framework Convention on Climate Change, opened for signature, June 4, 1992, *reprinted in* 31 ILM 849 (1992), at art. 4(2) & Annex I; Kyoto Protocol, *supra* note 128, at art. 3.

159. "Though projections vary substantially with assumptions about rates of economic growth, it appears that sometime between 2010 and 2020 China will overtake the United States as the world's largest greenhouse gas emitter, and that sometime thereafter emissions in the developing world will exceed those of the developed world." William L. Thomas, *The Kyoto Protocol: History, Facts, Figures and Projections*, 137(8) PUBLIC UTILITIES FORTNIGHTLY, April 15, 1999, at 48. Two-thirds of greenhouse gas emissions between 1995-2025 may be produced by developing countries. *Carbon Emissions to Rise Over Next 20 Years*, MODERN POWER SYSTEM, January 31, 1999, at 3 (LEXIS, World Library).

160. Clare Breidenich et al., *The Kyoto Protocol to the United Nations Framework Convention on Climate Change*, 92 AM. J. INT'L L. 315, 331 (1998); *Kyoto Protocol: The Unfinished Agenda*, 39 CLIMATIC CHANGE (1998), at 9.

ments under the Protocol. However, Argentina's effort to place the issue of voluntary commitments on the agenda was rejected by the G77, "most of which declined even to join in informal talks on the topic."[163]

Ultimately, the major greenhouse gas producing nations may find it cost-beneficial to enact more stringent measures to reduce emissions.[164] As Cline suggests, under a "moderate-central" assumption of the damages caused by climate change, benefits ultimately begin to exceed abatement costs on a global scale in 2050, with the benefits of damage avoidance rising to about 1 per cent of world gross domestic product (GDP) by 2050, and 5 per cent of GDP by 2275.[165]

Unfortunately, by that time most small island states may have paid a terrible price, and some may even have ceased to exist. If the FCCC's mandate to "prevent dangerous anthropogenic interference with the climate system" is to be complied with equitably, then major greenhouse emitting nations should not wait to act until they deem it expedient. However, given the record of industrialized states in confronting climate change over the past few decades, it is difficult to be sanguine about the future. In a world governed more by *realpolitik* than international law, small island states will likely remain hostages to forces far beyond their control.

161. William K. Stevens, *Climate Talks Enter Harder Phase of Cutting Back Emissions*, N.Y. TIMES, April 11, 1995, at 4; Davies, *supra* note 132, at 457. "20 per cent of the world's population is responsible for 63 per cent of carbon dioxide emissions, while another 20 per cent is responsible for only 2 per cent of these emissions." Robert Engelman, *Population, Consumption and Equity*, TIEMPO, December, 1998, at 5. *See also* MICHAEL GRUBB, CHRISTIAAN VROLIJK & DUNCAN BRACK, THE KYOTO PROTOCOL. A GUIDE AND ASSESSMENT 265-269 (1999).

162. *Id.*

163. GRUBB, *supra* note 145, at 251-52. Argentina subsequently announced its intention to enter into a binding commitment to reduce greenhouse emissions.

164. For analysis of how nations' cost-benefit analyses have hindered development of an effective emissions reduction regime, *see* Sean Fox, *Responding to Climate Change: The Case for Unilateral Trade Measures to Protect the Global Atmosphere*, 84 GEO. L.J. 2499, 2509 (1996); Martin J. LaLonde, *The Role of Risk Analysis in the 1992 Framework Convention on Climate Change*, 15 MICH. J. INT'L L. 215, 236 (1993).

165. William R. Cline, *Socially Efficient Abatement of Carbon Emissions*, in CLIMATE CHANGE & THE AGENDA FOR RESEARCH 102 (Ted Hanisch ed., 1994). However, even if the major emitters of greenhouse gases ultimately decide to make a serious commitment to emissions reductions, the task of stabilizing emissions will be extremely imposing. As Hoffert et al. recently concluded: "Stabilizing atmospheric CO_2 at twice pre-industrial levels while meeting the economic assumptions of 'business as usual' implies a massive transition to carbon-free power, particular [sic] in developing nations. There are no energy systems technologically ready at present to produce the required amounts of carbon-free power." Stabilizing atmospheric CO_s at double pre-industrial levels will require about 15 terawatts of carbon-free power by 2050, and even more thereafter. By comparison, total world energy consumption is currently only 13.5 terawatts. Martin I. Hoffert, et al., *Energy Implications of Future Stabilization of Atmospheric CO_2 Content*, 395 NATURE, October 29, 1998, at 884. *But see*, David G. Victor, *Strategies for Cutting Carbon*, 395 NATURE, October 29, 1998, at 837 ("if policies to improve energy efficiency could accelerate the 1 per cent annual decline in energy use per unit of economic output to 1.5 per cent, then the carbon-free energy required would be cut in half").

13. PAROCHIALISM AND EMPOWERMENT: RESPONDING TO ECOCOLONIALISM AND GLOBALISATION IN THE SOUTHWEST PACIFIC

MICHAEL EDWARDS
Institute for Sustainable Futures
University of Technology
Sydney, PO Box 123, Broadway, NSW 2007, Australia

1. Introduction

The water lapped at the shore with timeless repetition; the insects sang in unison from the trees that clung to the cliff like moss to a wall. A seagull circled on silent wings and headed out to sea. A crab scuttled over rocks and sought refuge in a small pool, iridescent in the evening light. A fish leaped from the sea and then, with a splash, returned to its coral world. The palm trees swayed gently in the breeze as the sun dipped below the horizon, signalling the end of another perfect day.

The beauty and serenity of Pacific islands has captured the attention of artists, writers, poets and discerning travellers for many centuries and will continue to do so in the future. However, the halcyonic image of palm-fringed beaches and azure seas belies a less palatable reality. Pacific islanders face many threats to their security; some unique to their region, common to those undermining human well-being in other parts of the world. The paradise described in the writings of Robert Louis Stevenson and depicted in the paintings of Paul Gauguin still exists. However, it will become increasingly imperiled in the years ahead.

Throughout the world, degradation of the environment is increasing at an alarming rate. Depletion of the ozone layer, loss of biodiversity, soil erosion, water and air pollution (witness the recent smogs in Southeast Asia) are beginning to undermine the security of the human race and the animal and plant species that share this planet. Although some academics believe classifying environmental problems as threats to security is counter-productive,[1] there is mounting evidence to suggest that environmental problems may

1. *See* D. Deudney, *The Mirage of Eco-War: The Weak Relationship Among Global Environmental Change, National Security and Interstate Violence* in GLOBAL ENVIRONMENTAL CHANGE AND INTERNATIONAL RELATIONS (I.H. Rowlands & M. Greene eds., 1992)

A. Gillespie and W. C. G. Burns (eds.), Climate Change in the South Pacific: Impacts and Responses in Australia, New Zealand, and Small Island States, 251–268.

have implications for national and international security.[2] Climate change, resulting from the enhanced anthropogenic greenhouse effect, is one environmental threat currently receiving attention. Although changes in climate have the potential to undermine human security worldwide,[3] people living on low-lying island states in the Southwest Pacific are thought to be particularly vulnerable. While there has been considerable research into the possible impacts of climate change in this region, few have bothered to investigate, or tried to understand, the links between climate change and security in this geographical and cultural setting. Over a three-year period, the author has explored these links and has made some discoveries that suggest that the implications of climate change threat are not as straight-forward as it may first appear.

Until recently, scholars interested in links between climate change and security have focused their attention on biophysical impacts and how these will aggravate the economic and political problems that can create and exacerbate tensions both within and between nations.[4] While biophysical impacts are important and must be examined, it is essential to realise that the discourse which is used to construct the threat, and the geographical scale at which the threat is seen to exist, also has implications for human security. To date, climate change discourse has been dominated by actors in rich and powerful nations, thus marginalising the concerns of people who have little or no control over the nature, magnitude and timing of the threat - the real victims of climate change. In this regard, climate change dis-

2. Readers interested in the links between environmental degradation and security should refer to: N. BROWN, THE STRATEGIC REVOLUTION: THOUGHTS FOR THE TWENTY-FIRST CENTURY (1992); S. Dalby, *Ecopolitical Discourse: 'Environmental Security' and Political Geography*, 16(4) PROGRESS IN HUMAN GEOGRAPHY (1992), at 503-522; S. Dalby, *Security, Modernity, Ecology: The Dilemmas of Post-Cold War Security Discourse*, 17 ALTERNATIVES (1992), at 95-134; M.J. EDWARDS, 1(1) DEFINITIONS, THREATS AND PYRAMIDS: THE CHANGING FACES OF ENVIRONMENT AND SECURITY (1996), at 96-123; P. Ehrlich & A. Ehrlich, *The Environmental Dimensions of National Security*, in GLOBAL PROBLEMS AND COMMON SECURITY (J. Rotblat & V.I. Goldanski eds., 1989); T.F. Homer-Dixon, *On the Threshold: Environmental Changes as Causes of Acute Conflict*, 16(2) INT'L SECURITY (1991), at 76-116; T.F. Homer-Dixon, J.F. Boutwell & G.W. Rathjens, *Environmental Change and Violent Conflict*, SCI. AM., Feb. 1993, 38-45; M.V. Soroos, *Global Change, Environmental Security, and the Prisoner's Dilemma*, 31(3) J. PEACE RES., at 317-332; A.H. Westing, *Environmental Security and its Relation to Ethiopia and Sudan*, 20(5) AMBIO (1991), at 168-171.

3. Views concerning how climate change will affect both national and international security vary. Some scholars believe it will merely exacerbate existing resource, demographic and economic pressures, T.F. Homer-Dixon, 19 ENVIRONMENTAL CHANGE AND VIOLENT CONFLICT: EVIDENCE FROM CASES' INTERNATIONAL SECURITY (1994), 5-40, 7-8, while others believe changes in climate are " . . . likely to cause unpredictable consequences." D.A. Wirth, *Catastrophic Climate Change* in WORLD SECURITY: TRENDS AND CHALLENGES AT CENTURY'S END (M.T. Klare & D.C. Thomas eds., 1991) at 388. Until there is greater certainty regarding the nature, timing, likelihood and intensity of regional impacts of climate change, it will be difficult to predict the implications for security *See* P.H. Gleick, *Global Climatic Change and International Security*, 1(1) COLO. J. INT'L ENVTL. L. & POL'Y (1990), at 41-56. Although this creates uncertainty, it is imperative that research is facilitated because, either way, climate change will have implications for security through the exacerbation of existing threats, the creation of new threats or both.

course needs to be viewed as a contemporary manifestation of other processes that have disempowered Pacific islanders since their islands were first colonised by Europeans.[5]

'Climate change discourse threat' (CCDT) has two primary strands. Both are important and have already begun to undermine human security in the Southwest Pacific. This is significant because it would appear that the impacts of the CCDT are occurring well before the most serious biophysical impacts of human-induced, or anthropogenically enhanced, climate change will be experienced in the region.[6] The first strand of the CCDT (to be discussed in part II of this chapter) has arisen out of the doomsday/worst-case scenarios of climate change that, understandably, dominate climate change discourse.[7] The second strand of the CCDT concerns the scale at which the threat of climate change has been constructed. As part III will demonstrate, there are economic, political and ideological reasons for constructing climate change as a global threat. It would appear that the threat has been globalised not simply because the impacts of an anthropogenic greenhouse effect will have global implications but, more importantly, because industrialised nations derive power from the globalisation of environmental problems.

---

4. *See* for example, P.H. Gleick, *The Implications of Global Climatic Changes for International Security*, 15(1/2) CLIMATIC CHANGE, at 309-325; P.H. Gleick, *How Will Climatic Changes and Strategies for the Control of Greenhouse-Gas Emissions Influence International Peace and Global Security?* in LIMITING THE GREENHOUSE EFFECT: OPTIONS FOR CONTROLLING ATMOSPHERIC CO_2 ACCUMULATION (G.I. Pearman ed., 1992); S.C. Lonergan & B. Kavanagh, *Climate Change, Water Resources and Security in the Middle East*, GLOBAL ENVIRONMENTAL CHANGE, September, 1991, at 272-290; L. Lunde, *North/South and Global Warming - Conflict or Cooperation?*, 22(2) BULLETIN PEACE PROPOSALS (1996), at 199-210; I.H. Rowlands, *Ozone Layer Depletion and Global Warming*, 16(3) NEW SOURCES FOR ENVTL. DISPUTES' PEACE & CHANGE (1991), at 260-284.

5. Since European colonisation, the island states of the Southwest Pacific have been assimilated into the global economic system. Pacific islanders have no power in a system that is designed to bring the greatest benefits to those with the political and economic clout to manipulate it. Assimilation has led to disempowerment. Climate change discourse may have much the same effect, albeit at a much smaller scale. At present, climate change discourse is constructed around the interests and concerns of scientists and politicians in industrialised countries. Those who have some control over the threat - the major greenhouse gas emitters - dominate the discourse. The concerns of Pacific islanders are marginalised in a discourse where the focus is primarily on how to reduce greenhouse gas emissions without jeopardising economic growth in industrialised nations. Unless Pacific islanders can reconstruct the threat of climate change in terms that are relevant to their 'local' lives they will find their engagement in climate change discourse, as presently constructed, will merely lead to further disempowerment.

6. This is a difficult assertion to make because some commentators would claim that anomalous weather events occurring around the world are the first evidence of changes in climate resulting from an enhanced anthropogenic greenhouse effect. However, there is no compelling proof that anomalous weather events in the Southwest Pacific are a direct result of an enhanced anthropogenic greenhouse effect.

7. Part I of this chapter is a revised and updated version of: M.J. Edwards, *Climate Change, Worst-Case Analysis and Ecocolonialism in the Southwest Pacific*, 8(1) PAC. REV. (1996), at 63-80.

It is important to state, at the outset, that climate change is an extremely serious threat to security in the island states of the Southwest Pacific.[8] Action must be taken in industrialised countries to reduce emissions of greenhouse gases. Indeed, the author believes that people in these countries have a moral obligation to change behaviour that directly threatens those who inhabit regions of the world that are most vulnerable to the impacts of climate change. It would be a mistake to construe what is written here as the work of a 'greenhouse sceptic.' There is no doubt in the author's mind that climate change needs to be addressed with great urgency. The only reason Pacific islanders are advised not to engage in climate change discourse (as it is presently constructed) is because this will do virtually nothing to reduce their vulnerability to the impacts of a threat over which they currently have little control.

2. Planning for the Worst Creates the Worst

Security in the Southwest Pacific region is undermined by a multiplicity of threats - some endogenous and some exogenous. In the years ahead, Pacific islanders are likely to face similar problems to those they face today, including environmental degradation and resource exploitation, dependence on foreign aid, rapid urbanisation, the Westernisation of their cultures, political corruption, external insensitivity to their sovereignty and indigenous militarisation.[9] In addition, they are likely to face threats associated with climate change arising from the enhanced anthropogenic greenhouse effect. In many respects, climate change is the 'wild card' in the future threat profile of the Southwest Pacific region. Not only may changes in climate exacerbate threats already faced by the region but, perhaps most importantly,

---

8. In the Southwest Pacific, climate change and sea-level rise could render atoll nations uninhabitable; inundate low-lying parts of larger/higher islands, *see* R. Chase & J. Veitayaki, *Implications of Climate Change and Sea Level Rise for Western Samoa,* Report of a Preparatory Mission, 59 SPREP REPORTS AND STUDIES SERIES, South Pacific Regional Environment Programme (1992); increase the frequency and intensity of extreme weather events, storm surges and high tide, *see* J.C. Pernetta, *Impacts of Climate Change and Sea-Level Rise on Small Island States: National and International Responses*, GLOBAL ENVIRONMENTAL CHANGE, March, 1992, at 19-31; lead to declines in agricultural and fisheries production, *see* J. Connell, *Climatic Change: A New Security Challenge for the Atoll States of the Southwest Pacific*, 31(2) J. COMMONWEALTH & COMPARATIVE POLITICS (1993), at 185; damage coral reef and mangrove ecosystems, *see* R.W. Buddemeier & S.V. Smith, *Coral Reef Growth in an Era of Rapidly Rising Sea Level: Predictions and Suggestions for Long-Term Research*, 7 CORAL REEFS (1988), at 51-56; J.C. Ellison & D.R. Stoddart, *Mangrove Ecosystem Collapse During Predicted Sea-Level Rise: Holocene Analogues and Implications*, 7(1) J. Coastal RES. (1991), at 151-165; increase the incidence of vector borne diseases such as malaria, filariasis and dengue in islands affected by these diseases. *See* ASPEI Task Team Members, *Overview of Potential Impacts of Climatic Change in the SPREP Region*, in IMPLICATIONS OF EXPECTED CLIMATE CHANGES IN THE SOUTH PACIFIC REGION: AN OVERVIEW, United Nations Environment Programme Seas Reports and Studies, No. 128, (J.C. Pernetta and P.J. Hughes, eds. 1990); and contaminate freshwater resources. *See* P. Roy and J. Connell, *Climatic Change and the Future of Atoll States*, 7(4) J. COASTAL RESEARCH (1991), at 1057-1075; R.W. Buddemeier & J.U. Oberdorfer, *Climate Change and Island Groundwater Resources* in Pernetta & Hughes, *id.*

they may alter the environmental and socio-cultural conditions of the localities where impacts may be experienced.

Uncertainty underpins most aspects of the climate change threat. As such, it is impossible to make any definitive statements about the nature, timing and magnitude of climate change impacts.[10] Whereas uncertainty has translated into a rather complacent approach to climate change in the industrialised world, in the Southwest Pacific it has increased the level of fear and apprehension surrounding the threat. Indeed, in a survey carried out by the author, climate change was considered by many to be the most pressing threat to regional security.[11] One respondent stated:

> I want to emphasise that climate change and sea-level rise are major threats to the island states of the Southwest Pacific, especially the low-lying islands. Whilst the economic threat, in particular dependence on foreign aid and vulnerability to external markets have been ongoing, those are not as urgent as the threat climate change poses to regional security. It is important to educate people that we in the Southwest Pacific are at great risk and that climate change will be our greatest challenge in the years ahead.

Most respondents to the survey said they feared the worst from climate change impacts. For example, it was stated that:

> . . . climate change and sea-level rise will bring about disastrous and permanent impacts on islands where generations of people have lived and died.

and:

> Climate changes and the rising seas pose the highest threat to regional security. Greenhouse effects must be controlled at all costs. Until appropriate measures are taken to control the greenhouse effect the human-race is bound for extinction. The last day is around

---

9. It is important to remember that the Southwest Pacific is a geographically and culturally diverse region and, therefore, threats to human security vary both in nature and in intensity between islands. Although all islands in the region face common threats such as vulnerability to natural disasters, environmental degradation and dependence on foreign aid, each island faces a number of unique threats that could undermine human security. In Fiji, for example, there is tension between Fijians and Indo-Fijians; in Solomon Islands, foreign logging companies are destroying the livelihoods of people who depend on forests for food, building materials and spiritual fulfillment; in Papua New Guinea, political corruption has sparked civil unrest; and in French Polynesia, security may be undermined as a result of the long-term health and environmental consequences of nuclear testing.
10. Gleick, *supra* note 3, at 45.
11. The author undertook a survey in 1995 to ascertain the nature of threats facing island states in the Southwest Pacific. The aim of the survey was to gauge how important the threat of climate change was perceived in comparison to other threats facing the region. The results of the survey can be obtained from the author.

> the corner . . . Do as you please and die as a result!

In addition, much of what has been written on climate change impacts in the Southwest Pacific has emphasised worst-case/doomsday scenarios.[12]

There would appear to be at least two reasons why the worst-case impacts of climate change are emphasised by the Pacific islanders who voice their concerns both nationally and internationally. First, they are genuinely concerned that their islands are threatened by an environmental disaster over which they have little control. Second, they hope that a doomsday scenario will ensure their plight receives international attention, perhaps forcing substantive changes in climate change policy in industrialised countries.

However, while a worst-case view of the future is an important diplomatic tool at the international level, it is creating problems at the domestic level. By emphasising the potentially catastrophic impacts of climate change and associated sea-level rise, politicians striving to save their island nations are actually contributing to the destruction of the region's environment by encouraging short-term planning.[13]

Short-term planning is not always by choice. There are many endogenous and exogenous factors at work, particularly in the less-developed world, which influence the planning process and make short-term planning a necessity. Poverty, aid dependency and debt are just three important factors that influence the time horizons over which decisions are made. In an attempt to address pressing economic problems, governments are forced to utilise natural resources at rates that simply cannot be sustained.[14] The contributions made by natural resources to global biodiversity and local livelihoods are of secondary importance in an economic system which, in its present form, grows as biophysical systems are undermined.

There are many scholars, bureaucrats and activists expending time and energy exploring ways to uncouple economic growth and development from resource pillage and environmental degradation. Although this author would argue that economic growth cannot occur without environmental damage, others believe the concept of sustainable development is the key to environmentally benign growth. Regardless of whether sustainable development proves to be an oxymoron or a

---

12. H. Brookfield, *Global Change and the Pacific: Problems for the Coming Half-Century*, THE CONTEMPORARY PACIFIC, Spring/Fall, 1989, at 16; A. Buckley, *The Proof We Fear Could Kill Us*, PAC. ISLANDS MONTHLY, June, 1992 at 35; J. Connell & J. Lea, *'My Country will not be There': Global Warming , Development and the Planning Response in Small Island States*, CITIES, November, 1992, at 295; W. Laban, *Keeping Our Heads Above Water: The Effects of Climate Change on Tuvalu*, 46 TOK BLONG PASIFIK (1994), at 19; P. Roy & J.Connell, *The Greenhouse Effect: Where have all the Islands Gone?*, PAC. ISLANDS MO., April/May, 1989, at 16.
13. "Short-term planning" is used here to refer to the time horizon over which decisions are made at a national and/or regional level.
14. P. CHATTERJEE & M. FINGER, THE EARTH BROKERS: POWER, POLITICS AND WORLD DEVELOPMENT (1994), at 148.

panacea to the world's environmental ills, it is a concept that should not be ignored. Most importantly, it emphasises the need for long-term planning to ensure intergenerational equity.[15]

At present, a conflicting message is being sent to Pacific islanders and their governments which will influence, to some extent, the degree to which policies geared towards sustainable development will be implemented. On the one hand, Pacific islanders are being urged to protect the environment and to think about the future;[16] on the other hand, they are being warned that some of their islands may cease to exist in 100 years.[17] The prospect that islands may be rendered uninhabitable may encourage politicians to promote unsustainable development. Indeed, the threat of climate change may force the governments of island states to adopt higher discount rates. This could have serious implications for the environment. According to Pearce et al.:

> The choice of the discount rate has a particular effect on the rate of exploitation of natural resources. The basic decision with regard to such resources is how much to consume now and how much to hold in store for future consumption . . . The higher the discount rate, the faster is the rate of depletion of the resource in the earlier years and the shorter is the interval before which the resource is exhausted. With a higher discount rate, a lower value is placed on future consumption relative to present consumption.[18]

The future will be discounted because it will be assumed that the impacts of climate change and associated sea-level rise will render some islands completely or partly uninhabitable. Even if human-induced changes in climate do not prove regionally or locally catastrophic, planning on the assumption that they will be may have much the same effect. The unfortunate thing is, those people who emphasise the potentially disastrous impacts of climate change because they are worried about the future of Pacific islanders (usually concerned academics, representatives of environmental NGOs and various politicians from island governments) may be inadvertently and unintentionally ensuring a more rapid rate of resource depletion and greater environmental degradation in the Southwest Pacific.

---

15. *See* R.B. NORGAARD, DEVELOPMENT BETRAYED: THE END OF PROGRESS AND A COEVOLUTIONARY REVISIONING OF THE FUTURE (1994).

16. According to the Programme of Action for Small Island States: " . . . small island developing States should, in accordance with their own priorities, endeavour to achieve the goals of sustainable development by, inter alia, formulating and implementing policies, strategies and programmes that take into account development, health and environmental goals, strengthening national institutions, and mobilising all available resources, all of which are aimed at improving the quality of life." *Programme of Action for Small Island States*, Earth Summit (Bridgetown, Barbados, Global Conference on the Sustainable Development of Small Island Developing States, 26 April-6 May, 1994, at 5.

17. *Id.* at 10.

18. D. PEARCE, A. MARKANDYA & E.B. BARBIER, BLUEPRINT FOR A GREEN ECONOMY (1989), at 144.

Although the unintentional outcomes of planning for climate change using a worst-case approach are important, perhaps more significant are the problems that could arise if worst-case analysis and the threat of climate change become integral components of a new genre of colonialism - ecocolonialism.[19] Indeed, research carried out in Fiji suggests that worst-case analysis could be used by foreign powers to increase the dependency of Pacific island states and to provide new opportunities for resource exploitation.

Although there have been few investigations into the role played by climate change discourse in ecocolonialism, Hay et al. have looked at the climate change issue from the perspective of "scientific neocolonialism."[20] Their findings suggest that the way monitoring is conducted and information controlled and interpreted by researchers and scientific organisations from outside the region is tantamount to exploitation. In this regard, some respondents to the questionnaire survey carried out by the authors, suggested that monitoring equipment is installed by donor countries in an attempt to fend off criticism for not taking action to halt climate change. Some respondents took a cynical view of monitoring programmes, suggesting they are nothing more than a cheap way of creating and maintaining a circle of inaction. Another important aspect of 'scientific neocolonialism,' identified by Hay et al., is the way Pacific islanders are forced to partake in, and operate at, international meetings according to an agenda which rarely allows their concerns to be addressed. While they are given a 'forum' to express their views, this is frequently in response to decisions that have already been made.[21]

In addition to the threat posed by 'scientific neocolonialism,' the threat of climate change may undermine Pacific Islanders' right to self-determination and increase their level of dependence on industrialised countries. Even though decolonisation has led to the emergence of self-governing political entities in the Southwest Pacific,[22] it has not translated Pacific islanders' control over their own affairs or freedom from outside influences.[23] This is important, because without control,[24] Pacific islanders will be unable to ". . . decide for themselves the dimensions of their political, economic, cultural and social conditions."[25] The threat of climate change is important in this regard because it effectively transfers control from

19. Ecocolonialism is the process by which industrialised nations manipulate concerns about the environment in order to maintain their political, economic and ideological hegemony.

20. J.E. Hay, C. Kaluwin & N. Koop, *Implications of Climate Change and Sealevel Rise for Small Island Nations of the South Pacific: A Regional Synthesis*, 15(2) WEATHER AND CLIMATE (1995), at 49-50.

21. *Id.*

22. *See* G. Bertram, *The Political Economy of Decolonisation and Nationhood in Small Pacific Societies*, in CLASS AND CULTURE IN THE SOUTH PACIFIC, Centre for Pacific Studies, University of Auckland and Institute of Pacific Studies, University of the South Pacific, (A. Hooper eds., 1987).

23. E. Hau'ofa, *The New South Pacific Society: Integration and Independence*, *id.*, at 1 & 9-10.

24. According to Knight, the word 'control' is particularly important in any discussion and understanding of the rights of indigenous people to self-determination. D.B. Knight, *Self-Determination for Indigenous Peoples: The Contest for Change* in NATIONAL SELF-DETERMINATION AND POLITICAL GEOGRAPHY (R.J. Johnston, D.B. Knight & E. Kofman eds., 1988), at 126.

Pacific islanders to politicians in industrialised countries who have a certain degree of influence over the nature and timing of future climate change impacts. So as not to relinquish further control, Pacific islanders must respond to the threat of climate change in ways that allow them to maintain their independence and their rights to self-determination. Possible ways of doing this are discussed in part 4.

So far this chapter has attempted to demonstrate that the threat of climate change could lead to short-term planning and increased rates of resource depletion. In addition, the threat is likely to create further dependency in the Southwest Pacific and, in the future, may be used to fulfil the neocolonial aspirations of those governments who seek to economically control weaker states. Considering what the author has discovered about climate change in the Southwest Pacific, it would appear that the threat is not being responded to in the most appropriate way. Of course, politicians from island states should continue to lobby the governments of the industrialised world to reduce their emissions of greenhouse gases. However, more attention needs to be focused on indigenous responses to climate change. These will, by definition, require action at the local scale. The recommendation to address climate change at the local level is a departure from prevailing orthodoxy and is likely to rile those who have political and economic reasons for constructing climate change as a global threat. The globalisation of climate change discourse is an important aspect of eco-colonialism and forms the basis of the second strand of the CCDT.

3. What Does the 'Global' Mean in Climate Change Discourse?

Contrary to popular belief, globalisation is not restricted to cultural, economic or institutional phenomena; it is becoming an important aspect of environmentalism.[26] As Aart Scholte points out, "Ecologically, globalization has taken place in the shape of planetary climate change, atmospheric ozone depletion, worldwide epidemics and the decline of Earth's biodiversity . . ."[27] Climate change is a global problem because it will have an impact on global biophysical processes and will be experienced (to different degrees) by every nation in the world. In another sense, climate change is global because all human beings contribute (albeit very unequally) to the problem and it will only be effectively confronted by widely replicated actions. Although climate change is a global problem, it needs to be remembered that the global prefix is not merely a descriptive term. Rather, it is often used to extend the power and influence of industrialised nations and hide the conflict that exists between rich and poor states, states and social movements, and the empowered and disempowered.[28]

---

25. G.T. Morris, *International Law and Politics: Toward a Right to Self-Determination for Indigenous Peoples*, in THE STATE OF NATIVE AMERICA: GENOCIDE, COLONIZATION AND RESISTANCE (M.A. James ed., 1992), at 56.
26. J. Aart Scholte, *Beyond the Buzzword: Towards a Critical Theory of Globalization*, IN GLOBALIZATION: THEORY AND PRACTICE (E. Kofman & G.Youngs eds., 1996), at 46.
27. *Id.* at 46.

Climate change can only be solved if knowledge that has been generated in a multiplicity of cultural and environmental settings is incorporated into climate change discourse. Diversity in the way the threat is perceived and interpreted will provide new insights into how to deal with the problem. Particularly important in this regard is the knowledge of indigenous people who frequently have a comprehensive understanding of their local environments and, therefore, skills and knowledge that they could use to reduce their vulnerability to climate change impacts. Indigenous people have learnt through direct experience to cope with extremes of climate. As such, agricultural practices/techniques, food storage systems, water conservation schemes and building strategies have all been developed in order to reduce the adverse impacts of inclement weather events and could be used in the future as the basis for adaptation strategies to cope with the impacts of climate change.[29] To ensure this knowledge is never lost,[30] it is important that archives of 'indigenous knowledge' are developed on a nation-by-nation basis. Ownership of these archives would rest with indigenous people, but could be accessed by anyone interested in local responses to global environmental threats.

Before indigenous knowledge can be incorporated into climate change discourse, in any form that is more than a token gesture, there will need to be a radical shift in the way the threat is constructed. At present, climate change discourse is based on parochial 'Western' knowledge.[31] The threat has been constructed by scientists and politicians in industrially developed nations and transnational NGOs dominated by campaigners in these countries.[32] Shiva uses the term 'dominant local' to describe the individuals and institutions in industrially developed nations with the power to determine which environmental problems become globally significant threats to human security.[33] In the Southwest Pacific, for example, fears about climate change are not based on local experiences or knowledge, but rather are based

28. A. Hawkins, *Contested Ground: International Environmentalism and Global Climate Change*, in THE STATE & SOCIAL POWER IN GLOBAL ENVIRONMENTAL POLITICS (R. Lipschutz & K. Conca eds., 1993), at 231-232.

29. *See* J. E. Hay, *Regional Assessment of the Vulnerability of Pacific Islands to the Impacts of Global Climate Change and Accelerated Sealevel Rise*, Report prepared for the South Pacific Regional Environment Programme (1996).

30. If their cultures are eradicated and their land is taken from them, the skills and knowledge of indigenous peoples will be lost forever. *See* J.H. BODLEY, VICTIMS OF PROGRESS (1990). Industrialisation that has caused climate change is also eradicating people who may be able to help industrialised nations find solutions to many environmental problems, including climate change.

31. V. Shiva, *Conflicts of Global Ecology: Environmental Activism in a Period of Global Reach*, 19 ALTERNATIVES (1994), at 195-207. *See also*, M.A. Hajer, *Ecological Modernisation as Cultural Politics*, in RISK, ENVIRONMENT AND MODERNITY: TOWARDS A NEW ECOLOGY (S. Lash, B. Szerszynski & B. Wynne eds., 1996), at 258; V. Shiva, *Does the New World Order Have Trees?*, 24 HOURS SUPPLEMENT, Feb., 1992, at 39; V. Shiva, *The Greening of the Global Reach*, in GLOBAL ECOLOGY: A NEW ARENA OF POLITICAL CONFLICT (W. Sachs ed., 1993).

32. A. Jamison, *The Shaping of the Global Environmental Agenda: The Role of Non-Governmental Organisations* in RISK, ENVIRONMENT & MODERNITY (Lash et al., eds.), at 224; V. Shiva, *The Greening of the Global Reach*, in GLOBAL VISIONS: BEYOND THE NEW WORLD ORDER (J. Brecher, J.B. Childs & J. Cutler eds., 1993).

on information generated by the 'dominant local.' The issue here is not whether information is correct or not, rather it is about the 'dominant local' and how it seeks global control through the globalisation of environmental problems such as climate change.[34]

There appears to be a problematic dualism in the adjective 'global.' On the one hand, global refers to the scale at which the threat of climate change will manifest itself. In this sense, the adjective is used in its true sense: ". . .worldwide; affecting, or taking into consideration, the whole world or all peoples. . ."[35] However, if the discourse rather than the impacts are examined, then the term 'global' becomes misleading. This is because it refers to a relatively small number of people who have power and influence at a global scale. In climate change discourse:

> . . . the concept of the "global" is not an expression of universal humanism, nor about a planetary consciousness . . . In the dominant discourses, the "global" is the political space in which the dominant local seeks global control and frees itself from local, national, and global control. The "global" does not represent any universal human interest; it represents a particular local and parochial interest that has been globalized through its reach and control.[36]

For people who are marginalised by the dominant discourse, attempts must be made to reconstruct the threat at a scale that is more appropriate to their needs and concerns. This means removing the 'global' from climate change discourse and reconstructing the threat at a local scale. The objective in the context of climate change is to disengage Pacific islanders from a discourse that does not address local concerns. The fact that the biophysical impacts of climate change will be experienced at the global scale has little bearing on the fact that Pacific islanders should find local solutions to problems they will experience at a local scale.

A critique of the notions of a 'global we'[37] and a 'global present'[38] are particularly important if the negative aspects of climate change discourse are to be addressed. The 'global we,' for example, creates the impression that there is an homogeneous group of human beings inhabiting the earth who are essentially equal.[39] The 'global we' says nothing about the distribution of power and the fact that within the

33. *Id.* at 53. *See also*, Hawkins, *supra* note 28, at 227; A. Hurrell, *A Crises of Ecological Viability?, Global Environmental Change and the Nation State*, XLII POLITICAL STUDIES (1994), at 161; D. Slater, *Other Contexts of the Global: A Critical Geopolitics of North-South Relations*, in Kofman & Youngs, *supra* note 26, at 281.

34. Shiva, *supra* note 31.

35. THE CHAMBERS DICTIONARY (C. Schwarz, ed. 1993), at 710.

36. Shiva, *supra* note 31, at 195-196.

37. B. Adam, *Re-Vision: The Centrality of Time for an Ecological Social Science Perspective*, in Lash, *supra* note 32, at 89.

38. *Id.*. Adam uses the phrase 'global present' to refer to the immediacy of information dissemination as a consequence of technological advancement. She suggests that 'global simultaneity' means the local becomes inseparably tied to the global.

'we' is a minority who should bear the most responsibility for environmental degradation. As Esteva and Prakesh suggest, the notion of a 'global we' and the global solutions predicated on this vision are the views of a small number of powerful hegemonic groupings of people who have very narrow views of what is good for humankind.[40] The notion of a 'global we' separates the symptoms of environmental degradation from their causes. Cultural globalisation has meant that cultural phenomena have been detached from the social units from which they arise,[41] and in much the same way, globalisation of environmental problems and the generation of a global consciousness have uncoupled environmental problems from their underlying causes. There are unpleasant ridges and troughs of inequality hidden in the 'global we.'

The notion of a 'global present' is particularly important with regard to the threat of climate change in the Southwest Pacific.[42] The 'global present' is created by the instantaneous or rapid transmission of information. In the past, information may have taken weeks, if not months, to circumnavigate the world; today, it is disseminated almost instantaneously so that people in one country become bystanders, or active participants, in events occurring in other parts of the world.[43] This sense of interconnectedness has both positive and negative implications for people's lives. For instance, the threat of climate change has become part of the 'local present' for many Pacific islanders as scientists in other parts of the world have generated and communicated information about future threats that have engendered fear in the region. It certainly is important for Pacific islanders to use this information in order to prepare for the impacts of climate change. However, the possibility exists that remotely generated information will cause fear and panic that will prevent Pacific islanders from living their 'local present.' This inability to live for today, because the future is purportedly so bleak, will render Pacific islanders more vulnerable to the impacts of climate change. Measures need to be taken today to reduce human vulnerability to biophysical impacts; however, adaptation strategies will not be developed or planned if doomsday scenarios of climate change prevail. The only way out of the conundrum of using remotely generated information which may undermine the ability of people to live their 'local present,' while at the same time preventing it from generating fear, is to construct the threat of climate change in terms that are more relevant to the cultural and geographical setting of the region where biophysical impacts will

---

39. *See* J. Saurin, *International Relations, Social Ecology and the Globalisation of Environmental Change* in THE ENVIRONMENT AND INTERNATIONAL RELATIONS (J. Vogler & M.F. Imber eds., 1996).
40. G. Esteva & M.S. Prakesh, *From Global to Local Thinking*, 24(5) THE ECOLOGIST (1994), at 163.
41. U. Hannerz, *Cultural Complexity: Studies in the Social Organisation of Meaning*, in ENVIRONMENTALISM: THE VIEW FROM ANTHROPOLOGY (K. Milton ed., 1993), at 9.
42. Shiva takes a slightly different view concerning the 'global present.' She believes ignoring the present shifts responsibility away from current polluters so that future polluters, such as India and China, can be implicated as major contributors. This "... creates a moral base for green imperialism." (Shiva, *supra* note 31, at 198).
43. Adam, *supra* note 37; Slater, *supra* note 33, at 227.

be experienced.

4. Think Locally, Act Locally

This chapter ends on a positive note suggesting that by 'thinking locally', Pacific islanders will be in a better position to reduce their vulnerability to the impacts of climate change. As was evident from interviews conducted by the author, many Pacific islanders are angry about the bleak picture commonly painted of their region. In an attempt to ameliorate this negativity, the following ideas are offered as a way to empower people who currently have little control over threats constructed at a global scale.

Perhaps the most important aspect of environmental globalisation is the effect it has on the way people view their position in the world. Global imagery isolates people from their local, 'lived' environments and subsumes them into an amorphous, vacuous space where their actions are controlled by processes over which they have little control.[44] Those who do not operate at a local scale soon become minor actors trapped in ". . . global arenas constructed by global thinking . . ."[45] To combat environmental globalisation:

> . . . active human solidarity is needed more than ever - the solidarity that comes through people who act *and* think locally, forging links to support, in whatever small way they can, the struggles of other local thinkers and actors who share their concerns, outlook and dilemmas. Growing coalitions of local thinkers-activists are learning to counteract the damage of global thinking through a shared rejection of 'global masterplans' and 'universal solutions.'[46]

Coalitions of like-minded people operating at the local scale are a force for change. They have the potential to create a different world,[47] where strength and power are derived from diverse ways of seeing, interpreting and responding to globally pervasive environmental problems. At present, environmental globalisation is being used to create new opportunities for powerful industrialised countries to maintain their current levels of economic activity and extend their political control. In the future, local activists will need to find ways to counter this process by universalis-

---

44. For Ingold, the alienation of people from their 'lived' environment is the cause of many globally pervasive environmental problems. He suggests: ". . .what is perhaps most striking about the contemporary discourse of global environmental change is the immensity of the gulf that divides the world as it is lived and experienced by the practitioners of this discourse, and the world of which they speak under the rubric of 'the globe.' No one, of course, denies the seriousness of the problems they address; there is good reason to believe, however, that many of these problems have their source in the very alienation of humanity from the world of which the notion of the global environment is a conspicuous expression." T. Ingold, *Globes and Spheres: The Topology of Environmentalism*, in Milton, *supra* note 41.
45. Esteva & Prakesh, *From Global to Local Thinking*, *supra* note 40, at 162.
46. *Id.* at 163.
47. Shiva, *The Greening of the Global Reach*, in Brecher, *supra* note 32.

ing values that promote human respect for the environment. Falk believes this can be achieved through a process known as "globalization-from-below." This is characterised by the emergence of new social movements whose agendas are constructed in response to the process of "globalization-from-above."[48]

According to Brecher *et al.*:

> Globalization-from-above extracts resources from the natural world and from local communities in order to increase the wealth and power of the wealthy and powerful. It concentrates that wealth and power in organizations which use their control of people and resources to expand their domination and to fight each other. It transfers power and resources from the natural world to human domination, from communities to elites, and from local societies to national and transnational power centres.[49]

In contrast, globalization-from-below:

> . . . aims to restore to communities the power to nurture their environments; to enhance the access of ordinary people to the resources they need; to democratize local, national, and transnational political institutions; and to impose pacification on conflicting power centres.[50]

Environmental globalisation is globalization-from-above. When problems such as climate change are globalised they take power away from local communities, making them more reliant upon industrialised countries. This chapter has emphasised the fact that Pacific islanders are vulnerable to both biophysical impacts of climate change and problems which result from constructing the threat at a global scale. While Pacific islanders can do little to mitigate climate change, they can reduce their vulnerability to these impacts to ensure they do not become powerless victims of globalisation. Some of the strategies open to Pacific islanders to respond to the threat of climate change are discussed below.

As suggested earlier in this chapter, Pacific islanders need to de-link their present existence from future threats constructed at a global scale. To do this, it is important to separate existence from discourse - if this is possible. Unless Pacific islanders isolate themselves from the discourse of globalisation, they may make decisions today based on perceptions of uncertain threats in the future. The threat of climate change could force a short-term perspective and increase resource exploitation in Pacific island states. People will be less concerned about environmental deterioration and the protection of the land if the impacts of climate change are predicted to render islands uninhabitable.[51] To ensure Pacific islanders con-

48. R. Falk, *The Making of Global Citizenship* in Brecher, *supra* note 32, at 39-50. *See also*, Slater, *supra* note 33, at 284-285.
49. J. Brecher, J.B. Childs & J. Cutler, *Introduction: Globalization-from-Below*, in Brecher, *supra* note 32, at xi.
50. Brecher et al., *Introduction: Globalization-from-Below*, Brecher, *supra* note 32, at xv.

tinue to take care of the environment, Atanraoi believes they must "... live for the present as best . . . they can."[52] People must live a 'local present,' not a 'global present'. While the argument could be made that living for the present will also ensure rapid rates of resource exploitation, the opposite could also be true if people have a positive view of the future.

The globalisation of climate change discourse means a great deal of attention is being focused on the Alliance of Small Island States (AOSIS)[53] as the best (perhaps only) way for Pacific islanders (and citizens of other low-lying island states) to respond to the threat of climate change.[54] While AOSIS has an important role to play, it is operating under a pre-determined agenda and will only be successful if politicians in industrialised countries are willing to set stringent and legally binding greenhouse gas emission targets. AOSIS is unlikely to have the power or influence to enforce the changes it feels are required in industrialised nations to reduce the severity of potential climate change impacts in its member states. This does not mean AOSIS does not have an important role to play. Rather, instead of relying solely on one organisation operating at the international scale, local level initiatives should be implemented to reduce vulnerability to climate change impacts.[55] Local

51. Anecdotal evidence collected by the author suggests that some islanders are less concerned about the environmental impacts of their actions because they believe their islands will be rendered uninhabitable by sea-level rise and other impacts of climate change.

52. P. Atanraoi, *Customary Tenure and Sustainability in an Atoll Nation: The Case of Kiribati*, in CUSTOMARY LAND TENURE AND SUSTAINABLE DEVELOPMENT: COMPLEMENTARITY OR CONFLICT, South Pacific Commission and Suva: Institute of Pacific Studies, University of the South Pacific, in (R.G. Crocombe ed., 1995), at 72.

53. The Alliance of Small Island States (AOSIS) came into existence in 1990 at the second World Climate Conference. According to Ambassador Tuiloma Neroni Slade: "It has since become established as an ad hoc lobby and negotiating voice for small island developing states (SIDS) within the UN system." AOSIS aims to find solutions to a number of shared environmental and development problems that threaten human security in small island developing states. To date, AOSIS has focused most of its attention on ways to reduce the severity of the threat of climate change to its member states - it has proposed a 20 per cent cut in CO_2 emissions from industrialised countries by 2005. In the future, AOSIS is set to focus increased attention on measures to reduce resource degradation and promote sustainable development in its member states. Address by Ambassador Tuiloma Neroni Slade, Chairman of the Alliance of Small Island States, Caribbean Ministerial Meeting on the Implementation of the Barbados Programme of Action for the Sustainable Development of Small Island States, Barbados, 10-14 November, 1997.

54. As of of April 1994, the following states comprised AOSIS:
Atlantic: Cape Verde; Guinea-Bissau; Sao Tome and Principe.
Caribbean: Antigua and Barbuda; Bahamas; Barbados; Belize; Cuba; Dominica; Grenada; Guyana; Jamaica; St. Kitts and Nevis; St. Lucia; St. Vincent and the Grenadines; Suriname; Trinidad and Tobago.
Indian Ocean: Comoros; Maldives; Mauritius; Seychelles.
Mediterranean: Cyprus; Malta.
Pacific: Cook Islands; Federated States of Micronesia; Fiji; Kiribati; Marshall Islands; Nauru; Papua New Guinea; Samoa; Solomon Islands; Tonga; Tuvalu; Vanuatu.
South China Sea: Singapore.
Observers: American Samoa; Guam; Netherlands Antilles; Niue; US Virgin Islands.

scale action will provide a means to counter the globalising discourse of climate change by providing "... space for indigenous strategies for self-reliance and empowerment."[56]

5. Conclusion

To conclude this chapter, four areas of research are identified that will help reduce the extent to which climate change will undermine security in the Southwest Pacific. This research needs to be organised and implemented, as far as possible, by Pacific islanders; if this is not possible, research must reflect their concerns.

First, despite this critique of environmental globalisation, international negotiations on the issue of climate change are desirable and must continue as long as they do not solely reflect the concerns of scientists, politicians and proponents of economic development in industrialised nations. Research needs to look at ways of raising the concerns of poor and marginalised people at the international scale and specifically, with regard to the Southwest Pacific, ways to assist Pacific island representatives to be more effective at voicing parochial concerns and forcing action to address them. At present, Pacific islanders have adopted an approach which emphasises their vulnerability to climate change impacts in the hope that other countries, motivated by altruism, will change their behaviour. This approach will have very limited success. This can be illustrated by events that recently took place at the South Pacific Forum in the Cook Islands. The Australian Prime Minister, John Howard, refused to sign a Forum communiqué which specified the need for legally binding greenhouse gas emission targets.[57] Despite the fact that Tuvalu's Prime Minister, Bikenibeu Paeniu, stated that his islands may no longer exist above sea-level if industrialised countries do not curb their emissions of greenhouse gases, John Howard refused to sign anything that had implications for Australia's fossil fuel dependent economy.[58] In situations where politicians from industrialised nations consciously place a lower value on the lives of people who inhabit low-lying island states, compared to the higher value they place on the economies of their countries, attempts by Pacific island leaders to force action on the basis of vulnerability are destined to fail.[59] To be fair, it is difficult to see how politicians from island states can approach

---

55. Local scale initiatives should use the knowledge and skills of the indigenous people who know their islands best. According to Trask: "No one knows how better to care for . . . our island home, than those of us who have lived here for thousands of years. On the other side of the world from us, no people understand the desert better than those who inhabit her do. And so on, throughout the magnificently varied places of the earth." H.K. Trask, *Malama 'Aina: Take Care of the Land*, in Brecher, *supra* note 32, at 128.

56. T. Skelton, *Globalization, Culture and Land: The Case of the Caribbean*, in Kofman & Youngs, *supra* note 26, at 326.

57. J. Short, *Howard Splits Forum over Greenhouse*, THE WEEKEND AUSTRALIAN, Sept. 20-21, 1997, at 2.

58. S. McKenzie, *PM Rejects Pacific Claims on Global Warming*, THE DAILY TELEGRAPH, Sept. 19, 1997, at 17.

the threat of climate change differently. Nevertheless, they need to find ways to influence decision making at the international level without marginalising pressing local and regional issues. AOSIS could be instrumental in this process.[60] Research, therefore, needs to explore ways of empowering Pacific island representatives so that they become more effective bargainers at the international level without becoming active participants in environmental globalisation. If action by industrialised nations is not forthcoming as a result of lobbying, the countries comprising AOSIS could investigate ways to use their combined voting power at international meetings in order to force the industrialised world to take action to address the global warming threat.[61]

Second, there needs to be more research into ways to adapt to climate change impacts in the Southwest Pacific. Although Hay identifies both mitigation and adaptation as possible response strategies in island states,[62] Pacific islanders can do little to mitigate the problem. Adaptation strategies on the other hand provide more opportunities to respond to the threat of climate change.[63] Importantly, they focus attention at the local scale and can be used to reduce dependence on industrialised countries. They also comprise 'no regrets' policies; that are ". . . the same as those which constitute sound environmental management . . ." and will be ". . . beneficial even if the climate does not change as anticipated."[64]

Third, it is necessary to look at how Pacific islanders have adapted to climate-related disasters in the past. Although the warming predicted by the Intergovern-

---

59. In 1996, the Director of the Australian Bureau of Agricultural and Resource Economics, Dr. Fisher, was reported as having suggested that an evacuation of small island states might be more efficient (cheaper) than forcing industrialised countries to cut greenhouse gas emissions. N. Bita, *Island Evacuation a Greenhouse Solution,* THE WEEKEND AUSTRALIAN, June 8-9, 1996, at 24. Fisher was obviously using cost-benefit analysis to decide the fate of Pacific islanders; he was clearly placing a lower value on the lives and cultures of inhabitants of the island states of the Southwest Pacific, compared to the higher value he was placing on the economies of industrialised nations. Economists of Working Group 3 of the Intergovernmental Panel on Climate Change (IPCC) have also been criticised for work they have carried out that places a higher value on the lives of people in developed countries compared to the lives of people in developing countries. C. Raghavan, *Climate Body Economists Asked to Redo their Work*, South-North Development Monitor, Third World Network (1995).

60. *See* R. Taplin, *International Policy on the Greenhouse Effect and the Island South Pacific*, 7(3) PAC. REV. (1994), at 271-281.

61. According to Ambassador Tuiloma Neroni Slade: "Thirty-six AOSIS countries are member States of the UN, or almost 20 per cent of the UN membership." This suggests that the combined voting power of AOSIS members could be used to influence international policies in favour of small island developing states. Address by Ambassador Tuiloma Neroni Slade, Caribbean Ministerial Meeting on the Implementation of the Barbados Programme of Action for the Sustainable Development of Small Island States.

62. Hay, *supra* note 29, at 72.

63. Research currently being undertaken by the author suggests that agricultural practices, building techniques and land tenure systems could all be adapted to reduce the vulnerability of Pacific islanders to climate change. *See also*, Hay, *supra* note 29.

64. Hay, *supra* note 29, at 72.

mental Panel on Climate Change (IPCC) would be greater than any seen in the past 10,000 years,[65] Pacific islanders have survived climate-related threats in the past and the cultures of the Pacific embody centuries of adaptation to a wide range of environmental (including climatic) factors.[66] In the future, more attention needs to be focused on identifying the nature of these traditional adaptation strategies and how modern strategies can be used to complement rather then undermine them.[67] However, it is important to emphasise that the ability to adapt is not an excuse for the neoliberal factions of industrialised countries to do nothing to reduce greenhouse gas emissions; rather it is a message to researchers, island governments and non-governmental organisations to concentrate their efforts on the identification of local response strategies to climate change. If more is known about how Pacific islanders addressed climate and environmental threats in the past, it may be possible to build robust adaptation strategies to counteract the impacts of climate change in the future. As Thaman points out, Pacific islanders will be better placed to respond to natural disasters (including climate change related disasters) if they "... return to effective traditional means of dealing with periodic disasters, such as shifts in food-system emphasis, activating intra- or inter-island trade networks, and encouraging self-help activities in general . . ."[68] Traditional systems are frequently more sustainable and will also allow Pacific islanders a greater degree of control over their own affairs.

Finally, research needs to consider ways of incorporating traditional skills and knowledge into the formal education system of Pacific island states. The knowledge that has helped Pacific islanders protect the cultural and environmental diversity of their island nations should be considered equal to, if not more important than, Western knowledge. Local knowledge, accumulated "... by breathing air, drinking water, tilling soil, harvesting forest produce, or fishing rivers, lakes and oceans"[69] will be an invaluable resource for designing and implementing strategies to respond to climate change.

---

65. Intergovernmental Panel on Climate Change, *The Science of Climate Change: Summary for Policymakers and Technical Summary of the Working Group 1 Report*, World Meteorological Organisation; United Nations Environment Programme, 1996, at 11.

66. R.C. Kiste, *Pre-Colonial Times*, in TIDES OF HISTORY (K.E. Howe, R.C. Kiste & B.V. Lal eds., 1994).

67. R. Thaman, *Environmental Issues in the Pacific Islands: Constraints to Sustainable Island Ecodevelopment,* 1 PAC. ISSUES (1988). *See also* J. Hay, *Regional Assessment of the Vulnerability and Resilience of Pacific Islands*, at 68.

68. Thaman, *supra* note 67, at 49.

69. S. Breyman, *Knowledge as Power: Ecology Movements and Global Environmental Problems*, in Lipschutz & Conca, *supra* note 28, at 131.

14. CLIMATE CHANGE IN THE PACIFIC: SCIENCE-BASED INFORMATION AND UNDERSTANDING

JOHN HAY
Woodward-Clyde Professor of Environmental Science
School of Environmental and Marine Sciences
Room 733.319, Tamaki Campus, Private Bag 92019
University of Auckland, New Zealand

1. Introduction

Regional and national manifestations of global climate change and accelerated sea level rise are a continuing and growing concern of the small island developing states (SIDS) of the Pacific.[1] For the Pacific island states, this concern is evinced in the technical reports prepared by the South Pacific Regional Environment Programme (SPREP)[2] and other regional intergovernmental organizations, and from the communiques issued by the South Pacific Forum and other political bodies.

Pittock[3] notes that, since impacts of climate change manifest themselves at a local or regional level, they will not be correctly anticipated by considering only globally averaged changes in the atmosphere or marine climates. For some variables, it may be that globally averaged changes may well be the antithesis of those occurring in a specific location or region. Pittock goes on to identify several climate change related issues that are unique to the South Pacific region. These include:

- tropical cyclones, which cannot be adequately simulated in global climate models as they are sub-grid scale;
- the El Niño-Southern Oscillation phenomenon which is a major source of atmospheric and oceanic variability in the South Pacific, but is not well modelled by the current generation of global climate models;

1. L.A. Nurse et al., *Small island states*, in THE REGIONAL IMPACTS OF CLIMATE CHANGE: AN ASSESSMENT OF VULNERABILITY (R. Watson, M. Zinyowera, R. Moss and D. Dokken eds., 1998), Intergovernmental Panel on Climate Change, at 331-354.
2. C. Kaluwin & J.E. Hay, 1998: *Climate Change and Sea Level Rise in the South Pacific Region, Proceedings of the Third SPREP Meeting*, Noumea, New Caledonia, August, 1997, South Pacific Regional Environment Programme (SPREP), Apia, Western Samoa.
3. A.B. Pittock , *Regional climate change scenarios for the South Pacific*, in Climate Change and Sea Level Rise in the South Pacific Region: Proceedings of the Second SPREP Meeting, (J. Jay & C. Kaluwin eds., 1993), at 50 –57.

A. Gillespie and W. C. G. Burns (eds.), Climate Change in the South Pacific: Impacts and Responses in Australia, New Zealand, and Small Island States, 269–287.

- variations in the Intertropical and South Pacific Convergence Zones which have a major influence on the seasonal and interannual patterns of rainfall in the region, but for which many global models are incapable of reproducing these variations under present conditions, let alone those associated with elevated levels of greenhouse gases;
- in addition to temporal variations in global mean sea level, relative sea level, which varies locally due to tectonic movements, changes in ocean currents, and regional variations in sea water density due to salinity and temperature changes;
- impacts on coastal zones which depend critically on any local changes in ocean currents and wave climatologies;
- regional and local variations in sea surface temperatures which have large potential effects on fisheries and on coral reef ecosystems;
- the sensitivity to changes in carbon dioxide and stratospheric ozone concentrations of natural and managed ecosystems of importance to the South Pacific, both terrestrial and marine, which vary locally and regionally; and
- changes in the magnitude and frequency of extreme or threshold events that produce significant impacts; since such changes are a composite effect of the above factors, they too will be specific to a given region or locality.

While awaiting the development of global climate models with appropriately increased spatial resolution, Pittock[4] argues that high resolution, limited area models offer a useful interim approach, especially when they are driven by or nested within global models. Various statistical approaches can also be used to interpolate to spatial scales finer than those of the current grid spacings in global climate models.

Thus, even now, the capability exists to prepare climate change and sea-level rise scenarios with an appropriate regional focus and spatial resolution that is tailored to the needs and priorities of the countries concerned. Pittock[5] warns that, until this is done, there is a danger that planning and other decisions will be based on inappropriate, poorly substantiated and possibly outdated scenarios

Despite the serious limitations of current global climate models for prediction of changes in regional climate, some benefits arise from studying the results of recent efforts. What follows is a review of progress in our science-based understanding

4. A.B. Pittock, *Developing regional climate change scenarios for the South Pacific*, 12 WEATHER & CLIMATE 17-31 (1992).
5. A.B. Pittock, *Climate scenario development*, in MODELLING CHANGE IN ENVIRONMENTAL SYSTEMS, (A.J. Wakeman, M.B. Beck & M.J. McAleer, eds. 1993), at 481-503.

of variations and changes in climate and sea level for the Pacific region and an attempt to identify gaps in the relevant information and knowledge bases. The approach taken is to use examples to illustrate the extent of our understanding of past climates (both atmospheric and oceanic), with reference to instrumental, proxy, interpreted and anecdotal records. Present understanding is also considered by way of examples, with emphasis on processes, fluxes and exchanges. Finally, our science-based extrapolation of future climates and sea levels is assessed using examples of both the methods employed by scientists, and the predictions that result.

While it is not possible in a contribution of this length to discuss every relevant topic and every pertinent study, the approach adopted here is to examine a representative cross-section of key studies that have been conducted in the past decade, and to assess the extent to which they contribute to our understanding of climate change in the Pacific.

2. Past Climates - Both Atmospheric and Oceanic

2.1 INSTRUMENTAL RECORD

The instrumental record for the Pacific is notable for its relative shortness and poor geographical coverage. As Salinger and Collen[6] note, the Apia (Western Samoa) record which dates from 1890 is one of the longest for any island in the South Pacific, and it is unreliable before 1902. Analyses of the instrumental records for both temperature and rainfall[7] show that the South Pacific Convergence Zone (SPCZ) is a key pivotal zone in terms of trends and variability of both temperature and rainfall. This century all areas of the South Pacific appear to have warmed at about 0.2 C per decade. South of the SPCZ, the increase has been steady, but north of the SPCZ there has been a rapid increase since the 1970s, after cooling from 1940s. This may well be related to changes in temporal patterns of the El Niño/ Southern Oscillation (ENSO), since to the north of the SPCZ warm/wet anomalies occur in El Niño years while areas to the south experience cool/dry anomalies. ENSO is one of the major sources of interannual variability in both atmospheric and marine climates of the Pacific. It manifests itself as persistent and large-scale anomalies in ocean currents and surface temperatures, and in wind patterns and convective activity.[8]

6. M.J. Salinger & B. Collen, *Climate trends in the South Pacific*, in CLIMATE CHANGE AND SEA LEVEL RISE IN THE SOUTH PACIFIC REGION, PROCEEDINGS OF THE SECOND SPREP MEETING (J.E. Hay & C. Kaluwin eds., 1993), South Pacific Regional Environment Programme (SPREP), Apia, Western Samoa, at 34-46.
7. M.J Salinger et al., *Climate trends in the Southwest Pacific*, 15 INT'L J. CLIMATOLOGY (1995), at 285-302; J.E. Hay, J. Salinger, B. Fitzharris & R. Basher, *Climatological seesaws in the Southwest Pacific*, 13 WEATHER & CLIMATE (1993), at 9-21.
8. UNEP, *The El Nino phenomenon*, United Nations Environment Programme (UNEP), Global Environmental Monitoring System (GEMS), UNEP/GEMS Environment Library No. 8, UNEP (1992).

The instrumental record of oceanic variability shows, among other things, that sea level changes are strongly influenced by local land movement. For tectonically stable areas, a regionally representative sea level increase is about 2 mm per year, averaged over the past 50 years. There is some evidence that regional sea level change in the central Pacific is decoupled from global changes. This needs close examination, given the implications of a proven decoupling. Any decoupling may be associated with ENSO, as large interannual variations in sea level occur as a consequence of this phenomenon.[9]

2.2 PROXY DATA

Proxy data are indirect sources of information – that is, information that is not based on direct measurement, but is inferred from analyses of other phenomena. In the Pacific, the proxy record takes on added significance due to both the short duration of instrumental records and to the sparseness of the current, conventional monitoring networks. Relevant sources of proxy data for the Pacific include tree rings, sediment cores and micro-atolls. The information they yield has been verified through cross- referencing to the instrumental record and to other proxy records. These sources have provided invaluable data, especially with respect to prehistoric changes in climate and other environmental conditions. For example, for Niuatopatapu (Tonga) the rate of tectonic uplift is estimated to be 5 mm per year,[10] far in excess of the rise in regional sea level (estimated to be 1 mm per year by Wyrtki.[11] Kirch[12] concludes that, consequently, the island has increased in area by 60% since initial settlement some 3000 years BP.

2.3 INTERPRETED RECORD

Interpretations of instrumental and other records will be discussed in this section in the context of ENSO, tropical cyclones and coastal changes.

In the Pacific, ENSO is the primary factor in climate variability on the 2-5 year time scale. Release of latent heat associated with El Niño episodes affects global temperature[13] and associated changes in oceanic upwelling influence atmospheric CO_2 levels.[14] The IPCC[15] notes that the behaviour of the ENSO has been unusual since the mid-1970s and especially since 1989. Since the mid-1970s warm episodes (El Niño) have been relatively more frequent or persistent than the opposite

9. S. G. Rintoul et al., *Ocean processes, climate and sea level*, IN GREENHOUSE: COPING WITH CLIMATE CHANGE (W.J. Bouma, G.I. Pearman & M.R. Manning eds., 1996), at 127-144; M.J. Salinger et al., *Observed variability and change in climate and sea level in Australia, New Zealand and the South Pacific*, in GREENHOUSE: COPING WITH CLIMATE CHANGE, (W.J. Bouma, G.I. Pearman and M.R. Manning, eds. 1996), at 100-127.

10. F.W.Taylor et al., *Coseismic and Quaternary vertical tectonic movements, Santo and Malekula Islands, New Hebrides island arc*, 85 J. GEOPHYSICAL RESEARCH (1980), at 5367-5381.

11. K. Wyrtki, *Sea level rise: the facts and the future*, 44 PAC. SCI. (1990), at 1-16.

12. P.V. Kirch, *The Lapitoid period in west Polynesia: excavations and survey in Niuatoputapu, Tonga*, 5 J. FIELD ARCHAEOLOGY (1978), at 1-13.

13. N.E. Graham, *Simulation of recent global temperature trends*, 267 SCI. 666-671 (1995).

phase (La Niña). The recent ENSO behaviour, and especially the consistent negative values of the Southern Oscillation Index[16] since 1989, appears unusual in the context of the instrumental record that spans the past 120 years. However, Allan and D'Arrigo[17] present evidence that suggests that such unusual patterns have occurred prior to the period of instrumental record.

Recent variations in precipitation over the tropical Pacific Ocean and surrounding land areas are related to this behaviour in ENSO. It has also affected the pattern and magnitude of surface temperatures. Graham[18] used satellite data, a course Tropical Pacific Ocean data set and numerical modelling based on observed sea surface temperatures, to demonstrate that tropical Pacific precipitation had increased in conjunction with the increased frequency of El Niño in the mid 1970s. Salinger et al.[19] analysed trends in precipitation observed at tropical Pacific atolls and islands. They found that central and eastern equatorial Pacific rainfall increased during the mid-1970s while that in the southwest Pacific decreased. These findings are consistent with linkages between spatial distribution in tropical Pacific rainfall anomalies and ENSO events.[20]

Since 1979, warm season sea surface temperature anomalies of 1 C or more have been associated with coral bleaching events in tropical areas. Bleaching was more prevalent in 1983, 1987 and 1991, coinciding with El Niño (warm) events, but was scarce in 1992, as a result of the cooling following the eruption of Mt. Pinatubo. Glynn[21] suggests that the extent of bleaching experienced since 1979 has not been observed previously.

In the southwest and southeast Pacific, the number of tropical cyclones appears to have increased.[22] In the latter case, this may be related to the increased frequency of El Niño events. However, such conclusions must be qualified in light of the quality (especially the lack of consistency) of the long-term cyclone database.[23]

---

14. C.D. Keeling, et al., *A three-dimensional model of atmospheric CO2 transport based on observed winds: Analysis of observational data*, in ASPECTS OF CLIMATE VARIABILITY IN THE PACIFIC AND THE WESTERN AMERICAS, (D.H. Peterson ed., 1989), at 55, 165-236
15. IPCC, *Climate Change 1995 – The Science of Climate Change: Contribution of Working Group I to the Second Assessment Report of the Intergovernmental Panel on Climate Change*, (J.J. Houghton eds., 1996).
16. J.E. Hay, J. Salinger, B. Fitzharris & R. Basher, *Climatological Seesaws in the Southwest Pacific*, 13 WEATHER & CLIMATE 9-21 (1993).
17. R.J. Allan & R.D. D'Arrigo, R.D., *'Persistent' ENSO Sequences: How unusual was the 1990-1995 El Nino?*, 9 HOLOCENE 101-118 (1999).
18. Graham, *supra* note 13.
19. Salinger, et al., *supra* note 7.
20. Hay, et al., *supra* note 7.
21. P.W. Glynn, *Coral reef bleaching: Ecological perspectives*, 12 CORAL REEFS 1-17 (1993).
22. C. Thompson, S. Ready & X. Zheng, *Tropical Cyclones in the Southwest Pacific: November 1979 to May 1989*, New Zealand Meteorological Service (1992).

Tropical cyclones can either add to island size by depositing lagoon sediment or ocean rubble onto land, or decrease it by eroding coastal landforms. For example, in 1972, Hurricane Bebe deposited a 19 km long bank of boulders up to 4 m high on the ocean side of Funafuti Atoll, Tuvalu. This increased the island size by about 20%.[24] Had a second hurricane occurred before sediment was replenished offshore, it is likely that the large waves would have been erosive as there would be little additional material to deposit and wave energy would not be dissipated by such sediment transport. Thus the frequency of tropical cyclones, a variable that might change with global warming, can determine whether such an extreme event will add to, or reduce, an island's land mass.

Many Pacific islands have long coastlines per unit area of land, making even small changes in the coastline of considerable significance. This is particularly so given the cultural importance of land to most island societies. The shorelines of the true atoll motu (sand islands) are especially dynamic, even in the absence of climate change. For example, since 1945 the west and northwest coasts of Betio, the western islet of South Tarawa, Kiribati, has undergone a maximum of 75-100 m of accretion while the southwest corner has receded by 40-50 m.[25] Should global warming bring about systematic changes in atmospheric and nearshore circulation patterns it is highly likely that atoll shorelines would become even more dynamic. Importantly, Cowell et al.[26] show that the processes most important for beach erosion may not be those directly associated with mean sea-level rise.

2.4 ANECDOTAL RECORD

The dearth of long-term and spatially-detailed climatic data in the Pacific also leads to greater emphasis being placed on anecdotal sources of information – material gleaned from oral and written records. Despite its typically more qualitative and subjective nature, such information often helps fill a data void and hence contributes to understanding. An example is provided by Nunn[27] who undertook a study designed to demonstrate changes in sea level since 1900, to record and analyse important environmental changes associated with recent sea level changes and

23. S.C.B. Raper, *Observational data on the relationships between climatic change and the frequency and magnitude of severe tropical storms*, in: CLIMATE AND SEA LEVEL CHANGE: OBSERVATIONS, PROJECTIONS AND IMPLICATIONS (R.A Warrick, E.M. Barrow & T.M.L. Wigley eds., 1993), at 192-212.
24. G.B.K. Baines & R.F. McLean, *Sequential studies of hurricane deposit evolutiona Funafuti Atoll*, 21 MAR. GEOLOGY (1976), at 1-8.
25. R. Howarth & B. Radke, *Investigation of historical evidence for shoreline changes: Betio, Tarawa Atoll, Kiribati; and Fongafale, Funafuti Atoll, Tuvalu*. Workshop on Coastal Processes in the South Pacific Island Nations, 7 SOPAC TECHNICAL BULLETIN (1991), at 91-98.
26. P.J. Cowell, P.S. Roy, T.Q. Zeng & B.G. Thom, *Practical relationships for predicting coastal geomorphic impacts of climate change* in PROCEEDINGS OF OCEAN AND ATMOSPHERE PACIFIC, (T.H. Aung ed., 1996), National Tidal Facility, The Flinders University of South Australia, at 16-21.
27. P. Nunn, *Geomorphological evidence of climate change in the Pacific*, in, Proceedings of the Ocean and Atmosphere Pacific International Conference (T. Aung ed., 1996) National Tidal Facility, The Flinders University of South Australia, at 265-275.

to summarize likely future impacts of global warming on coastal areas in the South Pacific. The basic data for the study were obtained by students who questioned elderly inhabitants of coastal settlements in the Cook Islands, Fiji, Kiribati, the Solomon Islands, Tonga, Tuvalu, Vanuatu and Western Samoa in order to obtain information about recent changes in sea level and other indicators of environmental quality and variability. Part of the interview encouraged elderly interviewees to indicate the approximate location of low tide in their youth. The researchers then measured the horizontal and vertical distances between the old and present low tide marks. Results for 48 tectonically-stable coastal settlements from the Solomon Islands to the Cook Islands show that shoreline inundation has occurred at the rate of around 10 cm/year.

Nunn acknowledges the imprecision of the information obtained through such procedures. He argues that discussion of the results based on national and regional averages will minimize the effects of any errors while still providing a generalized picture of recent shoreline changes in the Pacific.

2.5 DISCUSSION

The foregoing examples are intended to demonstrate the deeper and more comprehensive understanding that comes from interpretation of composite records - those based on a combination of instrumental, proxy and anecdotal information. Such an approach is especially beneficial in the Pacific where the extent of data of any one kind is often very limited.

3. Present Climates

Studies undertaken by the Centre ORSTOM (L' Institut Français de Recherche Scientifique pour le Developpement en Cooperation) in Noumea and Tahiti provide examples of comprehensive and integrated investigations of climate change related issues in the Pacific. Motivation for much of the research comes from the common understanding that irregularities in the Earth's climate are mainly due to interactions between tropical oceans and the global atmosphere, and particularly to the tropical Pacific El Niño Southern Oscillation (ENSO) phenomenon. The purpose of the research, conducted mainly in the context of the international TOGA (tropical ocean and global atmosphere) programme and its COARE (coupled ocean-atmosphere response experiment) sub-programme in the western Pacific, is to observe, study, understand, model and ultimately forecast these climatic variations.

For example, investigations are being conducted into the variations in planktonic biomass as a consequence of water mass changes. These changes are associated with interannual variations in the equatorial current system and the close coupling between the ocean and the atmosphere. The data set gathered since 1984 includes both El Niño and La Niña events. It shows a common response of chemical and biological parameters to variations in water mass composition along the

165 E meridian between 10 N and 20 S.

In turn, the observed variations in planktonic production have an impact on the structure of the pelagic ecosystem. System efficiency - measured with respect to fish production and the amount of matter (mainly carbon) being transferred to the deeper layers of the ocean - are two major consequences of the changes in ecosystem structure.

ORSTOM has also hosted several scientific meetings of relevance to climate change studies in the Pacific. For example, in 1995 international scientists gathered to assess the results of their investigations into the carbon cycle of the equatorial Pacific. This body of ocean is thought to play a major role in two aspects of the global carbon cycle: through the flux of CO_2 to the atmosphere and through the export of organic carbon to the deep ocean. New data obtained recently as part of the Joint Global Oceans Flux Study (JGOFS) compares the distributions and fluxes of selected parameters for 140 W and 155 - 165 E. The variables studied include winds, ocean temperatures, ocean currents, CO_2 flux, carbon fluxes and the food web. The results show that the boundary between the central and western equatorial Pacific occurs almost as a discontinuity, or step function, at about the international dateline. Primary, new and export production are all normally lower west of the dateline. The gradient of CO_2 across the air-sea interface is smaller and the food web is more oligotrophic (i.e., low in nutrients) in the west. These chemical and biological differences are largely the result of differences in physical forcing. The central equatorial Pacific is dominated by local winds that favour upwelling – the movement of deeper ocean water towards the surface. The nutrient content of the upwelled water, however, depends on remote westerly winds in the western Pacific. In the western equatorial Pacific winds are weaker and more variable. A low salinity surface layer lies over an isothermal barrier layer that blocks the vertical transport of nutrients, thereby making the near-surface waters less biological productive.

4. Future Climates

This section considers the "future" in terms of both the methods for predictive assessments and characterizations, and the information that comes from the explicit application of those methods. Methods which will be considered are modelling, including global climate models (GCMs), nested regional models and integrated impact assessment models, scenario development, vulnerability assessment, and, finally, "speculation." Results arising from the application of some of these methods will be described and discussed.

4.1 METHODS

4.1.1 *Modelling*

Global climate models use the fundamental thermodynamic equations to simulate the temporal evolution of climate (atmospheric or, in the case of coupled models, both atmospheric and oceanic). They can now provide useful predictions of climate at both global and continental scales, but little confidence can be placed in predictions at sub-continental scales.[28] Pittock et al.[29] and Whetton et al.[30] have reviewed the evolution of approaches being taken to improve the reliability of predictions at scales of relevance in the South Pacific.

Pittock et al.[31] highlight the uncertainty in characterizing present day rainfall in oceanic regions, let alone making predictions about changes in rainfall regimes as a consequence of global warming. There are significant differences between two observed data sets for the South Pacific.[32] However, both show high annual rainfall, associated with the South Pacific Convergence Zone (SPCZ), stretching from New Guinea across the Pacific towards the southeast.

The predictive power of a GCM is indicated, in part, by its ability to simulate critical features of the present climate – in so called "control runs." In such a control run of the CSIRO Mark 2 GCM (a global climate model developed by Australian scientists), the SPCZ is aligned more east-west than it is for the observations. Such a result is common to many other GCMs. This produces errors in the estimated precipitation for many island nations under present conditions, and brings into question the utility of GCM-based predictions of changes in rainfall associated with global warming.

Despite the serious constraints on the current use of global climate models for prediction of changes in regional climate, some benefits arise from studying the results of recent efforts. Zillman et al.[33] present anticipated changes in temperature and rainfall for the Pacific region. These are based on the work of McAvaney et al. (1992).[34] They determined the seasonal climate response to an effective doubling

---

28. IPCC, *Climate Change 1995 - The Science of Climate Change: Contribution of Working Group I to the Second Assessment Report of the Intergovernmental Panel on Climate Change* (J.J. Houghton et al., eds. 1996).
29. A.B. Pittock, et al., *Progress towards climate change scenarios for the southwest Pacific*, 15 WEATHER & CLIMATE 21-46 (1995).
30. P.H. Whetton et al., *Global comparison of the regional rainfall results of enhanced greenhouse coupled and mixed layer ocean experiments: implications for climate change scenario development*, 33 CLIMATE CHANGE (1996), at 497-519.
31. Pittock, *supra* note 29.
32. L. Jaeger, *Monthly and areal patterns of mean global precipitation*, in VARIATIONS IN THE GLOBAL WATER BUDGET (A. Street-Perrott, M. Beran & R. Ratcliffe eds., 1983), at 129-139; D.R. Legates & C.J. Willmott 1990, Mean seasonal and spatial variability in gauge corrected global precipitation, 19 INT'L J. CLIMATOLOGY 111-127 (1993).
33. J.W. Zillman, et al., *Current prognoses of climate change and its effects in the Pacific region, Symposium on Sustainable Development: Energy and Mineral Resources in the Circum-Pacific Region and the Environmental Impact of their Utilization* (1992).

of CO_2 in an equilibrium simulation using the Australian Bureau of Meteorology Research Centre's (BMRC) atmospheric general circulation model coupled with a mixed layer ocean model. The ocean hemisphere exhibits minor sensitivity to an effective doubling of CO_2 relative to adjacent continental areas, the sensitivity generally increasing with latitude and in winter relative to summer. One notable response in the Pacific region is a cooling of up to 1.5 C over northern Australia in the southern summer, with a general increase in rainfall over the Timor Sea. While changes in seasonal rainfall produce somewhat complex patterns, the annual pattern is much simpler - mean annual rainfall is generally slightly higher for the Pacific, except for mid-latitude areas.

Increased confidence may be placed in predictions produced by a global climate model if these are in agreement with those of other models. Nevertheless, this confirmation can never be absolute, as the various models have much in common. Nevertheless, it is educational to undertake such comparative analyses. For 14 representative locations over the Pacific region Zillman et al.[35] compared the BMRC model results (outlined above) with those for comparable experiments using four other models (Canadian Climate Centre, Geophysical Fluid Dynamics Laboratory, UK Meteorological Office and CSIRO). For the South Pacific, the BMRC estimates of increase in seasonal temperature are consistently lower than those provided by the other models, typically by at least one degree. For precipitation, the BMRC model is also somewhat more conservative, at least for the South Pacific locations that are analysed. There is no consensus amongst the models as to whether precipitation will increase or decrease at a given location.

On the other hand, there is more consistency in model simulations of heavy rainfall intensity. A general increase is indicated.[36] Results indicate that heavy rainfall events with a return period of at least a year will occur 2-4 times more often. In the UKMO and CSIRO9 GCMs, there is an approximate 50% increase in the frequency of rainfall exceeding the $1xCO_2$ 90th percentile in the southwest Pacific. Despite the greater confidence associated with these findings, practical constraints will likely limit the usefulness of approaches other than high resolution, nested modelling, such as undertaken recently by Hennessy et al.[37]

Recent variations in precipitation over the tropical Pacific Ocean and surrounding land areas are related to the fact that since the mid-1970s warm episodes (El Niño) have been relatively more frequent or persistent that the opposite phase (La

---

34. B.J. McAvaney, R.A. Colman, J.R. Fraser & R.R. Dahni, *The seasonal climatic response to a doubling of CO_2: BMRC control model results*, in MODELLING WEATHER & CLIMATE, Papers presented at the Third BMRC Workshop, November 1991, Australian Bureau of Meteorology Research Centre Research Report No. 33 (1992), at 401-20.
35. Zillman, *supra* note 33.
36. A.M. Fowler & K.J. Hennessy, *Potential impacts of global warming on the frequency and magnitude of heavy precipitation*, 11(3) NATURAL HAZARDS 283-303 (1995).
37. K.J. Hennessy, J.M. Gregory & J.F.B. Mitchell, *Changes in daily precipitation under enhanced greenhouse conditions*, 13 CLIMATE DYNAMICS 680 (1997).

Niña).[38] The El Niño Southern Oscillation (ENSO) phenomenon is the primary source of climate variability on the 2-5 year time scale. At present, the weight of evidence (from both observations and global GCM studies) is that there will be no significant changes to the amplitude or frequency of ENSO in the future. Thus, the current large interannual variability in rainfall associated with ENSO is likely to dominate over any effects attributable to global warming.

Integrated impact assessment models (CLIMPACTS and OzClim)[39] are in their infancy, but their development reflects the need to combine three aspects of climate change assessments:

- the sequence of changes, initially in atmospheric composition, but leading to subsequent changes in climate, biophysical systems and socioeconomic status;
- the linkages between systems, sectors and activities; and
- the interactions with system responses that are not climate induced.[40]

The predictive abilities of existing models are limited by the simplistic nature of the underlying assumptions and a lack of comprehensive information to support their use. However, the diagnostic capability they provide would be of immense use to Pacific island countries, either individually or, more appropriately, on a collective basis. Such diagnostic studies facilitate, for example, studies of the sensitivity and vulnerability of biophysical and socioeconomic systems to changes in climate and of the implications of selected response strategies.

4.2 SCENARIO DEVELOPMENT

At present, global climate models provide the most appropriate source of information for the development of climate change scenarios for the region. This is despite the notable differences in changes in rainfall derived from simulations using slab-ocean GCMs and the more comprehensive coupled-ocean models.[41] Such differences have increased doubts as to which of the methods provides the more appropriate basis for scenario development.

---

38. *Supra* note 7.
39. R.A.Warrick et al., *Integrated model system for national assessments of the effects of climate change: Applications in New Zealand and Bangladesh*, 19 J. WATER, AIR & SOIL POLLUTION (1996), at 215-227; CSIRO, *OzClim: A Climate Scenario Generator and Impacts Package for Australia*, November 1996.
40. J.Weyant et al., *Integrated assessment of climate change: an overview and comparison of approaches and results*, in CLIMATE CHANGE: ECONOMIC AND SOCIAL DIMENSIONS OF CLIMATE CHANGE, (J.P. Bruce, H. Lee & E.F. Haites eds., 1996), at 367-396.
41. Whetton, *supra* note 26.

While considerable effort has gone into developing scenarios for the Australasian region,[42] it must now be a priority to develop climate change scenarios for the Pacific island region. This should occur as soon as the above differences are resolved.

4.3. VULNERABILITY ASSESSMENT

The first assessments of the vulnerability of Pacific island countries, and the region as a whole, to climate change and to anticipated higher sea levels were undertaken by the UNEP/SPREP/ASPEI Task Team on Implications of Climate Change for the South Pacific (ASPEI).[43] The Team undertook its assessments assuming that low latitude temperatures would rise by 2 C by 2100, with an approximate warming rate of 0.3 C /decade. The case studies were also based on an assumed sea level rise of 1 m by about 2050, with the increase continuing beyond that date. The value of 1 m was considered conservative at the time, given that estimates ranged as high as 4.5 m by 2100. In a period as short as five years there were fundamental changes in assumptions and attitudes regarding the impact of climate change on small islands of the Pacific. Despite the lack of a more circumspect view, such as that subsequently provided by the IPCC assessments, the contributions of the various individuals working under the auspices of the UNEP/SPREP/ASPEI were substantial. They also laid an excellent foundation for the subsequent and more detailed country studies undertaken by SPREP with the assistance of UNEP and with the benefit of enhanced understanding provided by the IPCC assessments.

In recent years, SPREP, with the assistance of UNEP and regional experts, has undertaken studies of the potential impact of expected climate change on the national environment and socio-economic structure and activities. Ten countries in the region have been studied. The first study was for the Republic of Kiribati.[44] This served as a model for the subsequent investigations, by recognizing that changes in climate and sea level are only two, and in many cases not the most urgent, of a number of environmental problems facing small island nations in the Pacific. Thus the approach adopted in the Kiribati study, and followed to a large extent in subsequent assessments, was to focus on existing and anticipated environmental problems which are likely to be exacerbated by a change in climate or an increase in sea level. In their assessment of the vulnerability of Kiribati to climate change and sea-level rise, Sullivan and Gibson used a scenario of 2 C and 40 cm for the respective temperature and sea level increases by 2020. This was one of

42. Pittock, *supra* note 25.

43. ASPEI, *Potential Impacts of Greenhouse Gas Generated Climatic Change and Projected Sea-Level Rise on Pacific Island States of the SPREP Region,* Association off South Pacific Environmental Institutions (ASPEI), Prepared for the MEDU Joint Meeting of the Task Team on Implications of Climatic Change in the Mediterranean, Split, Yugoslavia, October 1988.

44. M. Sullivan & L. Gibson, *Environmental planning, climate change and potential sea level rise: report on a mission to Kiribati,* South Pacific Regional Environment Programme, SPREP Reports and Studies Series No. 50, Apia, Western Samoa (1991).

the earliest studies where the results of IPCC assessments and scenarios were incorporated explicitly in South Pacific vulnerability studies.

The subsequent nine studies were not full vulnerability assessments. Rather they were field and desk-top based scoping exercises designed to lay the foundation for more comprehensive assessments at a later date, and as such were called "preparatory missions." None of the studies used the IPCC's Common Methodology for Vulnerability Assessment. Some of the resulting reports do not even specify the changes in climate and sea level that are assumed in the preliminary studies. However, those that do tend to favour the projections provided by the IPCC,[45] viz. a temperature increase of 0.3 C per decade and a rise in sea level of 6 cm per decade.

The Coastal Zone Management Subgroup of IPCC Working Group II developed the "Seven Steps to the Assessment of the Vulnerability of Coastal Areas to Sea Level Rise - A Common Methodology."[46] It was published after a detailed process of review, comments and revisions. The Common Methodology is a guideline for assessing vulnerability to accelerated sea-level rise. In such an assessment, three levels of boundary conditions and scenarios are incorporated in the methodology: i) impacts on socio-economic development; ii) impacts on natural coastal systems; and iii) the implications of possible response strategies for adaptation.

In its present form, the Common Methodology focuses on the effects of accelerated sea-level rise and existing storm patterns. It is anticipated that more attention will be given in the future to the vulnerability of coastal areas with respect to other anticipated consequences of global warming, such as rainfall distribution and soil moisture and, possibly, changing storm patterns and intensities.

Kaluwin[47] has highlighted concerns about the applicability of the Common Methodology in the South Pacific and in comparable areas elsewhere. At issue is the biophysical framework for the assessment, the cost-benefit orientation of analyses of response options, and differences as to how governments and communities perceive the importance of climate change and sea-level rise relative to other regional planning and environmental issues. If a low priority is accorded, climate change and related response strategies could be biased towards reactive policies implemented only after erosional or storm events have caused damage.

45. IPCC, *Climate Change 1992: The Supplementary Report to the IPCC Scientific Assessment*, (J.T Houghton, B.A Callander & S.K. Varney eds., 1992), Intergovernmental Panel on Climate Change (IPCC); IPCC, *Climate Change - The IPCC Scientific Assessment*, (J.T. Houghton, G.J. Jenkins & J.J. Ephraums eds., 1990), Report prepared for the Intergovernmental Panel on Climate Change (IPCC) by Working Group I.

46. IPCC, *Assessment of the Vulnerability of Coastal Areas to Sea Level Rise: A Common Methodology*, Response Strategies Working Group, Intergovernmental Panel on Climate Change (1991).

47. C. Kaluwin, *The role of the South Pacific Regional Environment Programme in the IPCC work: appropriateness of the IPCC common methodology for the Pacific region*, in EASTERN HEMISPHERE WORKSHOP ON THE VULNERABILITY ASSESSMENT OF SEA LEVEL RISE AND COASTAL ZONE MANAGEMENT (N. Mimura & R. McLean eds., 1993), Intergovernmental Panel on Climate Change Regional Workshop, Tsukuba, Japan, at 5-9.

In response to the difficulties experienced in applying the Common Methodology to Pacific island countries, as outlined above, members of the SPREP/Japan Integrated Coastal Zone Management Programme for Western Samoa and Fiji built on and expanded the Common Methodology into a broader assessment and decision-making support framework appropriate to the South Pacific and even more widely applicable. The Stress-Response Methodology for the Assessment of Vulnerability and Resilience to Sea-Level Rise and Climate Change is described by Kay and Hay[48] and elaborated by Kay et al.,[49] and Nunn et al. and Yamada et al.[50] These publications also describe the application of the methodology to case study areas in both Fiji and Western Samoa.

A number of shortcomings in the Stress-Response Methodology are being examined. These include the subjectivity in assigning the vulnerability and resilience scores to individual coastal system components, the limitations engendered by working with only six levels of coastal system components, as opposed to aggregating the scores for sub-system elements, the failure to acknowledge explicitly the non-linear interactions between system components and the ongoing difficulty of quantifying intrinsic values (e.g., cultural importance of land in the Pacific) and valuing elements of subsistence societies (e.g., the gifting of food for festive occasions).

Experience gained during the First Assessment conducted by IPCC demonstrated the need for a more credible approach to the development of a global picture of the potential effects of climate change. The result was a study framework, which allows comparable assessments to be made of impacts in different regions, geographical areas, economic sectors and countries, the IPCC Guidelines for Assessing Climate Change Impacts and Adaptation Responses.[51]

---

48. R.C. Kay & J.E. Hay, *Possible future directions for integrated coastal zone management in the Eastern Hemisphere: a discussion paper*, in Eastern Hemisphere Workshop on the Vulnerability Assessment of Sea Level Rise and Coastal Zone Management (N. Mimura & R. McLean, eds. 1993), Intergovernmental Panel on Climate Change Regional Workshop, Tsukuba, Japan, at 181 194.

49. R.C. Kay et al., *Assessment of Coastal Vulnerability and Resilience to Sea-Level Rise and Climate Change. Case Study: 'Upolu Island, Western Samoa. Phase I: Concepts and Approach*, South Pacific Regional Environment Programme, Environment Agency, Government of Japan and Overseas Environmental Cooperation Center, Japan (1993).

50. P. Nunn et al., *Assessment of Coastal Vulnerability and Resilience to Sea-Level Rise and Climate Change,* Case Study: Yasawa Islands, Fiji. Phase 2: Development of Methodology. South Pacific Regional Environment Programme, Environment Agency, Government of Japan and Overseas Environmental Cooperation Center, Japan (1994).; P. Nunn et al., *Assessment of Coastal Vulnerability and Resilience to Sea-Level Rise and Climate Change*, Case Study: Savai'i Island, Western Samoa. Phase 2: Development of Methodology. South Pacific Regional Environment Programme, Environment Agency, Government of Japan and Overseas Environmental Cooperation Center, Japan, (1994); P. Nunn et al., *Assessment of Coastal Vulnerability and Resilience to Sea-Level Rise and Climate Change. Case Study: Viti Levu Island, Fiji, Phase I: Concepts and Approach.* South Pacific Regional Environment Programme, Environment Agency, Government of Japan and Overseas Environmental Cooperation Center, Japan (1993); K. Yamada et al., *Methodology for the assessment of vulnerability of South Pacific island countries to sea level rise and climate change*, 1 J. GLOBAL ENVTL. ENGINEERING (1995), at 101-125.

Until recently, there were no generally accepted procedures for formulating national and regional policies for adaptation to climate change.[52] The situation has changed as a result of initiatives by the IPCC to develop guidelines to enable estimations of impacts and adaptations that allow comparable assessments to be made for different regions/geographical areas, sectors and countries. The resulting "IPCC Technical Guidelines for Assessing Climate Change Impacts and Adaptations"[53] are intended to assist parties to the UNFCCC to meet, in part, commitments under Article 4 of the Convention. One such commitment of relevance to all Pacific island countries is the obligation to develop and elaborate appropriate and integrated plans for coastal zone management. The Guidelines for Development of an Adaptation Strategy[54] draw on the formal evaluation procedures used by many developed countries when formulating national policies and plans.

Empirical field observations are also critical to assessments of the vulnerability of terrestrial and marine areas to the consequences of climate change, including accelerated sea level rise. For example, field investigations are the principal method used by SOPAC to produce new information that will assist its member countries, and others, with coastal zone and EEZ resource assessment and management. The objective is to enhance understanding of all physical coastal processes and the way in which they interact with human activities.

4.4 SPECULATION

Regrettably, unsubstantiated and unwarranted speculation continues to be one of the methods used to predict the future states of the region's atmosphere and oceans. Culprits include some elements of the media, politicians, NGOs, and even some "scientists"! Examples are provided in Connell and Lea[55] and Garrett.[56]

4.4.1 *"Predictions" for the Future*

Changes in the regional climate expected as a consequence of a doubling of CO_2 include a 1 C rise in mean annual sea surface temperature[57] and substantial increases in mean rainfall intensity for central equatorial Pacific.[58] Several coupled ocean-atmosphere general circulation models are able to simulate ENSO-like

51. T.R. Carter et al., *IPCC Technical Guidelines for Assessing Climate Change Impacts and Adaptations*, Intergovernmental Panel on Climate Change (IPCC) (1994).
52. *Id.*
53. *Id.*
54. *Id.*
55. J. Connell & J. Lea, 1993: *Global warming: Meeting the planning challenge in island states*, in CLIMATE CHANGE AND SEA LEVEL RISE IN THE SOUTH PACIFIC REGION, PROCEEDINGS OF THE SECOND SPREP MEETING, Noumea, New Caledonia, April, 1992 (J. E. Hay & C. Kaluwin, April 1992 eds.). South Pacific Regional Environment Programme (SPREP), Apia, Western Samoa, at 49-156.
56. J. Garrett, *The media and global warming*, in HAY & KALUWIN, *supra* note 55, 217-221.
57. *Supra* note 25.
58. Whetton, *supra* note 27; Pittock, *supra* note 26.

sea surface temperature variability for present day climate, and also for climates associated with increased greenhouse gas concentrations.

Knowledge is currently insufficient to say whether there will be any changes in the occurrence or geographical distribution of severe storms, e.g., tropical cyclones. The formation of tropical cyclones depends not only on sea surface temperatures, but also on a number of atmospheric factors including the vertical lapse rate of temperature and vertical wind shear.[59] Although some models now represent tropical storms with some realism for present day climate,[60] the state of science does not allow conclusive assessment of future changes.[61] They suggest there are in fact no compelling reasons for expecting a major change in global tropical cyclone frequency, although substantial regional changes may occur. At present models are incapable of predicting the direction of such changes. This dismissal of the possibility that global climate models can provide useful information about the sensitivity of tropical cyclone climate to increased greenhouse gases is challenged by Broccoli et al.[62] They note that the climate modelling approach has much better potential for future improvement than do the empirical approaches offered as alternatives by Lighthill et al.[63]

The conclusions of Lighthill et al. are also questioned by Emanuel[64] (1995). He claims that basic physics coupled with the historic record of tropical cyclone intensities suggests that it is highly probable that increased thermodynamic disequilibrium between the tropical atmosphere and oceans as a result of global warming would be accompanied by an increase in the limiting intensity of actual tropical cyclones. Investigations by Holland[65] indicates that there is unlikely to be more intense tropical cyclones than the worst that occur at present, though there is some propensity for changes in cyclone frequency in regions where sea surface temperatures are between 26 and 29 C at present. Bengtsson et al.[66] found that under an enhanced greenhouse scenario, the geographical distribution of tropical cyclones was unchanged but the number decreased, especially in the Southern Hemisphere. The warmer surface waters associated with global warming were counteracted by

---

59. W.M. Gray, *Tropical Cyclone Genesis*, Dept. of Atmospheric Science Paper Number 234, Colorado State University (1975).
60. *See e.g.*, R.J. Haarsma, J.F.B. Mitchell & C.A. Senior, *Tropical disturbances in a GCM*, 8 CLIMATE DYNAMICS (1993), at 247-257.
61. J. Lighthill, et al., *Global climate change and tropical cyclones*, 75 BULLETIN AM. METEOROLOGICAL SOC'Y 2147-2157 (1994).
62. A.J Broccoli, *et al.*, *Comments on 'Global climate change and tropical cyclones,' Part II*, 76 BULLETIN AM. METEOROLOGICAL SOC'Y 2243-2245 (1995).
63. Lighthill, *supra* note 61.
64. K.A. Emanuel, *Comments on 'Global climate change and tropical cyclones,' Part I*, 76 BULLETIN AM. METEOROLOGICAL SOC'Y 2241-2243 (1995).
65. G.J. Holland, *The Maximum Intensity of Tropical Cyclones*, 54 J. ATMOSPHERIC SCI. 2519-2541 (1995).
66. L. Bengtsson, M. Botzet & M. Esch, *Hurricane-type vortices in a general circulation model*, 47 TELLUS 175-196 (1995).

generally weaker surface winds (reducing evaporation) and by changes in vertical stability.

The IPCC concludes that it is an open question as to whether the frequency, area of occurrence, time of occurrence, mean intensity or maximum intensity of tropical cyclones will change as a consequence of global warming.

Vulnerability assessments undertaken for Pacific island countries show that they:

- are highly vulnerable to changes in climate and sea level;
- have a large natural resilience impaired by human pressures;
- make insignificant contributions to global environmental changes; and
- have a low capacity to respond to changes. [67]

5. Conclusions

A common perception is that the dominant impact of global warming for Pacific island countries will be the drowning of atolls due to rising sea level, with the resultant creation of "environmental refugees." The truth is very different, but no less serious. Vulnerability and other assessments undertaken over the last few years, often in association with the Intergovernmental Panel on Climate Change (IPCC), reveal that climate change will impose diverse and significant impacts on Pacific island countries. While the form the global changes in climate will take in the Pacific is far from certain, the most significant and more immediate consequences are likely to be related to changes in rainfall regimes and soil moisture budgets, prevailing winds (both speed and direction) and in short-term variations in regional and local sea levels and patterns of wave action.

The large interannual variability in the climate signal in the Pacific presents challenges and opportunities, to both climate scientists and policy makers alike. While the present generation of global climate models fail to resolve the spatial scales of relevance to Pacific island countries, available evidence suggests that the predicted changes in the global climate may not manifest themselves in the Pacific, especially for temperature and sea level. Significantly, the findings recently reported by the IPCC suggest that the El Niño Southern Oscillation, a phenomenon of fundamental importance in the Pacific, may be relatively insensitive to climate change. However, there is also concern that key changes could occur should thresholds be exceeded. This is also the case in terms of the frequency and intensity of tropical cyclones. Their incidence is of critical importance – beneficial or detrimental, depending on circumstances – to most Pacific island countries. The conflicting evidence has led the IPCC to conclude that nothing definitive in terms of their

67. J.E. Hay, C. Kaluwin & N. Koop, *Implications of climate change and sealevel rise for small island nations of the South Pacific: A Regional Synthesis*, 15 WEATHER & CLIMATE (1995), at 5-20.

response to global warming can be stated at this time.

For Pacific island countries, the development of responses to global warming is further impeded by the fact that many regional and national vulnerability assessments have been undertaken using methodologies which are, at best, poorly harmonized with local conditions. Local applicability of methods is frequently in conflict with the legitimate desire to undertake studies using comparable methods that facilitate global intercomparisons and global assessments. However, in the Pacific there is little chance that assessments will have validity if they ignore the local dominance of subsistence economies, customary land ownership and village-based decision making.

Such equivocal findings must be considered along with the limited capacity of Pacific island countries. Perhaps of most relevance is limited adaptive capacity. Also of importance is limited local capacity to mitigate climate change, or even to characterize the anticipated regional changes in climate and identify appropriate response options.

Pacific island environments – both natural and human – are undeniably susceptible to extreme and anomalous persistent events occurring under present day conditions. Vulnerability and actual harm are enhanced by increased human pressure on natural systems. This sensitivity, and the consequences, leaves little doubt that should the changes predicted in the SAR manifest themselves in the future, the repercussions will threaten the life-supporting capacity of natural systems and the sustainability of human habitation.

Vulnerability assessments have revealed that not only the low islands of the Pacific are susceptible to the adverse effects of sea level rise. Human population, economic activity and infrastructural development are concentrated in the coastal areas of high islands. Thus, there are few viable options for retreat in face of inundation from rising sea levels or increased frequency and magnitude of storm waves and surges. Few landmasses in the Pacific are tectonically stable. Thus, systematic changes in sea level may be significantly offset or exacerbated by local uplift or subsidence of the land.

While there are areas of uncertainty, many of the anticipated changes may well be irreversible by the time there is certainty of outcome. Moreover, the momentum of change in the combined atmosphere-ocean system is such that the modifications of atmospheric composition taking place as a result of current human activity are already committing our children, and their children, to living in a world substantially different than the one we know today.

From the Pacific island perspective, dangerous anthropogenic interference is already occurring to the climate system. Concerns over threats and a perception that climate change related events are occurring with increasing frequency and intensity is taking a psycho-social toll on Pacific island communities even whilst the scientific debate continues as to how climate change may be manifested in the region.

In many instances there is heavy reliance, even by politicians, government officials and other senior level decision makers, on media reports for basic information related to climate change and sea-level rise. Furthermore, in the absence of reliable regional and national scale information, uncritical and imprudent use is commonly made of generalised scientific appraisals, impact assessments and response options that are applicable only at the global scale.

Change is often of less concern to those living in a naturally highly dynamic (variable) environment, leading to a feeling of powerlessness to control nature. In addition, there is a prevalent attitude that the current ability to cope with the devastating effects of tropical cyclones and other natural hazards is evidence of an aptitude to handle any future environmental threats. While this might have been the case in the past, many of today's natural systems have been degraded by human activity and are therefore more vulnerable to future stress, be it natural or human-induced. Moreover, changes in construction materials, methods and styles have all reduced the ability to make rapid and locally sourced repairs to homes and other buildings.

If policy development and the ensuing actions to address the anticipated impacts of climate change and accelerated sea level rise are to be effective they must be mainstreamed in both development planning and disaster management at the regional and national levels, with core initiatives being identified and implemented within an integrated environmental management framework. This is a current and continuing challenge for Pacific island countries where limitations on resources (human, financial, technical and information) and institutional capacities mean responses frequently fall short of the optimum approach. The cost is further stress on systems already under pressure. Thus, the diverse initiatives being undertaken in relation to both regional *and* international cooperation, and education, training *and* awareness raising can only be applauded. In regard to training, the activities being initiated in the region by CC:TRAIN, IGBP/START, PICCAP and VANDACLIM[68] are noteworthy.

68. R.A Warrick, G.J. Kenny, G.C. Sims, W. Ye & G. Sem, *VANDACLIM: A training tool for climate change vulnerability and adaptation assessment*, in CLIMATE CHANGE AND SEA LEVEL RISE IN THE SOUTH PACIFIC REGION, Proceedings of the Third SPREP Meeting, Noumea, New Caledonia, August, 1997, (C. Kaluwin & J.E. Hay eds., 1997). South Pacific Regional Environment Programme (SPREP), Apia, Western Samoa, at 147-156.

15. HOW SOUTH PACIFIC MANGROVES MAY RESPOND TO PREDICTED CLIMATE CHANGE AND SEA-LEVEL RISE

JOANNA C. ELLISON
School of Applied Science, University of Tasmania
P. O. Box 1214
Launceston, Tasmania 7250, Australia.

Abstract

In the Pacific islands, the total mangrove area is about 343,735 ha, with largest areas in Papua New Guinea, Solomon Islands, Fiji and New Caledonia. A total of 34 species of mangroves are found in the region, as well as 3 hybrids. These are of the Indo-Malayan assemblage (with one exception), and decline in diversity from west to east across the Pacific, reaching a limit at American Samoa. Mangrove resources are traditionally exploited in the Pacific islands, for construction and fuel wood, herbal medicines, and the gathering of crabs and fish.

There are two main environmental settings for mangroves in the Pacific, deltaic and estuarine mangroves of high islands, and embayment, lagoon and reef flat mangroves of low islands. It is indicated from past analogues that their close relationship with sea-level height renders these mangrove swamps particularly vulnerable to disruption by sea-level rise. Stratigraphic records of Pacific island mangrove ecosystems during sea-level changes of the Holocene Period demonstrate that low islands mangroves can keep up with a sea-level rise of up to 12 cm per 100 years. Mangroves of high islands can keep up with rates of sea-level rates of up to 45 cm per 100 years, depending on the supply of fluvial sediment. When the rate of sea-level rise exceeds the rate of accretion, mangroves experience problems of substrate erosion, inundation stress and increased salinity.

Rise in temperature and the direct effects of increased CO_2 levels are likely to increase mangrove productivity, change phenological patterns (such as the timing of flowering and fruiting), and expand the ranges of mangroves into higher latitudes.

Pacific island mangroves are expected to demonstrate a sensitive response to the predicted rise in sea level. A regional monitoring system is needed to provide data on ecosystem changes in productivity, species composition and sedimentation. This has been the intention of a number of programs, but none has yet been implemented.

1. Mangroves of the South Pacific Islands

1.1 MANGROVE AREAS AND DISTRIBUTIONS

Mangrove forests occur on low energy, sedimentary shorelines of the tropics, generally

A. Gillespie and W. C. G. Burns (eds.), Climate Change in the South Pacific: Impacts and Responses in Australia, New Zealand, and Small Island States, 289–301.

between mean tide and high tide elevations. Mangrove trees have special physiological and morphological adaptations to the environmental stresses of their intertidal habitat, of high salinity, low oxygen, poor nutrient availability and substrate mobility. Mangrove ecosystems are characterized by high faunal diversity of microbes (bacteria, Protozoa, fungi), invertebrate infauna (i.e., crabs), fish and birds.

In the Pacific islands, the total area of mangroves is about 343,735 ha, the largest areas occurring in Papua New Guinea, Solomon Islands, Fiji and New Caledonia (See the Appendix to this chapter). The world mangrove area is estimated at 14,197,635 ha;[1] hence, the Pacific islands have 2.42% of the world's mangroves. The mangrove area of the Pacific islands is small in global terms, but each island group has a unique community structure, and the mangrove forests provide significant uses for human populations.

The greatest diversity of mangroves in the world, with 33 species and 2 hybrids,[2] is southern Papua New Guinea, owing to its location at the centre of the Indo-Malayan mangrove center of diversity. Mangrove species diversity declines from Papua New Guinea east across the Pacific Islands,[3] and the eastern limit of mangroves is American Samoa. Mangroves do not occur further east due to difficulty of long-distance dispersal of propagules, and loss of habitats due to sea-level changes of the past.[4] Mangroves have been introduced in Hawaii and French Polynesia.

---

1. L.D. Lacerda et al., *Mangrove ecosystems of Latin America and the Caribbean: a summary*, in CONSERVATION AND SUSTAINABLE UTILIZATION OF MANGROVE FORESTS IN LATIN AMERICA AND AFRICA REGIONS, International Tropical Timber Organization and International Society for Mangrove Ecosystems (L. D. Lacerda ed., 1993), at 1-42.
2. J.C. Ellison, *The Pacific palaeogeography of Rhizophora mangle L. (Rhizophoraceae)*, 105 BOTANICAL J. LINNEAN SOC'Y (1991), at 271-284.
3. J. C. Ellison, *Systematics and distributions of Pacific Island mangroves*, in 1 MARINE AND COASTAL BIODIVERSITY IN THE TROPICAL ISLAND PACIFIC REGION (J.E. Maragos et al. eds., 1995), at 59-74.
4. J.C. Ellison & D. R. Stoddart, *Mangrove ecosystem collapse during predicted sea-level rise: Holocene analogues and implications*, 7 J. COASTAL RES. (1991), at 151-165; J.C. Ellison, *Palaeo-lake and swamp stratigraphic records of Holocene vegetational and sea-level changes, Mangaia, Cook Islands*, 48 PAC. SCI. (1994), at 1-15.

1.2 USES OF MANGROVE FORESTS

Mangrove resources have been traditionally exploited in the Pacific islands, for construction and fuel wood, herbal medicines, and the gathering of crabs and fish. In the last few centuries of economic development, significant areas have been lost to reclamation for settlement or agriculture, and others have been degraded by the dumping of rubbish. This continues today; for example, all significant mangrove areas in Tonga are now allocated for clearance. Despite the traditional use of mangroves in the Pacific islands, they are rarely a valued resource.

Legislation for the protection of mangroves exists in many Pacific island nations,[5] but enforcement of such laws is usually lax. More mangrove-protected areas need to be established to ensure that sufficient mangrove ecosystem biomass remains in each island nation for use by future generations. Mangrove conservation values can be enhanced by development of educational resource materials for distribution in schools and community fora.

Research needs to be carried out on forest community structure, primary productivity, ecology, food chains, fish, crustaceans, and other fauna. This would facilitate evaluation of the use of mangrove production by groups of organisms in adjacent systems, and bolster justification for mangrove protection. It would also provide guidelines for sustainable exploitation of the mangrove forest.

2. Impact of Climate Change on Mangrove Ecosystems

Concern over the capacity of intertidal wetlands to respond to climatic and sea level change has developed as a result of investigations of palaeo records[6] and mangrove ecophysiological response to climatic change.[7] These studies show that intertidal wetlands are likely to be one of the most sensitive ecosystems vis-à-vis predicted climate change due to their sensitivity to temperature,[8] salinity,[9] ambient CO_2,[10] and in particular because of the close relationship between the health of

5. D.A. SCOTT, A DIRECTORY OF WETLANDS IN OCEANIA (1993).

6. C.D. Woodroffe, *The impact of sea-level rise on mangrove shorelines*, 14 PROGRESS IN PHYSICAL GEOGRAPHY (1990), at 483-520; Ellison & Stoddart, *supra* note 4; R.W. Parkinson, R.D. DeLaune & J.R. White, *Holocene sea-level rise and the fate of mangrove forests within the wider Caribbean region*, 10 J. COASTAL RES. (1994), at 1077-1086; P.R. Bacon, *Template for evaluation of impacts of sea-level rise on Caribbean coastal wetlands*, 3 ECO. ENGINEERING (1994), at 171-186.

7. J.C. Pernetta, *Mangrove forests, climate change and sea-level rise*, IUCN (1993); V. Semeniuk, *Predicting the effect of sea-level rise on mangrove in northwestern Australia*, 10 J. COASTAL RES. (1994), at 1050-1076; J.C. Ellison, *Climate change and sea-level rise impacts on mangrove ecosystems, in Impacts of Climate Change on Ecosystems and Species: Marine and Coastal Ecosystems*, IUCN (J. Pernetta, R. Leemans, D. Elder, & S. Humphreys eds., 1994); S.C. Snedaker, *Mangroves and climate change in the Florida and Caribbean region: scenarios and hypotheses*, 295 HYDROBIOLOGIA (1995), at 43-49.

8. P. Saenger & J. Moverley, *Vegetative phenology of mangroves along the Queensland coastline*, 13 PROC. ECOL. SOC'Y. AUST. (1995), at 257-265; N.C. Duke, *Phenological Trends with Latitude in the Mangrove Tree Avicennia marina*, 78 J. ECOLOGY (1990), at 113-133.

intertidal vegetation communities and sea-level.[11]

The consequences of climate change for mangrove ecosystems are assessed below, identifying the processes that will respond to climate change, the direction of community response, the maximum rates of change that can be withstood, and implications for management and planning in mangrove areas.

2.1 MANGROVES AND SEA-LEVEL RISE

Large mangrove ecosystems develop on sedimentary shorelines of gentle gradient, between mean sea level (MSL) and the level of mean high water spring tides. Growing in the upper half of the tidal range, their close relationship with sea-level position renders mangrove swamps particularly vulnerable to disruption by sea-level rise. With most Pacific islands having a tidal range of less than 1 m, mangrove ecosystems will be disrupted by a sea-level rise of 0.3 m, and will retreat landwards with a sea-level rise of 1 m. However, as discussed later in this chapter, factors such as physiographic location, tidal range, species assemblage and sediment supply contribute to heterogeneity in mangrove response to rising sea-level, investigated below.

Research to investigate sea level rise impacts on mangroves falls into two areas. First, sea-level effects on an ecosystem can be reconstructed from the past, by analysis of stratigraphic deposits, provided the evidence is available from microfossils (i.e., pollen, diatoms) that can be used to indicate the vegetation and environmental conditions. This technique has the advantage of indicating long-term complex system response, though detail of short-term individual responses of species is usually only speculative. Second, present day case studies can be examined, such as areas of the world where sea-level rise is occurring, or areas where flooding of mangroves has been carried out for mosquito control. From these approaches, monitoring programmes can be developed for identification of changes in the mangrove ecosystem resulting from climate change and sea-level rise.

2.1.1 Past records of mangrove response to sea level rise

Within the intertidal habitat of mangroves, species have different preferences for elevation, salinity and frequency of inundation, resulting in species zones. Elevation of the ground surface can be raised under mangroves by accumulation of vegetative detritus to form a mangrove peat or mud, which may also contain matter brought in by the tides and by rivers. If the sedimentation rate keeps pace with rising sea level, then the salinity and frequency of inundation preferences of man-

---

9. M.C. Ball, *Salinity tolerance in mangroves Aegiceras corniculatum, and Avicennia marina I. Water use in relation to growth, carbon partitioning, and salt balance*, 15 AUSTRALIAN J. PLAN. (1988), at 447-464.
10. E.J. Farnsworth, M.A. Ellison & W.K. Gong, *Elevated CO_2 alters anatomy, physiology, growth and reproduction of red mangrove (Rhizophora mangle L.)*, 108 OECOLOGIA (1996), at 599-609.
11. Woodroffe, *supra* note 6; Ellison & Stoddart, *supra* note 4.

grove species zones will remain largely unaffected. If the rate of sea-level rise exceeds the rate of sedimentation, then mangrove species zones will migrate inland to their preferred elevation, and seaward margins will die back. The accumulation of sediment under mangroves will help to compensate for rising sea levels. However, expected rates of sedimentation must be established to assess the vulnerability of mangrove ecosystems. This requires analysis of Holocene stratigraphy.

Stratigraphy of Pacific island mangroves indicates that mangroves became established in expansive swamps during the middle of the Holocene Period, about 6500 years ago.[12] Before this time, they were limited by the interactive factors of rapid sea-level rise and the absence of sedimentary shorelines. Mangroves probably survived this period as individual trees, as seen today on shorelines with too steep of a gradient or that are too exposed for expansive swamp development.

There are two major environmental settings for mangroves of the Pacific, deltaic and estuarine mangroves of high islands, and embayment, lagoon and reef flat mangroves of low islands. Deltaic and estuarine mangroves occur on islands sufficiently high to develop a river system, and so deliver significant quantities of sediment to the coastal zone. These areas contain the most extensive mangroves of the Pacific islands, for example the Fly delta in Papua New Guinea, the Rewa delta in Viti Levu, the Dumbéa delta in New Caledonia, and the estuaries of Palau and Pohnpei. The mangrove areas receive fluvial sediment from the catchment, and also accumulate vegetative debris to form a mud. Stratigraphic reconstruction reveals fairly rapid accretion of up to 45 cm per 100 years.[13]

Embayment, harbour and lagoon mangroves of low islands can be extensive where fine sediment accumulates to form a broad intertidal slope. Low island mangroves do not have an external supply of sediment, and build up their substrate by accumulation of vegetative detritus to form a highly organic peat. Mangrove stratigraphy from low island mangrove ecosystems indicates rates of accumulation of up to 12 cm per 100 years.[14] This reflects the rate of peat production within the mangrove system, there being few sources of inorganic sediment in these environments.

Comparison of mangrove stratigraphy shows that low island mangroves are more susceptible to disruption by rising sea level owing to relatively low rates of sediment accretion. Stratigraphy from high islands and continental coastlines, with more sediment coming off the land into intertidal areas from rivers and longshore drift, indicates that mangrove ecosystems in these areas will be better able to keep pace with sea-level rise. Low island mangroves could keep up with sea-level rise of up to 12 cm per 100 years. High island mangroves could keep up with rates of

12. Ellison & Stoddart, *supra* note 4.

13. *Id.*

14. J.C. Ellison, *Pollen analysis of mangrove sediments as a sea level indicator: Assessment from Tongatapu, Tonga*, 74 PALAEOGEOGRAPHY, PALAEOCLIMATOLOGY, PALAEOECOLOGY (1989), at 327-341; J.C. Ellison, *Mangrove retreat with rising sea-level, Bermuda*, 37 ESTUARINE COASTAL & SHELF SCI. (1993), at 75-87.

45 cm per 100 years, provided the sediment supply is not restricted by construction, such as dams on rivers leading to mangrove deltas.

While the consequences of greenhouse-induced climate change are probably not as dramatic as feared in the late 1980's, sea-level rise remains an issue for coastal ecosystems such as mangroves. The anticipated magnitude of sea-level rise is based upon the findings of the WMO /UNEP Intergovernmental Panel on Climate Change[15] In summary, in the last 100 years there has been a rise of 10-25 cm in global eustatic sea-level and the IPCC working group predicted a future sea-level rise of 15-95 cm by 2100, (mid range 50 cm). This would imply mean rates of rise of 5 cm per decade over the next century. Comparison with the rates of mangrove accretion indicates that island mangroves will experience serious problems with rising sea level in the next 50 years, and low island mangroves could already be under stress. Examples of low islands in the Pacific with mangrove areas include Tongatapu (Tonga), the Marshall Islands, Kiribati and Tuvalu.

2.1.2 Present Case Studies

Some problems that mangroves in the Pacific may face as a consequence of climate change are demonstrated by mangrove dieback in Bermuda, a low island with a long tide gauge record of relatively rapid sea-level rise.[16] There are also other studies that indicate the problems that may develop in mangroves under such conditions.

Erosion. Bermuda is a low limestone island without rivers, and the most northerly location for mangroves. Tide gauge records since 1932 show sea-level rise at a rate of 28±18 cm per 100 years.[17] The largest mangrove area (6.26 acres) at Hungry Bay has for the last 2000 years been building peat at a rate of 8.5 to 10.6 cm per 100 years.[18] The rate of sea-level rise has exceeded the rate of sediment accretion, leading to retreat of the seaward margin and erosion. The substrate elevation of the seaward margin of mangroves is below mean sea level, the normal lower limit for mangroves. Retreat of the seaward edge has resulted in the loss of 2.24 acres of mangroves, commencing in the last few hundred years, with a second dieback between 1900 and 1947, and a third dieback in the last decade. Stratigraphy shows that beyond 4000 years ago sea-level rose at a rate of 25 cm per 100 years, from 4000 to 1000 years ago the rate of sea-level rise declined to 6 cm per 100 years, during which time mangroves were established, and in the last 1000 years there was an increase to 14.3 cm per 100 years, during which time the mangroves died back.

---

15. CLIMATE CHANGE 1996. THE SCIENCE OF CLIMATE CHANGE (J.T. Houghton et al. eds., 1996).
16. Ellison, *supra* note 14.
17. P.A. Pirazzoli, *Secular trends of relative sea-level (RSL) changes indicated by tide-gauge records*, 1 J. COASTAL RES. (1986), at 1-126.
18. Ellison, *supra* note 14.

The record from Bermuda demonstrates that mangrove sediment is subject to erosion by rising sea levels, with removal of mangrove substrate (above MSL) and with some deposition subtidally offshore of the mangroves.[19] This corresponds with the Bruun Rule, which has shown erosion of beach sediment to occur with sea-level rise.[20] Sheet erosion occurs at the peat surface, indicated by a 20 to 25 cm difference between the peat level above a small cliff and the former peat surface as indicated by exposed horizontal roots of Avicennia. As trees recede and loosen the sediment, more rapid erosion occurs and forms a 40-50 cm small cliff. Such erosion exacerbates the existing problem, for as the mangrove substrate surface is lowered and creeks widen, the differential between elevation and MSL increases.

Similar erosion patterns, with reversed succession as elevation declines, have been described by Semeniuk in N.W. Australia.[21] The effect of sheet erosion on mangrove zonation is migration of pioneer/seaward mangroves into more landward zones. The effect of cliffing on mangrove zonation is loss of the seaward zone, leading to truncated zonation and narrow fringes. The effect of tidal creek erosion is slumping of banks and loss of trees.

Erosion of mangrove substrate would seem to be a major problem in cases of sea level rise. It is indicated that the Bruun Rule of beach erosion with sea-level rise is also apposite for mangrove swamps, and may occur earlier owing to the finer texture of the sediment.

Increased Salinity. Increased salinity in mangroves leading to salt stress can result from a number of factors in addition to sea-level rise, such as groundwater depletion owing to reduced freshwater flux, groundwater extraction, and reduced rainfall. Two major physiological adaptations enable mangrove survival in saline ocean water,[22] salt exclusion in species of *Rhizophora* and *Laguncularia*, and salt excretion in species of *Aegialitis* and *Aegiceras*. Salt excluders also cease or diminish transpiration and photosynthesis when exposed to saline water. Salt secretors can continue photosynthesis, utilizing ocean water in transpiration, owing to salt glands in the leaves.

Stern and Voight[23] grew 200 seedlings of *Rhizophora* mangle under different salinities, finding that seedling survival and growth increase by dry weight and seedling height were all inversely related to salt concentrations of the growing solutions. Ball and Farquhar[24] studied gas exchange characteristics in *Aegiceras corniculatum* and *Avicennia marina* under different salinity and humidity conditions. They

19. *Id.*
20. P. Bruun, *Sea level rise as a cause of shore erosion*, 88 J. WATERWAYS & HARBOURS DIVISION, PROCEEDINGS OF THE AMERICAN SOCIETY OF CIVIL ENGINEERS (1962), at 117-130; M. Schwartz, *The Bruun theory of sea level rise as a cause of shore erosion*, 75 J. GEOLOGY (1967), at 76-92.
21. V. Semeniuk, *Mangrove zonation along an eroding coastline in King Sound, North-Western Australia*, 68 J. ECOLOGY (1980), at 789-812.
22. P.F. Scholander et al., *Salt balance in mangroves*, 37 PLANT PHYSIOLOGY (1962), at 722-729.
23. W.L. Stern & G.K. Voight, *Effect of salt concentration on growth of red mangrove in culture*, 131 BOTANICAL GAZETTE (1959), at 36-39.

concluded that photosynthetic capacity declined under conditions of increased salinity, with *Aegiceras* being the more sensitive. Ball and Farquhar[25] studied the gas exchange characteristics of *Avicennia marina* with increasing salinity, finding that CO_2 assimilation rate, stomatal conductance, intercellular CO_2 concentration and evaporation rate all decreased. Increased salinity has the effect of decreasing net primary productivity and results in reduced growth, with a differential effect on species. Reduced precipitation and a rise in sea levels could result in stress to and changing competition between mangrove species.

The shallow water table of S. Florida is susceptible to saline intrusion from sea-level rise and groundwater extraction owing to low topography and porous rock. Alexander[26] and Alexander and Crook[27] found evidence of pine stumps in salt-water mangrove areas of Key Largo and landward encroachment of mangroves into grasslands and cypress swamps. Sternberg and Swart[28] used stable isotopes in plant stem water to determine relative ocean water and fresh water utilization in plants. They concluded that mangroves can utilize water ranging from fresh to oceanic, indicating plasticity in the salt-tolerant physiological system. Thus, we can infer that mangroves gain a competitive edge over faster growing glycophylic plants in conditions of increasing salinity.

In the Northern Territory of Australia, Woodroffe and Mulrennan[29] have documented dramatic recent changes to the Lower Mary River floodplain, with saltwater intrusion and upstream expansion of the tidal creek network. This has resulted in the death of freshwater wetland communities with loss of 60 km2 of Melaleuca forest and upstream invasion of mangroves. There are a number of possible reasons for these events, including relative sea-level rise.[30] Similar, though less spectacular, extension of creeks has occurred on other river systems, such as the Alligator rivers.[31] Loss of freshwater wetlands with saline intrusion is documented in the Florida Keys,[32] where longer tide records have enabled researchers to attribute the cause to relative sea-level rise.

---

24. M.C. Ball & G.D. Farquhar, *Photosynthetic and stomatal responses of two mangrove species, Aegiceras corniculatum and Avicennia marina, to long term salinity and humidity conditions*, 74 PLANT PHYSIOLOGY (1984), at 1-6.
25. M.C. Ball & G.D. Farquhar, *Photosynthetic and stomatal responses of grey mangrove, Avicennia marina, to transient salinity conditions*, 74 PLANT PHYSIOLOGY (1984), at 7-11.
26. T. R. Alexander, *Evidence of recent sea-level rise derived from ecological studies on Key Largo, Florida*, in ENVIRONMENTS OF SOUTH FLORIDA: PRESENT AND PAST, Miami Geological Society Memoir 2, (P.J. Gleason ed., 1974), at 219-222.
27. T.R. Alexander & A.G.Crook, *Recent vegetational changes in southern Florida*, Gleason, *supra* note 26, at 61-72.
28. L. Sternberg & P.K. Swart, *Utilization of freshwater and ocean water by coastal plants of southern Florida*, 68 ECOLOGY (1987), at 1898-1905.
29. C. D. Woodroffe & M. E. Mulrennan, *Geomorphology of the Lower Mary River Plains, Northern Territory*, North Australia Research Unit (1993).
30. C.D. Woodroffe, *Response of tide dominated mangrove shorelines in Northern Australia to anticipated sea-level rise*, 20 EARTH SURFACE PROCESSES & LANDFORMS (1995), at 65-85.
31. *Id.*

Inundation. Impoundment of mangroves for mosquito control provides an analogy to the effects of increased inundation with sea-level rise. Harrington and Harrington[33] recorded extensive death of *Avicennia germinans* and *Rhizophora mangle* at India River, East Florida following 4 months of 30-45 cm depth of flooding of an impoundment. Inundation was sudden onset, then sustained over the period. The natural tidal range in this area is 10-20 cm.[34]

Naidoo[35] found that prolonged flooding resulted in lower ability of leaves to conduct water; an increase in stomatal closing; and, degeneration of chloroplasts in *Bruguiera gymnorrhiza*, leading to reduced rates of photosynthesis. When lenticels of aerial roots become inundated, oxygen concentrations in the plant fall dramatically.[36] If inundation is sustained, low oxygen conditions occur and mortalities follow. This is thought to have been the cause of widespread mortality of Avicennia germinans stands in Puerto Rico recorded by Jimenez et al.,[37] following permanent flooding caused by adjacent dredging.

Lahmann[38] found that rates of litterfall in an impounded mangrove forest in Florida were reduced relative to natural fringe forest during flooded months. This indicates that the above ground net productivity of *Rhizophora mangle* is reduced by flooding, though phenology (such as the timing of flowering and fruiting) was not changed. Over several years of flooding, survival of *Avicennia germinans* and *Laguncularia racemosa* seedlings was denuded, and *Rhizophora mangle* became more common, despite poor seedling establishment during flooded periods. The depth of flooding was not indicated in this study.

---

32. M.S Ross, J.J. O Brien & L.D. Sternberg, *Sea level rise and the reduction of pine forests in the Florida keys*, 4 ECOLOGICAL APPLICATIONS, (1994), at 144.
33. R.W. Harrington & E.S. Harrington, *Effects on fishes and their forage organisms of impounding a Florida salt marsh to prevent breeding by salt marsh mosquitos*, 32 BULL. MARINE SCI. (1982), at 523-531.
34. E.J. Lahmann, *Effects of different hydrological regimes on the productivity of Rhizophora mangle L. A case study of mosquito control impoundments at Hutchinson Island*, Saint Lucie County, Florida. Ph.D. dissertation, University of Miami (1988).
35. G. Naidoo, *Effects of flooding on leaf water potential and stomatal resistance in Bruguiera gymnorrhiza*, 93 NEW PHYTOLOGIST (1983), at 369-373.
36. P.F.L. Scholander, L. Van Dam & S.I. Scholander, *Gas exchange in roots of mangroves*, 42 AM. J. BOTANY, (1955), at 92-98.
37. J.A. Jimenez, R. Martinez & L. Encarnacion, *Massive tree mortality in a Puerto Rican mangrove forest*, 21(7) CARIBBEAN J. SCI. (1985), at 5-78.
38. Lahmann, *supra* note 34.

3. Impacts of Climate Change on Mangroves

3.1 CLIMATE WARMING

Climate warming should have a beneficial effect on mangrove ecosystems, with increased productivity and expansion of ranges of species. Mangrove forest ecosystems border herbaceous salt marsh ecosystems in subtropical latitudes, at the 16°C isotherm for air temperature of the coldest month, at the margins of incidence of ground frost, and where water temperatures never exceed 24°C. At these limits, species of *Avicennia* occur, *Avicennia marina* (grey mangrove) reaching the highest latitude in the South Pacific at Victoria in S.E. Australia (38°S) and 37°S in North Island, New Zealand.

With climate warming, these higher latitude marginal mangroves can be expected to increase in diversity, with expansion of the ranges of other mangrove species into mangrove margins only occupied currently by *Avicennia* species, and expansion of the ranges of mangroves into salt marsh environments. The heterogeneity of regional sea surface temperature limits of mangrove species ranges would indicate regional differences in mangrove response. Variable gradients of sea surface isotherms mean that, for example, the expansion of range in the southeastern United States will be limited relative to that in southeastern Australia or southeastern Africa.

Combined with higher atmospheric CO_2 levels (discussed below), climate warming can be expected to increase mangrove productivity, characterized by increased growth and litter production at all locations.

3.2 INCREASED CO_2

As well as its climate effects, increased CO_2 directly affects plant growth and development. Plants have different pathways of carbon fixation in photosynthesis, and mangroves operate in the C_3 pathway.[39] In this case, metabolic responses with increased atmospheric CO_2 are increased productivity and more efficient water use. Farnsworth et al.[40] grew seedlings of *Rhizophora mangle* in doubled levels of CO_2 and demonstrated significantly increased biomass, total stem length, branching activity and total leaf area compared with seedlings grown in normal levels of CO_2. Increases in atmospheric CO_2 can be expected to improve mangrove tree growth and litter production.

---

39. B.F. Clough, T.J. Andrews, & I.R. Cowan, *Physiological processes in mangroves*, MANGROVE ECOSYSTEMS IN AUSTRALIA, Australian Institute of Marine Science & Australian National (B.F. Clough ed., 1982), at 193-210.

40. *Supra* note 10.

3.3 PRECIPITATION CHANGES

Most of the world's mangroves are found in warm humid regions, with much more limited areas on drier coastlines. In equatorial and tropical summer rainfall regions, mangroves are tall, dense and floristically diverse. In subtropical dry regions, mangroves are low, scattered and sporadic. Mangrove diversity is a function of rainfall patterns, with the greatest abundance of species in areas of higher rainfall, owing to the benefits from fluvial runoff of sedimentation and nutrient supply.

The reasons for these patterns relate to salt stress. Under humid conditions, mangrove soils are almost continuously leached by heavy rains and fresh water is available from river discharge and groundwater outflow, which provides nutrients. Under arid conditions, evaporation from the intertidal mangroves at low tide leads to high concentrations of salt, in some cases resulting in unvegetated hypersaline flats around high tide level.

The IPCC found evidence of an increase in precipitation over the equatorial Pacific in the last few decades, with decreases to the north and south. Scenarios of future climate change all show an enhanced global mean hydrological cycle, but with uncertainties as to how this will affect rainfall patterns in the Pacific island area.

Snedaker[41] has postulated that changes in rainfall patterns will have a significant effect on mangrove ecosystems. Increased rainfall should result in reduced salinity and exposure to sulphate, and an increase in delivery of terrigenous nutrients. Mangrove park managers can expect decreased rainfall and increased evaporation to reduce the extent of mangrove areas, particularly with loss of the landward zone to unvegetated hypersaline flats. The number of mangrove zones and their diversity can be expected to decrease, and growth rates will also likely decline. In conditions of increased rainfall, the extent of mangrove areas can be expected to increase, with colonization of previously unvegetated areas of the landward fringe, and the diversity of mangrove zones and growth rates should increase.

These responses of Pacific island mangroves to changes in precipitation will occur in combination with response to climate warming, increased CO_2 and sea-level rise. The net response of mangroves at each location will also be combined with local factors and other impacts. Understanding and management of these changes will require data on the nature of mangrove changes in the region.

4. Monitoring of Climate Change Effects

Mangrove tidal wetlands are expected to show a sensitive response to predicted climate change and sea level rise.[42] The nature of this response is complex, and subject to factors related to environmental setting,[43] such as location on low or

41. *Supra* note 7.

42. *Supra* notes 4 & 5.

high islands. Sea-level position is central to the functional ecology of a mangrove swamp, and rise in sea level will perturb every aspect of the ecosystem. This combined with effects of climate change, and stresses from storms and human disturbances will cause mangroves to experience disruption and area losses in the next few decades.

Certain identification of climate change and sea-level rise effects on mangroves requires long-term monitoring of biological and physical parameters at a network of locations using standard techniques.[44] This would allow comparison of data in order to distinguish a regional trend in change from that resulting from local effects. Though aimed at distinguishing climate change effects, the monitoring system would also detect local effects and disturbances, which would provide environmental managers with ecological data to allow for informed management of mangrove ecosystems.

Several expert groups have identified the need for a global monitoring system of mangrove response to climate change,[45] but none to date has been implemented. Implementation of such a system may finally come to fruition with the UNEP-IOC-WMO-IUCN Long-Term Global Monitoring System of Coastal and Near Shore Phenomena Related to Climate Change,[46] which is to be established as part of the Global Ocean Observing System. In the South Pacific region, SPREP recently developed a Regional Wetland Action Plan, in which actions 3.3.1 and 3.3.5 call for development of a regional monitoring systems for mangrove ecosystem health.[47] Such regional monitoring networks would facilitate informed management with respect to climate change effects on mangroves.

---

43. *Supra* note 6 & 7.

44. UNEP/UNESCO, *Impact of Expected Climate Change on Mangroves*, 61 UNESCO REPORT IN MARINE SCI.

45. UNEP-IOC-WMO-IUCN, Meeting of Experts on a Long-Term Global Monitoring System of Coastal and Near Shore Phenomena Related to Climate Change. *Intergovernmental Oceanographic Commission Report of Meetings of Experts and Equivalent Bodies* 61 (1990); UNEP, *Assessment and Monitoring of Climatic Change Impacts on Mangrove Ecosystems*, UNEP Regional Seas Reports and Studies, No. 154 (1994).

46. UNEP-IOC-WMO-IUCN, *Meeting of Experts on a Long-Term Global Monitoring System of Coastal and Nearshore phenomena, Pilot projects on mangroves and coral reefs*, Intergovernmental Oceanographic Commission Report of Meetings of Experts and Equivalent Bodies, (1991) at 69.

47. N. Idechong, J. Ellison & R. Jaensch, *A Draft Regional Wetlands Action Plan for the Pacific Islands*, in *International Coral Reef Initiative Pacific Regional Workshop Report* (Suva, Fiji, 27 November-1 December 1995), South Pacific Regional Environment Programme, at 116-134.

Appendix I

TABLE 1. Mangrove areas and species diversity in the Pacific islands.

Island Group	**Mangrove species** (Hybrids are in parentheses)	**Mangrove Area (ha)**
Papua New Guinea	33 (2)	200,000
Palau	13	4,708
N. Mariana Islands	5	
Guam	11	70
Federated States of Micronesia	14	8,564
Marshall Islands	5	
Nauru	2	1
Solomon Islands	20 (2)	64,200
Vanuatu	14	2,750?
New Caledonia	14 (2)	20,250
Kiribati	4	
Tuvalu	2	40
Fiji	8 (1)	41,000
Tonga	8	1,000
Western Samoa	3	700
American Samoa	3	52
Niue	1	
French Polynesia	1	(introduced)
Hawaii	2	(introduced)
TOTAL	34 (3)	343,735

Acknowledgement

This contribution was written under an Australian Postdoctoral Research Fellowship for research into sea-level rise effects on mangroves.

Appendix I

UNITED NATIONS FRAMEWORK CONVENTION ON CLIMATE CHANGE

31 INTERNATIONAL LEGAL MATERIALS 849 (1992)

The Parties to this Convention,

Acknowledging that change in the Earth's climate and its adverse effects are a common concern of humankind,

Concerned that human activities have been substantially increasing the atmospheric concentrations of greenhouse gases, that these increases enhance the natural greenhouse effect, and that this will result on average in an additional warming of the Earth's surface and atmosphere and may adversely affect natural ecosystems and humankind,

Noting that the largest share of historical and current global emissions of greenhouse gases has originated in developed countries, that per capita emissions in developing countries are still relatively low and that the share of global emissions originating in developing countries will grow to meet their social and development needs,

Aware of the role and importance in terrestrial and marine ecosystems of sinks and reservoirs of greenhouse gases,

Noting that there are many uncertainties in predictions of climate change, particularly with regard to the timing, magnitude and regional patterns thereof,

Acknowledging that the global nature of climate change calls for the widest possible cooperation by all countries and their participation in an effective and appropriate international response, in accordance with their common but differentiated responsibilities and respective capabilities and their social and economic conditions,

Recalling the pertinent provisions of the Declaration of the United Nations Conference on the Human Environment, adopted at Stockholm on 16 June 1972,

Recalling also that States have, in accordance with the Charter of the United Nations and the principles of international law, the sovereign right to exploit their own resources pursuant to their own environmental and developmental policies, and the responsibility to ensure that activities within their jurisdiction or control do not cause damage to the environment of other States or of areas beyond the limits of national jurisdiction,

Reaffirming the principle of sovereignty of States in international cooperation to address climate change,

Recognizing that States should enact effective environmental legislation, that environmental standards, management objectives and priorities should reflect the environmental and developmental context to which they apply, and that standards applied by some countries may be inappropriate and of unwarranted economic and social cost to other countries, in particular developing countries,

Recalling the provisions of General Assembly resolution 44/228 of 22 December 1989 on the United Nations Conference on Environment and Development, and resolutions 43/53 of 6 December 1988, 44/207 of 22 December 1989, 45/212 of 21 December 1990 and 46/169 of 19 December 1991 on protection of global climate for present and future generations of mankind,

Recalling also the provisions of General Assembly resolution 44/206 of 22 December 1989 on the possible adverse effects of sea-level rise on islands and coastal areas, particularly low-lying coastal areas and the pertinent provisions of General Assembly resolution 44/172 of 19 December 1989 on the implementation of the Plan of Action to Combat Desertification,

Recalling further the Vienna Convention for the Protection of the Ozone Layer, 1985, and the Montreal Protocol on Substances that Deplete the Ozone Layer, 1987, as adjusted and amended on 29 June 1990,

Noting the Ministerial Declaration of the Second World Climate Conference adopted on 7 November 1990,

Conscious of the valuable analytical work being conducted by many States on climate change and of the important contributions of the World Meteorological Organization, the United Nations Environment Programme and other organs, organizations and bodies of the United Nations system, as well as other international and intergovernmental bodies, to the exchange of results of scientific research and the coordination of research,

Recognizing that steps required to understand and address climate change will be environmentally, socially and economically most effective if they are based on relevant scientific, technical and economic considerations and continually re-evaluated in the light of new findings in these areas,

Recognizing that various actions to address climate change can be justified economically in their own right and can also help in solving other environmental problems,

Recognizing also the need for developed countries to take immediate action in a flexible manner on the basis of clear priorities, as a first step towards comprehensive response strategies at the global, national and, where agreed, regional levels that take into account all greenhouse gases, with due consideration of their relative contributions to the enhancement of the greenhouse effect,

Recognizing further that low-lying and other small island countries, countries with low-lying coastal, arid and semi-arid areas or areas liable to floods, drought and desertification, and developing countries with fragile mountainous ecosystems are particularly vulnerable to the adverse effects of climate change,

Recognizing the special difficulties of those countries, especially developing countries, whose economies are particularly dependent on fossil fuel production, use and exportation, as a consequence of action taken on limiting greenhouse gas emissions,

Affirming that responses to climate change should be coordinated with social and economic development in an integrated manner with a view to avoiding adverse impacts on the latter, taking into full account the legitimate priority needs of developing countries for the achievement of sustained economic growth and the eradication of poverty,

Recognizing that all countries, especially developing countries, need access to resources required to achieve sustainable social and economic development and that, in order for developing countries to progress towards that goal, their energy consumption will need to grow taking into account the possibilities for achieving greater energy efficiency and for controlling greenhouse gas emissions in general, including through the application of new technologies on terms which make such an application economically and socially beneficial,

Determined to protect the climate system for present and future generations,

Have agreed as follows:

ARTICLE 1 - DEFINITIONS*

For the purposes of this Convention:

1. "Adverse effects of climate change" means changes in the physical environment or biota resulting from climate change which have significant deleterious effects on the composition, resilience or productivity of natural and managed ecosystems or on the operation of socio-economic systems or on human health and welfare.

2. "Climate change" means a change of climate which is attributed directly or indirectly to human activity that alters the composition of the global atmosphere and which is in addition to natural climate variability observed over comparable time periods.

3 "Climate system" means the totality of the atmosphere, hydrosphere, biosphere and geosphere and their interactions.

4. "Emissions" means the release of greenhouse gases and/or their precursors into the atmosphere over a specified area and period of time.

5. "Greenhouse gases" means those gaseous constituents of the atmosphere, both natural and anthropogenic, that absorb and re-emit infrared radiation.

6. "Regional economic integration organization" means an organization constituted by sovereign States of a given region which has competence in respect of matters governed by this Convention or its protocols and has been duly authorized, in accordance with its internal procedures, to sign, ratify, accept, approve or accede to the instruments concerned.
7. "Reservoir" means a component or components of the climate system where a greenhouse gas or a precursor of a greenhouse gas is stored.
8. "Sink" means any process, activity or mechanism which removes a greenhouse gas, an aerosol or a precursor of a greenhouse gas from the atmosphere.
9. "Source" means any process or activity which releases a greenhouse gas, an aerosol or a precursor of a greenhouse gas into the atmosphere.

* Titles of articles are included solely to assist the reader.

ARTICLE 2 - OBJECTIVE

The ultimate objective of this Convention and any related legal instruments that the Conference of the Parties may adopt is to achieve, in accordance with the relevant provisions of the Convention, stabilization of greenhouse gas concentrations in the atmosphere at a level that would prevent dangerous anthropogenic interference with the climate system. Such a level should be achieved within a time-frame sufficient to allow ecosystems to adapt naturally to climate change, to ensure that food production is not threatened and to enable economic development to proceed in a sustainable manner.

ARTICLE 3 - PRINCIPLES

In their actions to achieve the objective of the Convention and to implement its provisions, the Parties shall be guided, INTER ALIA, by the following:

1. The Parties should protect the climate system for the benefit of present and future generations of humankind, on the basis of equity and in accordance with their common but differentiated responsibilities and respective capabilities. Accordingly, the developed country Parties should take the lead in combating climate change and the adverse effects thereof.

2. The specific needs and special circumstances of developing country Parties, especially those that are particularly vulnerable to the adverse effects of climate change, and of those Parties, especially developing country Parties, that would have to bear a disproportionate or abnormal burden under the Convention, should be given full consideration.

3. The Parties should take precautionary measures to anticipate, prevent or minimize the causes of climate change and mitigate its adverse effects. Where there are threats of serious or irreversible damage, lack

of full scientific certainty should not be used as a reason for postponing such measures, taking into account that policies and measures to deal with climate change should be cost-effective so as to ensure global benefits at the lowest possible cost. To achieve this, such policies and measures should take into account different socio-economic contexts, be comprehensive, cover all relevant sources, sinks and reservoirs of greenhouse gases and adaptation, and comprise all economic sectors. Efforts to address climate change may be carried out cooperatively by interested Parties.

4. The Parties have a right to, and should, promote sustainable development. Policies and measures to protect the climate system against human-induced change should be appropriate for the specific conditions of each Party and should be integrated with national development programmes, taking into account that economic development is essential for adopting measures to address climate change.

5. The Parties should cooperate to promote a supportive and open international economic system that would lead to sustainable economic growth and development in all Parties, particularly developing country Parties, thus enabling them better to address the problems of climate change. Measures taken to combat climate change, including unilateral ones, should not constitute a means of arbitrary or unjustifiable discrimination or a disguised restriction on international trade.

ARTICLE 4 - COMMITMENTS

1. All Parties, taking into account their common but differentiated responsbilities and their specific national and regional development priorities, objectives and circumstances, shall:

   (a) Develop, periodically update, publish and make available to the Conference of the Parties, in accordance with Article 12, national inventories of anthropogenic emissions by sources and removals by sinks of all greenhouse gases not controlled by the Montreal Protocol, using comparable methodologies to be agreed upon by the Conference of the Parties;

   (b) Formulate, implement, publish and regularly update national and, where appropriate, regional programmes containing measures to mitigate climate change by addressing anthropogenic emissions by sources and removals by sinks of all greenhouse gases not controlled by the Montreal Protocol, and measures to facilitate adequate adaptation to climate change;

(c) Promote and cooperate in the development, application and diffusion, including transfer, of technologies, practices and processes that control, reduce or prevent anthropogenic emissions of greenhouse gases not controlled by the Montreal Protocol in all relevant sectors, including the energy, transport, industry, agriculture, forestry and waste management sectors;

(d) Promote sustainable management, and promote and cooperate in the conservation and enhancement, as appropriate, of sinks and reservoirs of all greenhouse gases not controlled by the Montreal Protocol, including biomass, forests and oceans as well as other terrestrial, coastal and marine ecosystems;

(e) Cooperate in preparing for adaptation to the impacts of climate change; develop and elaborate appropriate and integrated plans for coastal zone management, water resources and agriculture, and for the protection and rehabilitation of areas, particularly in Africa, affected by drought and desertification, as well as floods;

(f) Take climate change considerations into account, to the extent feasible, in their relevant social, economic and environmental policies and actions, and employ appropriate methods, for example impact assessments, formulated and determined nationally, with a view to minimizing adverse effects on the economy, on public health and on the quality of the environment, of projects or measures undertaken by them to mitigate or adapt to climate change;

(g) Promote and cooperate in scientific, technological, technical, socio-economic and other research, systematic observation and development of data archives related to the climate system and intended to further the understanding and to reduce or eliminate the remaining uncertainties regarding the causes, effects, magnitude and timing of climate change and the economic and social consequences of various response strategies;

(h) Promote and cooperate in the full, open and prompt exchange of relevant scientific, technological, technical, socio-economic and legal information related to the climate system and climate change, and to the economic and social consequences of various response strategies;

(i) Promote and cooperate in education, training and public awareness related to climate change and encourage the widest participation in this process, including that of non- governmental organizations; and

(j) Communicate to the Conference of the Parties information related to implementation, in accordance with Article 12.

2. The developed country Parties and other Parties included in Annex I commit themselves specifically as provided for in the following:

(a) Each of these Parties shall adopt nationall policies and take corresponding measures on the mitigation of climate change, by limiting its anthropogenic emissions of greenhouse gases and protecting and enhancing its greenhouse gas sinks and reservoirs. These policies and measures will demonstrate that developed countries are taking the lead in modifying longer-term trends in anthropogenic emissions consistent with the objective of the Convention, recognizing that the return by the end of the present decade to earlier levels of anthropogenic emissions of carbon dioxide and other greenhouse gases not controlled by the Montreal Protocol would contribute to such modification, and taking into account the differences in these Parties' starting points and approaches, economic structures and resource bases, the need to maintain strong and sustainable economic growth, available technologies and other individual circumstances, as well as the need for equitable and appropriate contributions by each of these Parties to the global effort regarding that objective. These Parties may implement such policies and measures jointly with other Parties and may assist other Parties in contributing to the achievement of the objective of the Convention and, in particular, that of this subparagraph;

(b) In order to promote progress to this end, each of these Parties shall communicate, within six months of the entry into force of the Convention for it and periodically thereafter, and in accordance with Article 12, detailed information on its policies and measures referred to in subparagraph (a) above, as well as on its resulting projected anthropogenic emissions by sources and removals by sinks of greenhouse gases not controlled by the Montreal Protocol for the period referred to in subparagraph (a), with the aim of returning individually or jointly to their 1990 levels these anthropogenic emissions of carbon dioxide and other greenhouse gases not controlled by the Montreal Protocol. This information will be reviewed by the Conference of the

Parties, at its first session and periodically thereafter, in accordance with Article 7;

(c) Calculations of emissions by sources and removals by sinks of greenhouse gases for the purposes of subparagraph (b) above should take into account the best available scientific knowledge, including of the effective capacity of sinks and the respective contributions of such gases to climate change. The Conference of the Parties shall consider and agree on methodologies for these calculations at its first session and review them regularly thereafter;

(d) The Conference of the Parties shall, at its first session, review the adequacy of subparagraphs (a) and (b) above. Such review shall be carried out in the light of the best available scientific information and assessment on climate change and its impacts, as well as relevant technical, social and economic information. Based on this review, the Conference of the Parties shall take appropriate action, which may include the adoption of amendments to the commitments in subparagraphs (a) and (b) above. The Conference of the Parties, at its first session, shall also take decisions regarding criteria for joint implementation as indicated in subparagraph (a) above. A second review of subparagraphs (a) and (b) shall take place not later than 31 December 1998, and thereafter at regular intervals determined by the Conference of the Parties, until the objective of the Convention is met;

(e) Each of these Parties shall :

(i) Coordinate as appropriate with other such Parties, relevant economic and administrative instruments developed to achieve the objective of the Convention; and

(ii) Identify and periodically review its own policies and practices which encourage activities that lead to greater levels of anthropogenic emissions of greenhouse gases not controlled by the Montreal Protocol than would otherwise occur;

(f) The Conference of the Parties shall review, not later than 31 December 1998, available information with a view to taking decisions regarding such amendments to the lists in Annexes I and II as may be appropriate, with the approval of the Party concerned;

(g) Any Party not included in Annex I may, in its instrument of ratification, acceptance, approval or accession, or at any time thereafter, notify the Depositary that it intends to be bound by subparagraphs (a) and (b) above. The Depositary shall inform the other signatories and Parties of any such notification.

3. The developed country Parties and other developed Parties included in Annex II shall provide new and additional financial resources to meet the agreed full costs incurred by developing country Parties in complying with their obligations under Article 12, paragraph 1. They shall also provide such financial resources, including for the transfer of technology, needed by the developing country Parties to meet the agreed full incremental costs of implementing measures that are covered by paragraph 1 of this Article and that are agreed between a developing country Party and the international entity or entities referred to in Article 11, in accordance with that Article. The implementation of these commitments shall take into account the need for adequacy and predictability in the flow of funds and the importance of appropriate burden sharing among the developed country Parties.

4. The developed country Parties and other developed Parties included in Annex II shall also assist the developing country Parties that are particularly vulnerable to the adverse effects of climate change in meeting costs of adaptation to those adverse effects.

5. The developed country Parties and other developed Parties included in Annex II shall take all practicable steps to promote, facilitate and finance, as appropriate, the transfer of, or access to, environmentally sound technologies and know-how to other Parties, particularly developing country Parties, to enable them to implement the provisions of the Convention. In this process, the developed country Parties shall support the development and enhancement of endogenous capacities and technologies of developing country Parties. Other Parties and organizations in a position to do so may also assist in facilitating the transfer of such technologies.

6. In the implementation of their commitments under paragraph 2 above, a certain degree of flexibility shall be allowed by the Conference of the Parties to the Parties included in Annex I undergoing the process of transition to a market economy, in order to enhance the ability of these Parties to address climate change, including with regard to the historical level of anthropogenic emissions of greenhouse gases not controlled by the Montreal Protocol chosen as a reference.

7. The extent to which developing country Parties will effectively implement their commitments under the Convention will depend on the effective implementation by developed country Parties of their commitments under the Convention related to financial resources and transfer of technology and will take fully into account that economic and social development and poverty eradication are the first and overriding priorities of the developing country Parties.

8. In the implementation of the commitments in this Article, the Parties shall give full consideration to what actions are necessary under the Convention, including actions related to funding, insurance and the transfer of technology, to meet the specific needs and concerns of developing country Parties arising from the adverse effects of climate change and/or the impact of the implementation of response measures, especially on:

   (a) Small island countries;

   (b) Countries with low-lying coastal areas;

   (c) Countries with arid and semi-arid areas, forested areas and areas liable to forest decay;

   (d) Countries with areas prone to natural disasters;

   (e) Countries with areas liable to drought and desertification;

   (f) Countries with areas of high urban atmospheric pollution;

   (g) Countries with areas with fragile ecosystems, including mountainous ecosystems;

   (h) Countries whose economies are highly dependent on income generated from the production, processing and export, and/or on consumption of fossil fuels and associated energy-intensive products; and

   (i) Land-locked and transit countries.

Further, the Conference of the Parties may take actions, as appropriate, with respect to this paragraph.

9. The Parties shall take full account of the specific needs and special situations of the least developed countries in their actions with regard to funding and transfer of technology.

10. The Parties shall, in accordance with Article 10, take into consideration in the implementation of the commitments of the Convention the situation of Parties, particularly developing country Parties, with economies that are vulnerable to the adverse effects of the implementation of measures to respond to climate change. This applies notably to Par-

ties with economies that are highly dependent on income generated from the production, processing and export, and/or consumption of fossil fuels and associated energy-intensive products and/or the use of fossil fuels for which such Parties have serious difficulties in switching to alternatives.

ARTICLE 5 - RESEARCH AND SYSTEMATIC OBSERVATION

In carrying out their commitments under Article 4, paragraph 1(g), the Parties shall:

(a) Support and further develop, as appropriate, international and intergovernmental programmes and networks or organizations aimed at defining, conducting, assessing and financing research, data collection and systematic observation, taking into account the need to minimize duplication of effort;

(b) Support international and intergovernmental efforts to strengthen systematic observation and national scientific and technical research capacities and capabilities, particularly in developing countries, and to promote access to, and the exchange of, data and analyses thereof obtained from areas beyond national jurisdiction; and

(c) Take into account the particular concerns and needs of developing countries and cooperate in improving their endogenous capacities and capabilities to participate in the efforts referred to in subparagraphs (a) and (b) above.

ARTICLE 6 - EDUCATION, TRAINING AND PUBLIC AWARENESS

In carrying out their commitments under Article 4, paragraph 1(i), the Parties shall:

(a) Promote and facilitate at the national and, as appropriate, subregional and regional levels, and in accordance with national laws and regulations, and within their respective capacities:

(i) The development and implementation of educational and public awareness programmes on climate change and its effects;

(ii) Public access to information on climate change and its effects;

(iii) Public participation in addressing climate change and its effects and developing adequate responses; and

(iv) Training of scientific, technical and managerial personnel.

(b) Cooperate in and promote, at the international level, and, where appropriate, using existing bodies:

   (i) The development and exchange of educational and public awareness material on climate change and its effects; and

   (ii) The development and implementation of education and training programmes, including the strengthening of national institutions and the exchange or secondment of personnel to train experts in this field, in particular for developing countries.

ARTICLE 7 - CONFERENCE OF THE PARTIES

1. A Conference of the Parties is hereby established.

2. The Conference of the Parties, as the supreme body of this Convention, shall keep under regular review the implementation of the Convention and any related legal instruments that the Conference of the Parties may adopt, and shall make, within its mandate, the decisions necessary to promote the effective implementation of the Convention. To this end, it shall:

   (a) Periodically examine the obligations of the Parties and the institutional arrangements under the Convention, in the light of the objective of the Convention, the experience gained in its implementation and the evolution of scientific and technological knowledge;

   (b) Promote and facilitate the exchange of information on measures adopted by the Parties to address climate change and its effects, taking into account the differing circumstances, responsibilities and capabilities of the Parties and their respective commitments under the Convention;

   (c) Facilitate, at the request of two or more Parties, the coordination of measures adopted by them to address climate change and its effects, taking into account the differing circumstances, responsibilities and capabilities of the Parties and their respective commitments under the Convention;

   (d) Promote and guide, in accordance with the objective and provisions of the Convention, the development and periodic refinement of comparable methodologies, to be agreed on by the Conference of the Parties, inter alia, for preparing inventories of greenhouse gas emissions by sources and removals by sinks, and for evaluating the effectiveness of measures to limit the emissions and enhance the removals of these gases;

(e) Assess, on the basis of all information made available to it in accordance with the provisions of the Convention, the implementation of the Convention by the Parties, the overall effects of the measures taken pursuant to the Convention, in particular environmental, economic and social effects as well as their cumulative impacts and the extent to which progress towards the objective of the Convention is being achieved;

(f) Consider and adopt regular reports on the implementation of the Convention and ensure their publication;

(g) Make recommendations on any matters necessary for the implementation of the Convention;

(h) Seek to mobilize financial resources in accordance with Article 4, paragraphs 3, 4 and 5, and Article 11;

(i) Establish such subsidiary bodies as are deemed necessary for the implementation of the Convention;

(j) Review reports submitted by its subsidiary bodies and provide guidance to them;

(k) Agree upon and adopt, by consensus, rules of procedure and financial rules for itself and for any subsidiary bodies;

(l) Seek and utilize, where appropriate, the services and cooperation of, and information provided by, competent international organizations and intergovernmental and non-governmental bodies; and

(m) Exercise such other functions as are required for the achievement of the objective of the Convention as well as all other functions assigned to it under the Convention.

3. The Conference of the Parties shall, at its first session, adopt its own rules of procedure as well as those of the subsidiary bodies established by the Convention, which shall include decision-making procedures for matters not already covered by decision- making procedures stipulated in the Convention. Such procedures may include specified majorities required for the adoption of particular decisions.

4. The first session of the Conference of the Parties shall be convened by the interim secretariat referred to in Article 21 and shall take place not later than one year after the date of entry into force of the Convention. Thereafter, ordinary sessions of the Conference of the Parties shall be held every year unless otherwise decided by the Conference of the Parties.

5. Extraordinary sessions of the Conference of the Parties shall be held at such other times as may be deemed necessary by the Conference, or at the written request of any Party, provided that, within six months of the request being communicated to the Parties by the secretariat, it is supported by at least one third of the Parties.

6. The United Nations, its specialized agencies and the International Atomic Energy Agency, as well as any State member thereof or observers thereto not Party to the Convention, may be represented at sessions of the Conference of the Parties as observers. Any body or agency, whether national or international, governmental or non- governmental, which is qualified in matters covered by the Convention, and which has informed the secretariat of its wish to be represented at a session of the Conference of the Parties as an observer, may be so admitted unless at least one third of the Parties present object. The admission and participation of observers shall be subject to the rules of procedure adopted by the Conference of the Parties.

ARTICLE 8 - SECRETARIAT

1. A secretariat is hereby established.

2. The functions of the secretariat shall be:

   (a) To make arrangements for sessions of the Conference of the Parties and its subsidiary bodies established under the Convention and to provide them with services as required;

   (b) To compile and transmit reports submitted to it;

   (c) To facilitate assistance to the Parties, particularly developing country Parties, on request, in the compilation and communication of information required in accordance with the provisions of the Convention;

   (d) To prepare reports on its activities and present them to the Conference of the Parties;

   (e) To ensure the necessary coordination with the secretariats of other relevant international bodies;

   (f) To enter, under the overall guidance of the Conference of the Parties, into such administrative and contractual arrangements as may be required for the effective discharge of its functions; and

   (g) To perform the other secretariat functions specified in the Convention and in any of its protocols and such other functions as may be determined by the Conference of the Parties.

3. The Conference of the Parties, at its first session, shall designate a permanent secretariat and make arrangements for its functioning.

ARTICLE 9 - SUBSIDIARY BODY FOR SCIENTIFIC AND TECHNOLOGICAL ADVICE

1. A subsidiary body for scientific and technological advice is hereby established to provide the Conference of the Parties and, as appropriate, its other subsidiary bodies with timely information and advice on scientific and technological matters relating to the Convention. This body shall be open to participation by all Parties and shall be multidisciplinary. It shall comprise government representatives competent in the relevant field of expertise. It shall report regularly to the Conference of the Parties on all aspects of its work.

2. Under the guidance of the Conference of the Parties, and drawing upon existing competent international bodies, this body shall:

   (a) Provide assessments of the state of scientific knowledge relating to climate change and its effects;

   (b) Prepare scientific assessments on the effects of measures taken in the implementation of the Convention;

   (c) Identify innovative, efficient and state-of-the-art technologies and know-how and advise on the ways and means of promoting development and/or transferring such technologies;

   (d) Provide advice on scientific programmes, international cooperation in research and development related to climate change, as well as on ways and means of supporting endogenous capacity-building in developing countries; and

   (e) Respond to scientific, technological and methodological questions that the Conference of the Parties and its subsidiary bodies may put to the body.

3. The functions and terms of reference of this body may be further elaborated by the Conference of the Parties.

ARTICLE 10 - SUBSIDIARY BODY FOR IMPLEMENTATION

1. A subsidiary body for implementation is hereby established to assist the Conference of the Parties in the assessment and review of the effective implementation of the Convention. This body shall be open to participation by all Parties and comprise government representatives who are experts on matters related to climate change. It shall report regularly to the Conference of the Parties on all aspects of its work.

2. Under the guidance of the Conference of the Parties, this body shall:

    (a) Consider the information communicated in accordance with Article 12, paragraph 1, to assess the overall aggregated effect of the steps taken by the Parties in the light of the latest scientific assessments concerning climate change;

    (b) Consider the information communicated in accordance with Article 12, paragraph 2, in order to assist the Conference of the Parties in carrying out the reviews required by Article 4, paragraph 2(d); and

    (c) Assist the Conference of the Parties, as appropriate, in the preparation and implementation of its decisions.

ARTICLE 11 - FINANCIAL MECHANISM

1. A mechanism for the provision of financial resources on a grant or concessional basis, including for the transfer of technology, is hereby defined. It shall function under the guidance of and be accountable to the Conference of the Parties, which shall decide on its policies, programme priorities and eligibility criteria related to this Convention. Its operation shall be entrusted to one or more existing international entities.

2. The financial mechanism shall have an equitable and balanced representation of all Parties within a transparent system of governance.

3. The Conference of the Parties and the entity or entities entrusted with the operation of the financial mechanism shall agree upon arrangements to give effect to the above paragraphs, which shall include the following:

    (a) Modalities to ensure that the funded projects to address climate change are in conformity with the policies, programme priorities and eligibility criteria established by the Conference of the Parties;

    (b) Modalities by which a particular funding decision may be reconsidered in light of these policies, programme priorities and eligibility criteria;

    (c) Provision by the entity or entities of regular reports to the Conference of the Parties on its funding operations, which is consistent with the requirement for accountability set out in paragraph 1 above; and

    (d) Determination in a predictable and identifiable manner of the amount of funding necessary and available for the implementation of this Convention and the conditions under which that amount shall be periodically reviewed.

4. The Conference of the Parties shall make arrangements to implement the above- mentioned provisions at its first session, reviewing and taking into account the interim arrangements referred to in Article 21, paragraph 3, and shall decide whether these interim arrangements shall be maintained. Within four years thereafter, the Conference of the Parties shall review the financial mechanism and take appropriate measures.

5. The developed country Parties may also provide and developing country Parties avail themselves of, financial resources related to the implementation of the Convention through bilateral, regional and other multilateral channels.

ARTICLE 12 - COMMUNICATION OF INFORMATION RELATED TO IMPLEMENTATION

1. In accordance with Article 4, paragraph 1, each Party shall communicate to the Conference of the Parties, through the secretariat, the following elements of information:

   (a) A national inventory of anthropogenic emissions by sources and removals by sinks of all greenhouse gases not controlled by the Montreal Protocol, to the extent its capacities permit, using comparable methodologies to be promoted and agreed upon by the Conference of the Parties;

   (b) A general description of steps taken or envisaged by the Party to implement the Convention; and

   (c) Any other information that the Party considers relevant to the achievement of the objective of the Convention and suitable for inclusion in its communication, including, if feasible, material relevant for calculations of global emission trends.

2. Each developed country Party and each other Party included in Annex I shall incorporate in its communication the following elements of information:

   (a) A detailed description of the policies and measures that it has adopted to implement its commitment under Article 4, paragraphs 2(a) and 2(b); and

   (b) A specific estimate of the effects that the policies and measures referred to in subparagraph (a) immediately above will have on anthropogenic emissions by its sources and removals by its sinks of greenhouse gases during the period referred to in Article 4, paragraph 2(a).

3. In addition, each developed country Party and each other developed Party included in Annex II shall incorporate details of measures taken in accordance with Article 4, paragraphs 3, 4 and 5.

4. Developing country Parties may, on a voluntary basis, propose projects for financing, including specific technologies, materials, equipment, techniques or practices that would be needed to implement such projects, along with, if possible, an estimate of all incremental costs, of the reductions of emissions and increments of removals of greenhouse gases, as well as an estimate of the consequent benefits.

5. Each developed country Party and each other Party included in Annex I shall make its initial communication within six months of the entry into force of the Convention for that Party. Each Party not so listed shall make its initial communication within three years of the entry into force of the Convention for that Party, or of the availability of financial resources in accordance with Article 4, paragraph 3. Parties that are least developed countries may make their initial communication at their discretion. The frequency of subsequent communications by all Parties shall be determined by the Conference of the Parties, taking into account the differentiated timetable set by this paragraph.

6. Information communicated by Parties under this Article shall be transmitted by the secretariat as soon as possible to the Conference of the Parties and to any subsidiary bodies concerned. If necessary, the procedures for the communication of information may be further considered by the Conference of the Parties.

7. From its first session, the Conference of the Parties shall arrange for the provision to developing country Parties of technical and financial support, on request, in compiling and communicating information under this Article, as well as in identifying the technical and financial needs associated with proposed projects and response measures under Article 4. Such support may be provided by other Parties, by competent international organizations and by the secretariat, as appropriate.

8. Any group of Parties may, subject to guidelines adopted by the Conference of the Parties, and to prior notification to the Conference of the Parties, make a joint communication in fulfilment of their obligations under this Article, provided that such a communication includes information on the fulfilment by each of these Parties of its individual obligations under the Convention.

9. Information received by the secretariat that is designated by a Party as confidential, in accordance with criteria to be established by the Conference of the Parties, shall be aggregated by the secretariat to protect its confidentiality before being made available to any of the bodies involved in the communication and review of information.

10. Subject to paragraph 9 above, and without prejudice to the ability of any Party to make public its communication at any time, the secretariat shall make communications by Parties under this Article publicly available at the time they are submitted to the Conference of the Parties.

ARTICLE 13 - RESOLUTION OF QUESTIONS REGARDING IMPLEMENTATION

The Conference of the Parties shall, at its first session, consider the establishment of a multilateral consultative process, available to Parties on their request, for the resolution of questions regarding the implementation of the Convention.

ARTICLE 14 - SETTLEMENT OF DISPUTES

1. In the event of a dispute between any two or more Parties concerning the interpretation or application of the Convention, the Parties concerned shall seek a settlement of the dispute through negotiation or any other peaceful means of their own choice.
2. When ratifying, accepting, approving or acceding to the Convention, or at any time thereafter, a Party which is not a regional economic integration organization may declare in a written instrument submitted to the Depositary that, in respect of any dispute concerning the interpretation or application of the Convention, it recognizes as compulsory ipso facto and without special agreement, in relation to any Party accepting the same obligation:
   (a) Submission of the dispute to the International Court of Justice, and/or
   (b) Arbitration in accordance with procedures to be adopted by the Conference of the Parties as soon as practicable, in an annex on arbitration.

A Party which is a regional economic integration organization may make a declaration with like effect in relation to arbitration in accordance with the procedures referred to in subparagraph (b) above.

3. A declaration made under paragraph 2 above shall remain in force until it expires in accordance with its terms or until three months after written notice of its revocation has been deposited with the Depositary.
4. A new declaration, a notice of revocation or the expiry of a declaration shall not in any way affect proceedings pending before the International Court of Justice or the arbitral tribunal, unless the parties to the dispute otherwise agree.

5. Subject to the operation of paragraph 2 above, if after twelve months following notification by one Party to another that a dispute exists between them, the Parties concerned have not been able to settle their dispute through the means mentioned in paragraph 1 above, the dispute shall be submitted, at the request of any of the parties to the dispute, to conciliation.

6. A conciliation commission shall be created upon the request of one of the parties to the dispute. The commission shall be composed of an equal number of members appointed by each party concerned and a chairman chosen jointly by the members appointed by each party. The commission shall render a recommendatory award, which the parties shall consider in good faith.

7. Additional procedures relating to conciliation shall be adopted by the Conference of the Parties, as soon as practicable, in an annex on conciliation.

8. The provisions of this Article shall apply to any related legal instrument which the Conference of the Parties may adopt, unless the instrument provides otherwise.

ARTICLE 15 - AMENDMENTS TO THE CONVENTION

1. Any Party may propose amendments to the Convention.

2. Amendments to the Convention shall be adopted at an ordinary session of the Conference of the Parties. The text of any proposed amendment to the Convention shall be communicated to the Parties by the secretariat at least six months before the meeting at which it is proposed for adoption. The secretariat shall also communicate proposed amendments to the signatories to the Convention and, for information, to the Depositary.

3. The Parties shall make every effort to reach agreement on any proposed amendment to the Convention by consensus. If all efforts at consensus have been exhausted, and no agreement reached, the amendment shall as a last resort be adopted by a three-fourths majority vote of the Parties present and voting at the meeting. The adopted amendment shall be communicated by the secretariat to the Depositary, who shall circulate it to all Parties for their acceptance.

4. Instruments of acceptance in respect of an amendment shall be deposited with the Depositary. An amendment adopted in accordance with paragraph 3 above shall enter into force for those Parties having accepted it on the ninetieth day after the date of receipt by the Depositary of an instrument of acceptance by at least three fourths of the Parties to the Convention.

5. The amendment shall enter into force for any other Party on the ninetieth day after the date on which that Party deposits with the Depositary its instrument of acceptance of the said amendment.

6. For the purposes of this Article, "Parties present and voting" means Parties present and casting an affirmative or negative vote.

ARTICLE 16 - ADOPTION AND AMENDMENT OF ANNEXES TO THE CONVENTION

1. Annexes to the Convention shall form an integral part thereof and, unless otherwise expressly provided, a reference to the Convention constitutes at the same time a reference to any annexes thereto. Without prejudice to the provisions of Article 14, paragraphs 2(b) and 7, such annexes shall be restricted to lists, forms and any other material of a descriptive nature that is of a scientific, technical, procedural or administrative character.

2. Annexes to the Convention shall be proposed and adopted in accordance with the procedure set forth in Article 15, paragraphs 2, 3 and 4.

3. An annex that has been adopted in accordance with paragraph 2 above shall enter into force for all Parties to the Convention six months after the date of the communication by the Depositary to such Parties of the adoption of the annex, except for those Parties that have notified the Depositary, in writing, within that period of their non-acceptance of the annex. The annex shall enter into force for Parties which withdraw their notification of non-acceptance on the ninetieth day after the date on which withdrawal of such notification has been received by the Depositary.

4. The proposal, adoption and entry into force of amendments to annexes to the Convention shall be subject to the same procedure as that for the proposal, adoption and entry into force of annexes to the Convention in accordance with paragraphs 2 and 3 above.

5. If the adoption of an annex or an amendment to an annex involves an amendment to the Convention, that annex or amendment to an annex shall not enter into force until such time as the amendment to the Convention enters into force.

ARTICLE 17 - PROTOCOLS

1. The Conference of the Parties may, at any ordinary session, adopt protocols to the Convention.

2. The text of any proposed protocol shall be communicated to the Parties by the secretariat at least six months before such a session.

3. The requirements for the entry into force of any protocol shall be established by that instrument.
4. Only Parties to the Convention may be Parties to a protocol.
5. Decisions under any protocol shall be taken only by the Parties to the protocol concerned.

ARTICLE 18 - RIGHT TO VOTE

1. Each Party to the Convention shall have one vote, except as provided for in paragraph 2 below.
2. Regional economic integration organizations, in matters within their competence, shall exercise their right to vote with a number of votes equal to the number of their member States that are Parties to the Convention. Such an organization shall not exercise its right to vote if any of its member States exercises its right, and vice versa.

ARTICLE 19 - DEPOSITARY

The Secretary-General of the United Nations shall be the Depositary of the Convention and of protocols adopted in accordance with Article 17.

ARTICLE 20 - SIGNATURE

This Convention shall be open for signature by States Members of the United Nations or of any of its specialized agencies or that are Parties to the Statute of the International Court of Justice and by regional economic integration organizations at Rio de Janeiro, during the United Nations Conference on Environment and Development, and thereafter at United Nations Headquarters in New York from 20 June 1992 to 19 June 1993.

ARTICLE 21 - INTERIM ARRANGEMENTS

1. The secretariat functions referred to in Article 8 will be carried out on an interim basis by the secretariat established by the General Assembly of the United Nations in its resolution 45/212 of 21 December 1990, until the completion of the first session of the Conference of the Parties.
2. The head of the interim secretariat referred to in paragraph 1 above will cooperate closely with the Intergovernmental Panel on Climate Change to ensure that the Panel can respond to the need for objective scientific and technical advice. Other relevant scientific bodies could also be consulted.
3. The Global Environment Facility of the United Nations Development Programme, the United Nations Environment Programme and the International Bank for Reconstruction and Development shall be the international entity entrusted with the operation of the financial mechanism referred to in Article 11 on an interim basis. In this connection, the Glo-

bal Environment Facility should be appropriately restructured and its membership made universal to enable it to fulfil the requirements of Article 11.

ARTICLE 22 - RATIFICATION, ACCEPTANCE, APPROVAL OR ACCESSION

1. The Convention shall be subject to ratification, acceptance, approval or accession by States and by regional economic integration organizations. It shall be open for accession from the day after the date on which the Convention is closed for signature. Instruments of ratification, acceptance, approval or accession shall be deposited with the Depositary.

2. Any regional economic integration organization which becomes a Party to the Convention without any of its member States being a Party shall be bound by all the obligations under the Convention. In the case of such organizations, one or more of whose member States is a Party to the Convention, the organization and its member States shall decide on their respective responsibilities for the performance of their obligations under the Convention. In such cases, the organization and the member States shall not be entitled to exercise rights under the Convention concurrently.

3. In their instruments of ratification, acceptance, approval or accession, regional economic integration organizations shall declare the extent of their competence with respect to the matters governed by the Convention. These organizations shall also inform the Depositary, who shall in turn inform the Parties, of any substantial modification in the extent of their competence.

ARTICLE 23 - ENTRY INTO FORCE

1. The Convention shall enter into force on the ninetieth day after the date of deposit of the fiftieth instrument of ratification, acceptance, approval or accession.

2. For each State or regional economic integration organization that ratifies, accepts or approves the Convention or accedes thereto after the deposit of the fiftieth instrument of ratification, acceptance, approval or accession, the Convention shall enter into force on the ninetieth day after the date of deposit by such State or regional economic integration organization of its instrument of ratification, acceptance, approval or accession.

3. For the purposes of paragraphs 1 and 2 above, any instrument deposited by a regional economic integration organization shall not be counted as additional to those deposited by States members of the organization.

ARTICLE 24 - RESERVATIONS

No reservations may be made to the Convention.

ARTICLE 25 - WITHDRAWAL

1. At any time after three years from the date on which the Convention has entered into force for a Party, that Party may withdraw from the Convention by giving written notification to the Depositary.

2. Any such withdrawal shall take effect upon expiry of one year from the date of receipt by the Depositary of the notification of withdrawal, or on such later date as may be specified in the notification of withdrawal.

3. Any Party that withdraws from the Convention shall be considered as also having withdrawn from any protocol to which it is a Party.

ARTICLE 26 - AUTHENTIC TEXTS

The original of this Convention, of which the Arabic, Chinese, English, French, Russian and Spanish texts are equally authentic, shall be deposited with the Secretary- General of the United Nations.

IN WITNESS WHEREOF the undersigned, being duly authorized to that effect, have signed this Convention.

DONE at New York this ninth day of May one thousand nine hundred and ninety-two.

ANNEX I AND ANNEX II COUNTRIES

Annex I
Australia
Austria
Belarus a/
Belgium
Bulgaria a/
Canada
Czechoslovakia a/
Denmark
European Economic Community
Estonia a/
Finland
France
Germany
Greece
Hungary a/
Iceland
Ireland
Italy
Japan

Latvia a/
Lithuania a/
Luxembourg
Netherlands
New Zealand
Norway
Poland a/
Portugal
Romania a/
Russian Federation a/Spain
Sweden
Switzerland
Turkey
Ukraine a/
United Kingdom of Great Britain and Northern Ireland
United States of America

a/ Countries that are undergoing the process of transition to a market economy.

Annex II
Australia
Austria
Belgium
Canada
Denmark
European Economic Community
Finland
France
Germany
Greece
Iceland
Ireland
Italy
Japan
Luxembourg
Netherlands
New Zealand
Norway
Portugal
Spain
Sweden
Switzerland
Turkey
United Kingdom of Great Britain and Northern Ireland
United States of America

Appendix II

Distr.

LIMITED

FCCC/CP/1997/L.7/Add.1

10 December 1997

Subject to technical revision

Original: ENGLISH

CONFERENCE OF THE PARTIES

Third session

Kyoto, 1-10 December 1997

Agenda item 5

KYOTO PROTOCOL TO THE UNITED NATIONS FRAMEWORK CONVENTION ON CLIMATE CHANGE

The Parties to this Protocol,

Being Parties to the United Nations Framework Convention on Climate Change, hereinafter referred to as "the Convention",

In pursuit of the ultimate objective of the Convention as stated in its Article 2,

Recalling the provisions of the Convention,

Being guided by Article 3 of the Convention,

Pursuant to the Berlin Mandate adopted by decision 1/CP.1 of the Conference of the Parties to the Convention at its first session,

Have agreed as follows:

ARTICLE 1

For the purposes of this Protocol, the definitions contained in Article 1 of the Convention shall apply. In addition:

1. "Conference of the Parties" means the Conference of the Parties to the Convention.

UKY.97

2. "Convention" means the United Nations Framework Convention on Climate Change, adopted in New York on 9 May 1992.

3. "Intergovernmental Panel on Climate Change" means the Intergovernmental Panel on Climate Change established in 1988 jointly by the World Meteorological Organization and the United Nations Environment Programme.

4. "Montreal Protocol" means the Montreal Protocol on Substances that Deplete the Ozone Layer, adopted in Montreal on 16 September 1987 and as subsequently adjusted and amended.

5. "Parties present and voting" means Parties present and casting an affirmative or negative vote.

6. "Party" means, unless the context otherwise indicates, a Party to this Protocol.

7. "Party included in Annex I" means a Party included in Annex I to the Convention, as may be amended, or a Party which has made a notification under Article 4, paragraph 2(g), of the Convention.

ARTICLE 2

1. Each Party included in Annex I in achieving its quantified emission limitation and reduction commitments under Article 3, in order to promote sustainable development, shall:

   (a) Implement and/or further elaborate policies and measures in accordance with its national circumstances, such as:

      (i) Enhancement of energy efficiency in relevant sectors of the national economy;

      (ii) Protection and enhancement of sinks and reservoirs of greenhouse gases not controlled by the Montreal Protocol, taking into account its commitments under relevant international environmental agreements; promotion of sustainable forest management practices, afforestation and-reforestation;

      (iii) Promotion of sustainable forms of agriculture in light of climate change considerations;

      (iv) Promotion, research, development and increased use of new and renewable forms of energy, of carbon dioxide sequestration technologies and of advanced and innovative environmentally sound technologies;

      (v) Progressive reduction or phasing out of market imperfections, fiscal incentives, tax and duty exemptions and subsidies in all greenhouse gas emitting sectors that run counter to the objective of the Convention and apply market instruments;

(vi) Encouragement of appropriate reforms in relevant sectors aimed at promoting policies and measures which limit or reduce emissions of greenhouse gases not controlled by the Montreal Protocol;

(vii) Measures to limit and/or reduce emissions of greenhouse gases not controlled by the Montreal Protocol in the transport sector;

(viii) Limitation and/or reduction of methane through recovery and use in waste management, as well as in the production, transport and distribution of energy;

(b) Cooperate with other such Parties to enhance the individual and combined effectiveness of their policies and measures adopted under this Article, pursuant to Article 4, paragraph 2(e)(i), of the Convention. To this end, these Parties shall take steps to share their experience and exchange information on such policies and measures, including developing ways of improving their comparability, transparency and effectiveness. The Conference of the Parties serving as the meeting of the Parties to this Protocol shall, at its first session or as soon as practicable thereafter, consider ways to facilitate such cooperation, taking into account all relevant information.

2. The Parties included in Annex I shall pursue limitation or reduction of emissions of greenhouse gases not controlled by the Montreal Protocol from aviation and marine bunker fuels, working through the International Civil Aviation Organization and the International Maritime Organization, respectively.

3. The Parties included in Annex I shall strive to implement policies and measures under this Article in such a way as to minimize adverse effects, including the adverse effects of climate change, effects on international trade, and social, environmental and economic impacts on other Parties, especially developing country Parties and in particular those identified in Article 4, paragraphs 8 and 9 of the Convention, taking into account Article 3 of the Convention. The Conference of the Parties serving as the meeting of the Parties to this Protocol may take further action, as appropriate, to promote the implementation of the provisions of this paragraph.

4. The Conference of the Parties serving as the meeting of the Parties to this Protocol, if it decides that it would be beneficial to coordinate any of the policies and measures in paragraph 1(a) above, taking into account different national circumstances and potential effects, shall consider ways and means to elaborate the coordination of such policies and measures.

ARTICLE 3

1. The Parties included in Annex I shall, individually or jointly, ensure that their aggregate anthropogenic carbon dioxide equivalent emissions of the greenhouse gases listed in Annex A do not exceed their assigned amounts, calculated pursuant to their quantified emission limitation and reduction commitments inscribed in Annex B and in accordance with the provisions of this Article, with a view to reducing their overall emissions of such gases by at least 5 per cent below 1990 levels in the commitment period 2008 to 2012.

2. Each Party included in Annex I shall, by 2005, have made demonstrable progress in achieving its commitments under this Protocol.

3. The net changes in greenhouse gas emissions from sources and removals by sinks resulting from direct human-induced land use change and forestry activities, limited to afforestation, reforestation, and deforestation since 1990, measured as verifiable changes in stocks in each commitment period shall be used to meet the commitments in this Article of each Party included in Annex I. The greenhouse gas emissions from sources and removals by sinks associated with those activities shall be reported in a transparent and verifiable manner and reviewed in accordance with Articles 7 and 8.

4. Prior to the first session of the Conference of the Parties serving as the meeting of the Parties to this Protocol, each Party included in Annex I shall provide for consideration by the Subsidiary Body for Scientific and Technological Advice data to establish its level of carbon stocks in 1990 and to enable an estimate to be made of its changes in carbon stocks in subsequent years. The Conference of the Parties serving as the meeting of the Parties to this Protocol shall, at its first session or as soon as practicable thereafter, decide upon modalities, rules and guidelines as to how and which additional human-induced activities related to changes in greenhouse gas emissions and removals in the agricultural soil and land use change and forestry categories, shall be added to, or subtracted from, the assigned amount for Parties included in Annex I, taking into account uncertainties, transparency in reporting, verifiability, the methodological work of the Intergovernmental Panel on Climate Change, the advice provided by the Subsidiary Body for Scientific and Technological Advice in accordance with Article 5 and the decisions of the Conference of the Parties. Such a decision shall apply in the second and subsequent commitment periods. A Party may choose to apply such a decision on these additional human-induced activities for its first commitment period, provided that these activities have taken place since 1990.

5. The Parties included in Annex I undergoing the process of transition to a market economy whose base year or period was established pursuant to decision 9/CP.2 of the Conference of the Parties at its second session, shall use that base year or period for the implementation of their commitments under this Article. Any other Party included in Annex I undergoing the process of transition to a market economy which has not yet submitted its first national communication under Article 12 of the Convention may also notify the Conference of the Parties serving as the meeting of the Parties to this Protocol that it intends to use a historical base year or period other than 1990 for the implementation of its commitments under this Article. The Conference of the Parties serving as the meeting of the Parties to this Protocol shall decide on the acceptance of such notification.
6. Taking into account Article 4, paragraph 6, of the Convention, in the implementation of their commitments under this Protocol other than those in this Article, a certain degree of flexibility shall be allowed by the Conference of the Parties serving as the meeting of the Parties to this Protocol to the Parties included in Annex I undergoing the process of transition to a market economy.
7. In the first quantified emission limitation and reduction commitment period, from 2008 to 2012, the assigned amount for each Party included in Annex I shall be equal to the percentage inscribed for it in Annex B of its aggregate anthropogenic carbon dioxide equivalent emissions of the greenhouse gases listed in Annex A in 1990, or the base year or period determined in accordance with paragraph 5 above, multiplied by five. Those Parties included in Annex I for whom land use change and forestry constituted a net source of greenhouse gas emissions in 1990 shall include in their 1990 emissions base year or period the aggregate anthropogenic carbon dioxide equivalent emissions minus removals in 1990 from land use change for the purposes of calculating their assigned amount.
8. Any Party included in Annex I may use 1995 as its base year for hydrofluorocarbons, perfluorocarbons and sulphur hexafluoride, for the purposes of the calculation referred to in paragraph 7 above.
9. Commitments for subsequent periods for Parties included in Annex I shall be established in amendments to Annex B to this Protocol, which shall be adopted in accordance with the provisions of Article 20, paragraph 7. The Conference of the Parties serving as the meeting of the Parties to this Protocol shall initiatc thc considcration of such commitments at least seven years before the end of the first commitment period mentioned in paragraph 7 above.

10. Any emission reduction units, or any part of an assigned amount, which a Party acquires from another Party in accordance with the provisions of Article 6 and of Article 16 bis shall be added to the assigned amount for that Party.

11. Any emission reduction units, or any part of an assigned amount, which a Party transfers to another Party in accordance with the provisions of Article 6 and of Article 16 bis shall be subtracted from the assigned amount for that Party.

12. Any certified emission reductions which a Party acquires from another Party in accordance with the provisions of Article 12 shall be added to the assigned amount for that Party.

13. If the emissions of a Party included in Annex I during a commitment period are less than its assigned amount under this Article, this difference shall, on request of that Party, be added to the assigned amount for that Party for subsequent commitment periods.

14. Each Party included in Annex I shall strive to implement the commitments mentioned in paragraph 1 above in such a way as to minimize adverse social, environmental and economic impacts on developing country Parties, particularly those identified in Article 4, paragraphs 8 and 9, of the Convention. In line with relevant decisions of the Conference of the Parties on the implementation of those paragraphs, the Conference of the Parties serving as the meeting of the Parties to this Protocol shall, at its first session, consider what actions are necessary to minimize the adverse effects of climate change and/or the impacts of response measures on Parties referred to in those paragraphs. Among the issues to be considered shall be the establishment of funding, insurance and transfer of technology.

ARTICLE 4

1. Any Parties included in Annex I that have agreed to jointly fulfil their commitments under Article 3 shall be deemed to have met those commitments provided that their total combined aggregate anthropogenic carbon dioxide equivalent emissions of the greenhouse gases listed in Annex A do not exceed their assigned amounts calculated pursuant to their quantified emission limitation and reduction commitments inscribed in Annex B and in accordance with the provisions of Article 3. The respective emission level allocated to each of the Parties to the agreement shall be set out in that agreement.

2. The Parties to any such agreement shall notify the secretariat of the terms of the agreement on the date of deposit of their instruments of ratification, acceptance, approval or accession. The secretariat shall in turn inform the Parties and signatories to the Convention of the terms of the agreement.

3. The agreement shall remain in operation for the duration of the commitment period specified in Article 3, paragraph 7.

4. If Parties acting jointly do so in the framework of, and together with, a regionaleconomic integration organization, any alteration in the composition of the organization after adoption of this Protocol shall not affect existing commitments under this Protocol. Any alteration in the composition of the organization shall only apply for the purposes of those commitments under Article 3 that are adopted subsequent to that revision.

5. In the event of failure by the Parties to such an agreement to achieve their totalcombined level of emission reductions, each Party to such an agreement shall be responsible for its own level of emissions set out in the agreement.

6. If Parties acting jointly do so in the framework of, and together with, a regional economic integration organization which is itself a Party to this Protocol, each member State of that regional economic integration organization individually, and together with the regional economic integration organization acting in accordance with Article 23, shall, in the event of failure to achieve the total combined level of emission reductions, be responsible for its level of emissions as notified in accordance with this Article.

ARTICLE 5

1. Each Party included in Annex I shall have in place, no later than one year prior to the start of the first commitment period, a national system for the estimation of anthropogenic emissions by sources and removals by sinks of all greenhouse gases not controlled by the Montreal Protocol. Guidelines for such national systems, which shall incorporate the methodologies specified in paragraph 2 below, shall be decided upon by the Conference of the Parties serving as the meeting of the Parties to this Protocol at its first session.

2. Methodologies for estimating anthropogenic emissions by sources and removals by sinks of all greenhouse gases not controlled by the Montreal Protocol shall be those accepted by the Intergovernmental Panel on Climate Change and agreed upon by the Conference of the Parties at its third session. Where such methodologies are not used, appropriate adjustments shall be applied according to methodologies agreed upon

by the Conference of the Parties serving as the meeting of the Parties to this Protocol at its first session. Based on the work of, *inter alia*, the Intergovernmental Panel on Climate Change and advice provided by the Subsidiary Body for Scientific and Technological Advice, the Conference of the Parties serving as the meeting of the Parties to this Protocol shall regularly review and, as appropriate, revise such methodologies and adjustments, taking fully into account any relevant decisions by the Conference of the Parties. Any revision to methodologies or adjustments shall be used only for the purposes of ascertaining compliance with commitments under Article 3 in respect of any commitment period adopted subsequent to that revision.

3. The global warming potentials used to calculate the carbon dioxide equivalence of anthropogenic emissions by sources and removals by sinks of greenhouse gases not controlled by the Montreal Protocol listed in Annex A shall be those accepted by the Intergovernmental Panel on Climate Change and agreed upon by the Conference of the Parties at its third session. Based on the work of, *inter alia*, the Intergovernmental Panel on Climate Change and advice provided by the Subsidiary Body for Scientific and Technological Advice, the Conference of the Parties serving as the meeting of the Parties to this Protocol shall regularly review and, as appropriate, revise the global warming potential of each such greenhouse gas, taking fully into account any relevant decisions by the Conference of the Parties. Any revision to a global warming potential shall apply only to those commitments under Article 3 in respect of any commitment period adopted subsequent to that revision.

ARTICLE 6

1. For the purpose of meeting its commitments under Article 3, any Party included in

   Annex I may transfer to, or acquire from, any other such Party emission reduction units resulting from projects aimed at reducing anthropogenic emissions by sources or enhancing anthropogenic removals by sinks of greenhouse gases in any sector of the economy, provided that:

   (a) Any such project has the approval of the Parties involved;

   (b) Any such project provides a reduction in emissions by sources, or an enhancement of removals by sinks, that is additional to any that would otherwise occur;

   (c) It does not acquire any emission reduction units if it is not in compliance with its obligations under Articles 5 and 7; and

(d) The acquisition of emission reduction units shall be supplemental to domestic actions for the purposes of meeting commitments under Article 3.

2. The Conference of the Parties serving as the meeting of the Parties to this Protocol may, at its first session or as soon as practicable thereafter, further elaborate guidelines for the implementation of this Article, including for verification and reporting.

3. A Party included in Annex I may authorize legal entities to participate, under its responsibility, in actions leading to the generation, transfer or acquisition under this Article of emission reduction units.

4. If a question of implementation by a Party included in Annex I of the requirements referred to in this paragraph is identified in accordance with the relevant provisions of Article 8, transfers and acquisitions of emission reduction units may continue to be made after the question has been identified, provided that any such units may not be used by a Party to meet its commitments under Article 3 until any issue of compliance is resolved.

ARTICLE 7

1. Each Party included in Annex I shall incorporate in its annual inventory of anthropogenic emissions by sources and removals by sinks of greenhouse gases not controlled by the Montreal Protocol, submitted in accordance with the relevant decisions of the Conference of the Parties, the necessary supplementary information for the purposes of ensuring compliance with Article 3, to be determined in accordance with paragraph 4 below.

2. Each Party included in Annex I shall incorporate in its national communication, submitted under Article 12 of the Convention, the supplementary information necessary to demonstrate compliance with its commitments under this Protocol, to be determined in accordance with paragraph 4 below.

3. Each Party included in Annex I shall submit the information required under paragraph 1 above annually, beginning with the first inventory due under the Convention for the first year of the commitment period after this Protocol has entered into force for it. Each such Party shall submit the information required under paragraph 2 above as part of the first national communication due under the Convention after this Protocol has entered into force for it and after the adoption of guidelines as provided for in paragraph 4 below. The frequency of subsequent submission of information required under this Article shall be determined by the Conference of the Parties serving as the meeting of the Parties to this Protocol, taking into

account any timetable for the submission of national communications decided upon by the Conference of the Parties.

4. The Conference of the Parties serving as the meeting of the Parties to this Protocol shall adopt at its first session, and review periodically thereafter, guidelines for the preparation of the information required under this Article, taking into account guidelines for the preparation of national communications by Parties included in Annex I adopted by the Conference of the Parties. The Conference of the Parties serving as the meeting of the Parties to this Protocol shall also, prior to the first commitment period, decide upon modalities for the accounting of assigned amounts.

ARTICLE 8

1. The information submitted under Article 7 by each Party included in Annex I shall be reviewed by expert review teams pursuant to the relevant decisions of the Conference of the Parties and in accordance with guidelines adopted for this purpose by the Conference of the Parties serving as the meeting of the Parties to this Protocol under paragraph 4 below. The information submitted under Article 7, paragraph 1, by each Party included in Annex I shall be reviewed as part of the annual compilation and accounting of emissions inventories and assigned amounts. Additionally, the information submitted under Article 7, paragraph 2, by each Party included in Annex I shall be reviewed as part of the review of communications.

2. Expert review teams shall be coordinated by the secretariat and shall be composed of experts selected from those nominated by Parties to the Convention and, as appropriate, by intergovernmental organizations, in accordance with guidance provided for this purpose by the Conference of the Parties.

3. The review process shall provide a thorough and comprehensive technical assessment of all aspects of the implementation by a Party of this Protocol. The expert review teams shall prepare a report to the Conference of the Parties serving as the meeting of the Parties to this Protocol, assessing the implementation of the commitments of the Party and identifying any potential problems in, and factors influencing, the fulfilment of commitments. Such reports shall be circulated by the secretariat to all Parties to the Convention. The secretariat shall list those questions of implementation indicated in such reports for further consideration by the Conference of the Parties serving as the meeting of the Parties to this Protocol.

4. The Conference of the Parties serving as the meeting of the Parties to this Protocol shall adopt at its first session, and review periodically thereafter, guidelines for the review of implementation by expert review teams taking into account the relevant decisions of the Conference of the Parties.

5. The Conference of the Parties serving as the meeting of the Parties to this Protocol shall, with the assistance of the Subsidiary Body for Implementation and, as appropriate, the Subsidiary Body for Scientific and Technological Advice, consider:

   (a) The information submitted by the Parties under Article 7 and the reports of the expert reviews thereon conducted under this Article; and

   (b) Those questions of implementation listed by the secretariat under paragraph 3 above, as well as any questions raised by Parties.

6. Pursuant to its consideration of the information referred to in paragraph 5 above, the Conference of the Parties serving as the meeting of the Parties to this Protocol shall take decisions on any matter required for the implementation of this Protocol.

ARTICLE 9

1. The Conference of the Parties serving as the meeting of the Parties to this Protocol shall periodically review this Protocol in the light of the best available scientific information and assessments on climate change and its impacts, as well as relevant technical, social and economic information. Such reviews shall be coordinated with pertinent reviews under the Convention, in particular those required by Article 4, paragraph 2(d), and Article 7, paragraph 2(a), of the Convention. Based on these reviews, the Conference of the Parties serving as the meeting of the Parties to this Protocol shall take appropriate action.

2. The first review shall take place at the second session of the Conference of the Parties serving as the meeting of the Parties to this Protocol. Further reviews shall take place at regular intervals and in a timely manner.

ARTICLE 10

All Parties, taking into account their common but differentiated responsibilities and their specific national and regional development priorities, objectives and circumstances, without introducing any new commitments for Parties not included in Annex I, but reaffirming existing commitments in Article 4, paragraph 1, of the Convention, and continuing to advance the implementation of these commitments in order to achieve sustainable development, taking into account Article 4, paragraphs 3, 5 and 7, of the Convention, shall:

(a) Formulate, where relevant and to the extent possible, cost-effective national, and where appropriate regional programmes to improve the quality of local emission factors, activity data and/or models which reflect the socio-economic conditions of each Party for the preparation and periodic updating of national inventories of anthropogenic emissions by sources and removals by sinks of all greenhouse gases not controlled by the Montreal Protocol, using comparable methodologies to be agreed upon by the Conference of the Parties, and consistent with the guidelines for national communications adopted by the Conference of the Parties;

(b) Formulate, implement, publish and regularly update national and, where appropriate, regional programmes containing measures to mitigate climate change and measures to facilitate adequate adaptation to climate change:

(i) Such programmes would, *inter alia*, concern the energy, transport and industry sectors as well as agriculture, forestry and waste management. Furthermore, adaptation technologies and methods for improving spatial planning would improve adaptation to climate change; and

(ii) Parties included in Annex I shall submit information on action under this Protocol, including national programmes, according to the guidelines laid down in Article 8; and other Parties shall seek to include in their national communications, as appropriate, information on programmes which contain measures that the Party believes contribute to addressing climate change and its adverse impacts, including the abatement of increase in greenhouse gas emissions, and enhancement of and removals by sinks, capacity building and adaptation measures.

(c) Cooperate in the promotion of effective modalities for the development, application and diffusion of, and take all practicable steps to promote, facilitate and finance, as appropriate, the transfer of, or access to, environmentally sound technologies, know-how, practices and processes pertinent to climate change, in particular to developing countries, including the formulation of policies and programmes for the effective transfer of environmentally sound technologies that are publicly owned or in the public domain and the creation of an enabling environment for the private sector, to promote and enhance access to, and transfer of, environmentally sound technologies;

(d) Cooperate in scientific and technical research and promote the maintenance and the development of systematic observation systems and development of data archives to reduce uncertainties related to the climate system, the adverse impacts of climate change and the economic and social consequences of various response strategies, and promote the development and strengthening of endogenous capacities and capabilities to participate in international and intergovernmental efforts, programmes and networks on research and systematic observation, taking into account Article 5 of the Convention;

(e) Cooperate in and promote at the international level, and, where appropriate, using existing bodies, the development and implementation of education and training programmes, including the strengthening of national capacity building, in particular human and institutional capacities and the exchange or secondment of personnel to train experts in this field, in particular for developing countries, and facilitate at the national level public awareness and public access to information on climate change. Suitable modalities should be developed to implement these activities through the relevant bodies of the Convention taking into account Article 6 of the Convention;

(f) Include in their national communications information on programmes and activities undertaken pursuant to this Article in accordance with relevant decisions of the Conference of the Parties; and

(g) Give full consideration, in implementing the commitments in this Article, to Article 4, paragraph 8, of the Convention.

ARTICLE 11

1. In the implementation of Article 10, Parties shall take into account the provisions of Article 4, paragraphs 4, 5, 7, 8 and 9 of the Convention.

2. In the context of the implementation of Article 4, paragraph 1, of the Convention, in accordance with the provisions of Article 4, paragraph 3, and Article 11 of the Convention, and through the operating entity or entities of the financial mechanism of the Convention, the developed country Parties and other developed Parties included in Annex II to the Convention shall:

(a) Provide new and additional financial resources to meet the agreed full costs incurred by developing country Parties in advancing the implementation of existing commitments under Article 4, paragraph 1(a), of the Convention that are covered in Article 10, subparagraph (a); and

(b) Also provide such financial resources, including for the transfer of technology, needed by the developing country Parties to meet the agreed full incremental costs of advancing the implementation of existing commitments in Article 4, paragraph 1, of the Convention that are covered by Article 10 and that are agreed between a developing country Party and the international entity or entities referred to in Article 11 of the Convention, in accordance with that Article.

The implementation of these existing commitments shall take into account the need for adequacy and predictability in the flow of funds and the importance of appropriate burden sharing among developed country Parties. The guidance to the financial mechanism of the Convention in relevant decisions of the Conference of the Parties, including those agreed before the adoption of this Protocol, shall apply *mutatis mutandis* to the provisions of this paragraph.

3. The developed country Parties and other developed Parties in Annex II to theConvention may also provide, and developing country Parties avail themselves of, financial resources for the implementation of Article 10, through bilateral, regional and other multilateral channels.

ARTICLE 12

1. A clean development mechanism is hereby defined.

2. The purpose of the clean development mechanism shall be to assist Parties not included in Annex I in achieving sustainable development and in contributing to the ultimate objective of the Convention, and to assist Parties included in Annex I in achieving compliance with their quantified emission limitation and reduction commitments under Article 3.

3. Under the clean development mechanism:

   (a) Parties not included in Annex I will benefit from project activities resulting in certified emission reductions; and

   (b) Parties included in Annex I may use the certified emission reductions accruing from such project activities to contribute to compliance with part of their quantified emission limitation and reduction commitments under Article 3, as determined by the Conference of the Parties serving as the meeting of the Parties to this Protocol.

4. The clean development mechanism shall be subject to the authority and guidance of the Conference of the Parties serving as the meeting of the Parties to this Protocol and be supervised by an executive board of the clean development mechanism.

5. Emission reductions resulting from each project activity shall be certified by operational entities to be designated by the Conference of the Parties serving as the meeting of the Parties to this Protocol, on the basis of:

   (a) Voluntary participation approved by each Party involved;

   (b) Real, measurable, and long-term benefits related to the mitigation of climate change; and

   (c) Reductions in emissions that are additional to any that would occur in the absence of the certified project activity.

6. The clean development mechanism shall assist in arranging funding of certified project activities as necessary.

7. The Conference of the Parties serving as the meeting of the Parties to this Protocol shall, at its first session, elaborate modalities and procedures with the objective of ensuring transparency, efficiency and accountability through independent auditing and verification of project activities.

8. The Conference of the Parties serving as the meeting of the Parties to this Protocol shall ensure that a share of the proceeds from certified project activities is used to cover administrative expenses as well as to assist developing country Parties that are particularly vulnerable to the adverse effects of climate change to meet the costs of adaptation.

9. Participation under the clean development mechanism, including in activities mentioned in paragraph 3(a) above and acquisition of certified emission reductions, may involve private and/or public entities, and is to be subject to whatever guidance may be provided by the executive board of the clean development mechanism.

10. Certified emission reductions obtained during the period from the year 2000 up to the beginning of the first commitment period can be used to assist in achieving compliance in the first commitment period.

ARTICLE 13

1. The Conference of the Parties, the supreme body of the Convention, shall serve as the eeting of the Parties to this Protocol.

2. Parties to the Convention that are not Parties to this Protocol may participate asobservers in the proceedings of any session of the Conference of the Parties serving as the meeting of the Parties to this Protocol.

When the Conference of the Parties serves as the meeting of the Parties to this Protocol, decisions under this Protocol shall be taken only by those that are Parties to it.

3. When the Conference of the Parties serves as the meeting of the Parties to this Protocol, any member of the Bureau of the Conference of the Parties representing a Party to the Convention but, at that time, not a Party to this Protocol, shall be substituted by an additional member to be elected by and from amongst the Parties to this Protocol.

4. The Conference of the Parties serving as the meeting of the Parties to this Protocol shall keep under regular review the implementation of this Protocol and shall make, within its mandate, the decisions necessary to promote its effective implementation. It shall perform the functions assigned to it by this Protocol and shall:

   (a) Assess, on the basis of all information made available to it in accordance withthe provisions of this Protocol, the implementation of this Protocol by the Parties, the overall effects of the measures taken pursuant to this Protocol, in particular environmental, economic and social effects as well as their cumulative impacts and the extent to which progress towards the objective of the Convention is being achieved;

   (b) Periodically examine the obligations of the Parties under this Protocol, giving due consideration to any reviews required by Article 4, paragraph 2(d), and Article 7, paragraph 2, of the Convention, in the light of the objective of the Convention, the experience gained in its implementation and the evolution of scientific and technological knowledge, and in this respect consider and adopt regular reports on the implementation of this Protocol;

   (c) Promote and facilitate the exchange of information on measures adopted by the Parties to address climate change and its effects, taking into account the differing circumstances, responsibilities and capabilities of the Parties and their respective commitments under this Protocol;

   (d) Facilitate, at the request of two or more Parties, the coordination of measures adopted by them to address climate change and its effects, taking into account the differing circumstances, responsibilities and capabilities of the Parties and their respective commitments under this Protocol;

(e) Promote and guide, in accordance with the objective of the Convention and the provisions of this Protocol, and taking fully into account the relevant decisions by the Conference of the Parties, the development and periodic refinement of comparable methodologies for the effective implementation of this Protocol, to be agreed on by the Conference of the Parties serving as the meeting of the Parties to this Protocol;

(f) Make recommendations on any matters necessary for the implementation of this Protocol;

(g) Seek to mobilize additional financial resources in accordance with Article 11, paragraph 2;

(h) Establish such subsidiary bodies as are deemed necessary for the implementation of this Protocol;

(i) Seek and utilize, where appropriate, the services and cooperation of, and information provided by, competent international organizations and intergovernmental and non-governmental bodies; and

(j) Exercise such other functions as may be required for the implementation of this Protocol, and consider any assignment resulting from a decision by the Conference of the Parties.

5. The rules of procedure of the Conference of the Parties and financial procedures of the Convention shall be applied *mutatis mutandis* under this Protocol, except as may be otherwise decided by consensus by the Conference of the Parties serving as the meeting of the Parties to this Protocol.

6. The first session of the Conference of the Parties serving as the meeting of the Parties to this Protocol shall be convened by the secretariat in conjunction with the first session of the Conference of the Parties that is scheduled after the date of the entry into force of this Protocol. Subsequent ordinary sessions of the Conference of the Parties serving as the meeting of the Parties to this Protocol shall be held every year and in conjunction with ordinary sessions of the Conference of the Parties unless otherwise decided by the Conference of the Parties serving as the meeting of the Parties to this Protocol.

7. Extraordinary sessions of the Conference of the Parties serving as the meeting of the Parties to this Protocol shall be held at such other times as may be deemed necessary by the Conference of the Parties serving as the meeting of the Parties to this Protocol, or at the written request of any Party, provided that, within six months of the request being communicated to the Parties by the secretariat, it is supported by at least one third of the Parties.

8. The United Nations, its specialized agencies and the International Atomic Energy Agency, as well as any State member thereof or observers thereto not party to the Convention, may be represented at sessions of the Conference of the Parties serving as the meeting of the Parties to this Protocol as observers. Any body or agency, whether national or international, governmental or non-governmental, which is qualified in matters covered by this Protocol and which has informed the secretariat of its wish to be represented at a session of the Conference of the Parties serving as the meeting of the Parties to this Protocol as an observer, may be so admitted unless at least one third of the Parties present object. The admission and participation of observers shall be subject to the rules of procedure, as referred to in paragraph 5 above.

ARTICLE 14

1. The secretariat established by Article 8 of the Convention shall serve as the secretariat of this Protocol.

2. Article 8, paragraph 2, of the Convention on the functions of the secretariat, and Article 8, paragraph 3, of the Convention on arrangements made for the functioning of the secretariat, shall apply *mutatis mutandis* to this Protocol. The secretariat shall, in addition, exercise the functions assigned to it under this Protocol.

ARTICLE 15

1. The Subsidiary Body for Scientific and Technological Advice and the Subsidiary Body for Implementation established by Articles 9 and 10 of the Convention shall serve as, respectively, the Subsidiary Body for Scientific and Technological Advice and the Subsidiary Body for Implementation of this Protocol. The provisions relating to the functioning of these two bodies under the Convention shall apply *mutatis mutandis* to this Protocol. Sessions of the meetings of the Subsidiary Body for Scientific and Technological Advice and the Subsidiary Body for Implementation of this Protocol shall be held in conjunction with the meetings of, respectively, the Subsidiary Body for Scientific and Technological Advice and the Subsidiary Body for Implementation of the Convention.

2. Parties to the Convention that are not Parties to this Protocol may participate as observers in the proceedings of any session of the subsidiary bodies. When the subsidiary bodies serve as the subsidiary bodies of this Protocol, decisions under this Protocol shall be taken only by the Parties to this Protocol.

3. When the subsidiary bodies established by Articles 9 and 10 of the Convention exercise their functions with regard to matters concerning this Protocol, any member of the Bureaux of those subsidiary bodies representing a Party to the Convention but, at that time, not a party to this Protocol, shall be substituted by an additional member to be elected by and from amongst the Parties to this Protocol.

ARTICLE 16

The Conference of the Parties serving as the meeting of the Parties to this Protocol shall, as soon as practicable, consider the application to this Protocol of, and modify as appropriate, the multilateral consultative process referred to in Article 13 of the Convention, in the light of any relevant decisions that may be taken by the Conference of the Parties. Any multilateral consultative process that may be applied to this Protocol shall operate without prejudice to the procedures and mechanisms established in accordance with Article 17.

ARTICLE 16 BIS

The Conference of the Parties shall define the relevant principles, modalities, rules and guidelines, in particular for verification, reporting and accountability for emissions trading. The Parties included in Annex B may participate in emissions trading for the purposes of fulfilling their commitments under Article 3 of this Protocol. Any such trading shall be supplemental to domestic actions for the purpose of meeting quantified emission limitation and reduction commitments under that Article.

ARTICLE 17

The Conference of the Parties serving as the meeting of the Parties to this Protocol shall, at its first session, approve appropriate and effective procedures and mechanisms to determine and to address cases of non-compliance with the provisions of this Protocol, including through the development of an indicative list of consequences, taking into account the cause, type, degree and frequency of non-compliance. Any procedures and mechanisms under this Article entailing binding consequences shall be adopted by means of an amendment to this Protocol.

ARTICLE 18

The provisions of Article 14 of the Convention on settlement of disputes shall apply *mutatis mutandis* to this Protocol.

ARTICLE 19

1. Any Party may propose amendments to this Protocol.

2. Amendments to this Protocol shall be adopted at an ordinary session of the Conference of the Parties serving as the meeting of the Parties to this Protocol. The text of any proposed amendment to this Protocol shall be communicated to the Parties by the secretariat at least six

months before the meeting at which it is proposed for adoption. The secretariat shall also communicate the text of any proposed amendments to the Parties and signatories to the Convention and, for information, to the Depositary.

3. The Parties shall make every effort to reach agreement on any proposed amendment to this Protocol by consensus. If all efforts at consensus have been exhausted, and no agreement reached, the amendment shall as a last resort be adopted by a three-fourths majority vote of the Parties present and voting at the meeting. The adopted amendment shall be communicated by the secretariat to the Depositary, who shall circulate it to all Parties for their acceptance.

4. Instruments of acceptance in respect of an amendment shall be deposited with the Depositary. An amendment adopted in accordance with paragraph 3 above shall enter into force for those Parties having accepted it on the ninetieth day after the date of receipt by the Depositary of an instrument of acceptance by at least three fourths of the Parties to this Protocol.

5. The amendment shall enter into force for any other Party on the ninetieth day after the date on which that Party deposits with the Depositary its instrument of acceptance of the said amendment.

ARTICLE 20

1. Annexes to this Protocol shall form an integral part thereof and, unless otherwise expressly provided, a reference to this Protocol constitutes at the same time a reference to any annexes thereto. Any annexes adopted after the entry into force of this Protocol shall be restricted to lists, forms and any other material of a descriptive nature that is of a scientific, technical, procedural or administrative character.

2. Any Party may make proposals for an annex to this Protocol and may propose amendments to annexes to this Protocol.

3. Annexes to this Protocol and amendments to annexes to this Protocol shall be adopted at an ordinary session of the Conference of the Parties serving as the meeting of the Parties to this Protocol. The text of any proposed annex or amendment to an annex shall be communicated to the Parties by the secretariat at least six months before the meeting at which it is proposed for adoption. The secretariat shall also communicate the text of any proposed annex or amendment to an annex to the Parties and signatories to the Convention and, for information, to the Depositary.

4. The Parties shall make every effort to reach agreement on any proposed annex or amendment to an annex by consensus. If all efforts at consensus have been exhausted, and no agreement reached, the annex or amendment to an annex shall as a last resort be adopted by a three-fourths majority vote of the Parties present and voting at the meeting. The adopted annex or amendment to an annex shall be communicated by the secretariat to the Depositary, who shall circulate it to all Parties for their acceptance.

5. An annex, other than Annex A or B, that has been adopted or amended in accordance with paragraphs 3 and 4 above shall enter into force for all Parties to this Protocol six months after the date of the communication by the Depositary to such Parties of the adoption or amendment of the annex, except for those Parties that have notified the Depositary in writing within that period of their non-acceptance of the annex or amendment to the annex. The annex or amendment to an annex shall enter into force for Parties which withdraw their notification of non-acceptance on the ninetieth day after the date on which withdrawal of such notification has been received by the Depositary.

6. If the adoption of an annex or an amendment to an annex involves an amendment to this Protocol, that annex or amendment to an annex shall not enter into force until such time as the amendment to this Protocol enters into force.

7. Amendments to Annexes A and B to this Protocol shall be adopted and enter into force in accordance with the procedure set out in Article 19, provided that any amendments to Annex B shall be adopted only with the written consent of the Party concerned.

ARTICLE 21

1. Each Party shall have one vote, except as provided for in paragraph 2 below.

2. Regional economic integration organizations, in matters within their competence, shall exercise their right to vote with a number of votes equal to the number of their member States which are Parties to this Protocol. Such an organization shall not exercise its right to vote if any of its member States exercises its right, and vice versa.

ARTICLE 22

The Secretary-General of the United Nations shall be the Depositary of this Protocol.

ARTICLE 23

1. This Protocol shall be open for signature and subject to ratification, acceptance or approval by States and regional economic integration organizations which are Parties to the Convention. It shall be open for signature at United Nations Headquarters in New York from 16 March 1998 to 15 March 1999. This Protocol shall be open for accession from the day after the date on which it is closed for signature. Instruments of ratification, acceptance, approval or accession shall be deposited with the Depositary.

2. Any regional economic integration organization which becomes a Party to this Protocol without any of its member States being a Party shall be bound by all the obligations under this Protocol. In the case of such organizations, one or more of whose member States is a Party to this Protocol, the organization and its member States shall decide on their respective responsibilities for the performance of their obligations under this Protocol. In such cases, the organization and the member States shall not be entitled to exercise rights under this Protocol concurrently.

3. In their instruments of ratification, acceptance, approval or accession, regional economic integration organizations shall declare the extent of their competence with respect to the matters governed by this Protocol. These organizations shall also inform the Depositary, who shall in turn inform the Parties, of any substantial modification in the extent of their competence.

ARTICLE 24

1. This Protocol shall enter into force on the ninetieth day after the date on which not less than 55 Parties to the Convention, incorporating Parties included in Annex I which accounted in total for at least 55 per cent of the total carbon dioxide emissions for 1990 of the Parties included in Annex I, have deposited their instruments of ratification, acceptance, approval or accession.

2. For the purposes of this Article, "the total carbon dioxide emissions for 1990 of the Parties included in Annex I" means the amount communicated on or before the date of adoption of this Protocol by the Parties included in Annex I in their first national communications submitted in accordance with Article 12 of the Convention.

3. For each State or regional economic integration organization that ratifies, accepts or approves this Protocol or accedes thereto after the conditions set out in paragraph 1 above for the entry into force have been fulfilled, this Protocol shall enter into force on the ninetieth day following the date of deposit of its instrument of ratification, acceptance, approval or accession.

4. For the purposes of this Article, any instrument deposited by a regional economic integration organization shall not be counted as additional to those deposited by States members of the organization.

ARTICLE 25

No reservations may be made to this Protocol.

ARTICLE 26

1. At any time after three years from the date on which this Protocol has entered into force for a Party, that Party may withdraw from this Protocol by giving written notification to the Depositary.
2. Any such withdrawal shall take effect upon expiry of one year from the date of receipt by the Depositary of the notification of withdrawal, or on such later date as may be specified in the notification of withdrawal.
3. Any Party that withdraws from the Convention shall be considered as also having withdrawn from this Protocol.

ARTICLE 27

The original of this Protocol, of which the Arabic, Chinese, English, French, Russian and Spanish texts are equally authentic, shall be deposited with the Secretary-General of the United Nations.

Done at Kyoto this tenth day of December one thousand nine hundred and ninety-seven.

Annex A

Greenhouse gases

Carbon dioxide (CO_2)
Methane (CH_4)
Nitrous oxide (N_2O)
Hydrofluorocarbons (HFCs)
Perfluorocarbons (PFCs)
Sulphur hexafluoride (SF_6)

Sectors/source categories

Energy
Fuel combustion
Energy industries
Manufacturing industries and construction
Transport
Other sectors
Other

Fugitive emissions from fuels
Solid fuels
Oil and natural gas
Other

Industrial processes
Mineral products
Chemical industry
Metal production
Other production
Production of halocarbons and sulphur hexafluoride
Consumption of halocarbons and sulphur hexafluoride
Other

Solvent and other product use

Agriculture
Enteric fermentation
Manure management
Rice cultivation
Agricultural soils
Prescribed burning of savannas
Field burning of agricultural residues
Other

Waste
Solid waste disposal on land
Wastewater handling
Waste incineration
Other

Annex B

Party Quantified emission limitation or reduction commitment
(percentage of base year or period)

Australia 108
Austria 92
Belgium 92
Bulgaria* 92
Canada 94
Croatia* 95
Czech Republic* 92
Denmark 92
Estonia* 92
European Community 92
Finland 92

France 92
Germany 92
Greece 92
Hungary* 94
Iceland 110
Ireland 92
Italy 92
Japan 94
Latvia* 92
Liechtenstein 92
Lithuania* 92
Luxembourg 92
Monaco 92
Netherlands 92
New Zealand 100
Norway 101
Poland* 94
Portugal 92
Romania* 92
Russian Federation* 100
Slovakia* 92
Slovenia* 92
Spain 92
Sweden 92
Switzerland 92
Ukraine* 100
United Kingdom of Great Britain and Northern Ireland 92
United States of America 93

* Countries that are undergoing the process of transition to a market economy.

Appendix III

Distr. LIMITED

A/AC.237/L.23/Add.1

27 September 1994

Original: ENGLISH

INTERGOVERNMENTAL NEGOTIATING COMMITTEE FOR A FRAMEWORK CONVENTION ON CLIMATE CHANGE

Eleventh session

New York, 6-17 February 1995

Item 7 (b) of the provisional agenda

MATTERS RELATING TO COMMITMENTS

REVIEW OF THE ADEQUACY OF COMMITMENTS IN ARTICLE 4, PARAS. 2(A) AND (B)

Letter dated 22 September 1994 from the Federal Ministry for the Environment, Nature Conservation and Nuclear Safety of Germany to the Executive Secretary of the interim secretariat, transmitting further elements of a protocol to the Convention

Note by the interim secretariat

Addendum

1. Document A/AC.237/L.23 contains the text of a proposed protocol to the Convention as submitted by Trinidad and Tobago, on behalf of the States Parties to the Convention that are members of the Alliance of Small Island States.

2. The Government of Germany has transmitted to the interim secretariat proposals for further elements of such a protocol. These proposals were sent on 27 September 1994 with a note verbale to all Permanent Missions to the United Nations in New York and at Geneva.

3. The Committee is invited to consider these proposals from Germany in addition to the proposals from Trinidad and Tobago at its eleventh session, in preparation for the first session of the Conference of the Parties, to be held in Berlin from 28 March to 7 April 1995.

Letter dated 22 September 1994 from the Federal Ministry for the Environment, Nature Conservation and Nuclear Safety of Germany to the Executive Secretary of the interim secretariat, transmitting further elements of a protocol to the Convention.

Germany very much welcomes that the Alliance of Small Island States has taken the initiative in proposing a "Draft Protocol to the United Nations Framework Convention on Climate Change on Greenhouse Gas Emission Reduction ". In order to further stimulate substantial consideration of this central political issue to be decided upon at COP 1 in Berlin, we ask you to kindly circulate the paper enclosed among the Parties to the Convention before 28 September 1994.

(Signed)

Cornelia Quennet-Thielen
Head of Division
International Co-operation
Environment and Development, International Legal Affairs

Elements for a comprehensive protocol to the FCCC

Reviewing the adequacy of commitments under the Framework Convention on Climate Change (FCCC) will be a central policy issue at the first Conference of Parties (COP). Assuming that COP 1 will conclude that the commitments of Annex I Parties under Article 4.2(a) and (b) of the FCCC are inadequate, it is essential that the commitments are further elaborated as soon as possible.

It is noted with interest that Trinidad and Tobago on behalf of the AOSIS have tabled a Draft protocol to the FCCC on 21 September 1994.

In this paper, Germany presents proposals for further elements of such a protocol, in particular in the field of policies and measures. The proposals are, in general, based on the German position paper presented at INC 10.

General Considerations

In order to reach the ultimate objective of the FCCC, it is necessary not only to regulate CO_2 but also to limit or reduce other relevant greenhouse gases.

Therefore a comprehensive protocol for greenhouse gases and their sources and sinks, as well as for all sectors, should be negotiated without delay, flexible enough to allow the gradual incorporation of relevant substances in line with the progress of scientific knowledge, and combining targets and timetables for the limitation and/or reduction commitments with coordinated policies and measures to be implemented.

The commitments relating to greenhouse gases laid down in the protocol should be reviewed at regular intervals to be determined and, if need be, further developed in the light of the ultimate objective enshrined in Article 2 of the FCCC, taking into account the best available scientific information and assessments on climate change and its impacts, as well as relevant technical, social and economic information.

In setting new, farther reaching commitments, attention must be paid to ensuring the principles in Art. 3 FCCC guiding the Parties in their endeavour to protect the climate system, including the principle of a balanced burden sharing in line with the common but differentiated responsibilities of Parties as well as their capabili-

ties and possibilities.

We should continue to work towards balanced commitments on the part of industrialized and developing countries, for example by means of further reporting commitments for non-Annex I Parties and commitments to limit the rise in emissions in the case of certain more advanced developing countries.

Carbon dioxide

I. Basic commitments

1. The Annex I Parties shall commit themselves to stabilising their CO2 emissions - individually or jointly - by the year 2000 at 1990 levels, i.e., to return their emissions to 1990 levels until 2000 and maintain them thereafter.

2. a) Moreover it is necessary to undertake ambitious CO2 reduction in the period after 2000. Therefore, specific targets and timetables for the reduction of CO2 emissions of Annex I Parties will have to be agreed upon in the protocol.

   b) The protocol should stipulate the launching of negotiations on additional reduction steps in the light of the ultimate objective laid down in Art. 2 of the FCCC.

3. Emission reduction should basically be carried out within each Party's own territory. With regard to the reduction commitments listed in No. 2a above a certain portion yet to be determined may be met by joint implementation 1/, whereby a significant part of the commitments must be met through measures within each Party's own territory.

II. Further policies and measures

1. Energy (energy supply, building-related issues, industry)

   1.1 Annex I Parties shall adopt national 2/ policies and take corresponding measures to improve the efficiency of large scale combustion plants taking account of Annex (...).

   1.2 The Annex I Parties shall adopt national 2/ policies and take corresponding measures to improve the efficiency of facilities providing low-temperature heat (e.g. heating boilers etc.) and other small scale combustion plants, taking account of Annex (...), as well as for the regular control of these installations, in principle.

1.3 The Annex I Parties shall adopt national 2/ policies and take corresponding measures to improve the efficiency of certain products and/or to provide energy efficiency standards and labelling giving information on energy consumption; this applies particularly to:

- household appliances,

- appliances in entertainment and communication technology,

- air conditioning systems and refrigeration facilities,

-the insulation properties of certain materials.

---

* In this context it is noted that the Draft Protocol tabled by Trinidad and Tobago proposes in its Art. 3 (1) that each of the Annex I Parties shall reduce its 1990 level of anthropogenic emissions of CO_2 by at least 20% by the year 2005.

1.4 The Annex I Parties shall adopt national 2/policies and take cor responding measures

- to reduce energy loss during distribution,

- to use waste heat from large industrial plants,

- to improve heating and insulation in buildings,

- to ensure consumption based accounting of costs for heating, air conditioning and hot water,

- to implement energy diagnostic assessments of areas of indus

- to step up the replacement of fuels with high CO_2 emissions with those which are low in CO_2 or CO_2-free,

-to utilize least-cost planning,

-to increase the use of contracting (planning, implementation, financing and operation of energy supply by third parties).

2. Renewable energies

The Annex I Parties shall adopt national 2/ policies and take corresponding measures for the development, generation and increased use of renewable energies.

3. Traffic and transport

3.1 The Annex I Parties shall adopt national 2/ policies and take corresponding measures to avoid and reduce unnecessary traffic and transport.

3.2 The Annex I Parties shall adopt national 2/ policies and take corresponding measures to shift to more environmentally sound means of transport.

3.3 The Annex I Parties shall adopt national 2/ policies and take corresponding measures to reduce gradually the average fuel consumption of newly licensed cars to 5 1/100 km, as far as possible, and increase the efficiency of other means of tranport.

3.4 The Annex I Parties shall commit themselves within the framework of international negotiations to work towards dismantling tax reliefs for air traffic, in particular

- the petrol tax exemption for aircraft fuel,
- the value added tax exemption for cross-border traffic.

4. Forests

4.1 The Annex I Parties shall adopt national policies and take corresponding measures with regard to management, conservation and sustainable development of forests in order to conserve and enhance, as appropriate, sinks and reservoirs.

4.2 The Annex I Parties shall work towards ensuring that internationally agreed criteria are determined and applied for the management, conservation and sustainable development of forests.

Methane

In order to limit CH_4 emissions, the Annex I Parties shall adopt national policies and take corresponding measures to

1. reduce CH_4 emissions in the extraction, transport and use of crude oil- and natural gas,
2. reduce emissions of pit gas from hard-coal mining (use of pit gas as energy),
3. avoid and/or utilize landfill gas,
4. reduce and utilize sewage gas,
5. utilize biogas.

Nitrous oxide

The Annex I Parties shall adopt national 2/ policies and take corresponding measures to reduce N_2O emissions

1. from industrial installations,
2. from agriculture, in particular by means of fertilization adapted to plant needs and site-related, and by improving fertilization dosage,
3. in the field of animal husbandry and the storage of animal waste products.

Fluorocarbons

1. The Annex I Parties shall commit themselves to reporting on the production and consumption of FCs and HFCs.
2. The Annex I Parties shall adopt national 2/ policies and take corresponding measures
    - 2.1 to deal with commercial and industrial refrigeration equipment and air conditioning equipment in such a way that such used substances may be recovered or disposed of in an environmentally sound way, as possible,
    - 2.2 to limit by precautionary measures the leakage of these substances
        - * during manufacture, installation, operation and servicing of commercial and industrial refrigeration and air conditioning equipment,
        - * when such substances are used as feedstocks in the manufacture of other chemicals and
        - * when such substances are inadvertently produced by the manufacture of other chemicals.

General policies and measures

1. The Annex I Parties shall in the aforementioned areas promote the increased use of economic instruments, for example charges such as taxes, fees, contributions and special levies, including a CO2/energy tax, promotional measures, tradeable permits, bubble of offset policies including joint implementation, advantages to those who act in an environmentally sound way, and voluntary agreements, product markings and the environmental label.

2. The Annex I Parties shall in principle dismantle fiscal and other benefits which promote conduct contrary to the objective of the protocol.

3. The Annex I Parties shall develop and implement education and training programmes in the aforementioned areas.

4. The Annex I Parties shall intensify research, if possible in the framework of international and intergovernmental programmes, extend scientific cooperation and ensure the processing, evaluation and transmission of scientific knowledge.

5. The Annex I Parties shall develop and implement information and advice programmes in the aforementioned areas.

6. The protocol should provide for further development of the commitments under the Convention entered into by Annex II Parties (Art. 4, para. 5) as to promote, facilitate and finance, as appropriate, the transfer of, or access to, environmentally sound technologies and know how to other Parties, in particular non Annex I Parties.

Notes

1/ This in principle also applies to possible commitments regarding other greenhouse gases.

2/ This also includes the policies and measures adopted by regional economic integration organizations.

Distr. LIMITED

A/AC.237/L.23

27 September 1994

Original: ENGLISH

INTERGOVERNMENTAL NEGOTIATING COMMITTEE FOR A FRAMEWORK CONVENTION ON CLIMATE-CHANGE

Eleventh session

New York, 6-17 February 1995

Item 7 (b) of the provisional agenda

MATTERS RELATING TO COMMITMENTS

REVIEW OF THE ADEQUACY OF COMMITMENTS IN ARTICLE 4, PARAS. 2 (A) AND (B)

Letter dated 20 September 1994 from the Permanent Representative of Trinidad and Tobago to the United Nations in New York to the Executive Secretary of the interim secretariat, transmitting a draft protocol to the United Nations Framework Convention on Climate Change on Greenhouse Gas Emissions Reduction.

Note by the interim secretariat

1. Article 17.1 of the Convention provides that "the Conference of the Parties may, at any ordinary session, adopt protocols to the Convention", and Article 17.2 provides that "the text of any proposed protocol shall be communicated to the Parties by the secretariat at least six months before such a session".
2. In accordance with these provisions, the Permanent Representative of Trinidad and Tobago to the United Nations, on behalf of the States Parties to the Convention that are members of the Alliance of Small Island States, transmitted the text of a proposed protocol to the interim secretariat. Consequently, on 21 September 1994, the interim secretariat sent a note verbale containing this text to all Permanent Missions to the United Nations in New York and at Geneva.
3. The Committee is invited to consider this draft protocol at its eleventh session, in preparation for the first session of the Conference of the Parties, to be held in Berlin from 28 March to 7 April 1995.

Letter dated 20 September 1994 from the Permanent Representative of Trinidad and Tobago to the United Nations in New York to the Executive Secretary of the interim secretariat, transmitting a draft protocol to the United Nations Framework Convention on Climate Change on Greenhouse Gas Emissions Reduction.

Please find attached a "Draft Protocol to the United Nations Framework Convention on Climate Change on Greenhouse Gas Emissions Reduction" which has been prepared by the Alliance of Small Island States.

I would be grateful if arrangements could be made for the Draft Protocol to be considered at the eleventh session of the Intergovernmental Negotiating Committee for a Framework Convention on Climate Change and at the first session of the Conference of the Parties in 1995.

(Signed)

Annette des Iles
Ambassador
Permanent Representative

DRAFT PROTOCOL TO THE UNITED NATIONS FRAMEWORK CONVENTION ON CLIMATE CHANGE ON GREENHOUSE GAS EMISSION REDUCTION

PREAMBLE

The Parties to this Protocol,

Being Parties to the 1992 United Nations Framework Convention on Climate Change (the Convention),

Acknowledging that the ultimate objective of the Convention and of this protocol is to achieve stabilisation of atmospheric greenhouse gas concentrations at a level which would prevent dangerous anthropogenic interference with the climate system within a time frame sufficient to allow ecosystems to adapt naturally to climate change, to ensure that food production is not threatened and to enable economic development to proceed in a sustainable manner,

Noting that Article 3 of the Convention requires developed country Parties to take the lead in combating climate change and the adverse effects thereof,

Conscious of the need for developed country Parties to adopt specific targets and time frames for reducing emission of greenhouse gases to achieve the Objective of the Convention,

Reaffirming that per capita emissions in developing countries are still relatively low and that the share of global emissions originating in developing countries will-grow to meet their social and development needs,

Aware of the advantages of coordinating relevant measures and strategies, including specific administrative and economic instruments to achieve the Objective of the Convention,

Acknowledging that in accordance with the principle of common but differentiated responsibility Parties to the Convention and this Protocol should in future re-examine the impact of global efforts to combat climate change and the adverse effects thereof,

HAVE AGREED AS FOLLOWS:

ARTICLE 1 - DEFINITIONS

For the purposes of this protocol:

1. "Annex I Parties" means the developed country Parties and other developed Parties included in Annex I of the Convention, that are also Parties to this Protocol.
2. "Conference of the Parties" means the Conference of the Parties to the Convention established pursuant to Article 7 of the Convention.
3. "Convention" means the United Nations Framework Convention on Climate Change adopted on 9 May 1992, and unless the text otherwise indicates, the terms defined in Article 1 of the Convention shall have the same meaning in this Protocol.
4. "Meeting of the Parties" means the Conference of the Parties established pursuant to Article 8 of this Protocol.
5. "Montreal Protocol" means the 1987 Montreal Protocol on Substances that Deplete the Ozone Layer, as subsequently adjusted and amended.
6. "Objective" means the ultimate objective stated in Article 2 of the Convention.
7. "Parties" means the Parties to the present Protocol.
8. "Parties to the Convention" means Parties for whom the Convention has legally entered into force in accordance with the Convention's provisions.
9. "Principles" means, unless the context otherwise requires, the Principles stated in Article 3 of the Convention.
10. "Secretariat" means the Secretariat established under Article 8 of the Convention.

ARTICLE 2 - BASIC COMMITMENT

In accordance with the Objective and Principles of the Convention, all Parties, taking into account their common but differentiated responsibilities and their specific national and regional development priorities, objectives and circumstances shall implement national and, where appropriate, regional programmes containing measures to mitigate climate change by addressing anthropogenic emissions by sources and removals by sinks of all greenhouse gases not controlled by the Montreal Protocol.

ARTICLE 3 - TARGETS FOR GREENHOUSE GAS REDUCTIONS

1. Each of the Annex I Parties shall:

   (a) Reduce its 1990 level of anthropogenic emissions of carbon dioxide by at least 20 percent by the year 2005; and

   (b) Adopt specific targets and timetables to limit or reduce other greenhouse gases not controlled by the Montreal Protocol, including, targets and timetables for methane, nitrous oxide and fluorocarbons, in accordance with a programme of additional commitments to be negotiated and adopted by the first Meeting of the Parties.

2. The Meeting of the Parties shall review and revise the commitments of the Annex I Parties contained in subparagraphs (a), and the commitments adopted pursuant to subparagraph (b) above, in accordance with the precautionary principle and the best available scientific information and assessment of climate change, not later than five years after the entry into force of this Protocol and thereafter at regular intervals to be determined by the Meeting of the Parties.

3. Any Party not included in Annex I of the Convention that has expressed its intention to be bound by Article 4.2(a) and (b) of the Convention in accordance with Article 4.2(g) of the Convention, may in its instrument of ratification, acceptance, approval or accession to this Protocol, or at any time thereafter, notify the Depositary that it intends to be bound by Articles 3 to 5 of this Protocol. The Depositary shall inform the other signatories and Parties of any such notification.

ARTICLE 4 - COORDINATION MECHANISM

1. A mechanism to facilitate Annex I Parties' coordination of measures developed to achieve the Objective of the Convention is hereby established to provide the Meeting of the Parties, and as appropriate, the institutions established by the Convention and other relevant international organizations, with timely advice on the coordination of such measures.

2. The mechanism shall provide advice on the full range of measures the coordination of which could assist Annex I Parties implement their commitments to combat climate change and the adverse effects thereof. These measures shall include, inter alia, the coordination of economic instruments such as taxes or subsidies, administrative instruments such as least cost or integrated resource planning, energy efficiency standards and recycling, and specific measures covering the industrial, energy, transportation, land use, agriculture, waste management and forestry sectors.

3. The mechanism shall be open to participation by all Parties to this Protocol and shall be multidisciplinary. It shall comprise governmental representatives competent in the relevant field of expertise. It shall report regularly to the Meeting of the Parties on all aspects of its work.

4. The functions, terms of reference, organization and operation of this mechanism shall be elaborated further at the first Meeting of the Parties.

ARTICLE 5 - REPORTING REQUIREMENTS

1. Each of the Annex I Parties shall communicate to the Meeting of the Parties, through the Secretariat, the following information:

   (a) A detailed description of the policies, programmes and measures it has undertaken to implement its commitments under Articles 2 to 4 above; and

   (b) A specific estimate of the effects that these policies, programmes and measures will have on anthropogenic emissions by its sources and removals by its sinks of greenhouses gases.

2. Each of the Annex I Parties shall also provide information on the full costs and benefits of the policies and measures described in subparagraphs (a) and (b), and indicate how such policies and measures form part of a least cost implementation strategy. At their first Meeting, Parties shall consider and agree on methodologies for Annex I Parties to undertake calculations of the full costs and benefits referred to above.

3. Each of the Annex I Parties shall make its initial communication within one year of the entry into force of the Protocol for that Party. The frequency of subsequent communications shall be determined by the first Meeting of the Parties.

ARTICLE 6 - INSTITUTIONAL ARRANGEMENTS

1. Decisions under this Protocol shall be taken only by Parties to this Protocol. Parties shall bear in mind that the Conference of the Parties, as the supreme body of the Convention, must also keep under regular review the implementation of any related legal instruments, such as this Protocol.

2. The Secretariat, financial mechanism and subsidiary bodies established by the Convention or by the Conference of the Parties shall be available for use by the Parties subject to the prior approval of such arrangements by the Conference of the Parties.

3. To avoid duplication, overlap and conflicts between the institutional structures and reporting requirements established by the Convention and those established by the Protocol, the first Meeting of the Parties shall seek guidance on these matters from the Conference of the Parties.

4. The first Meeting of the Parties shall adopt by consensus financial rules, in accordance with guidance received from the Conference of the Parties, to ensure that any additional funds for the operation of this Protocol are provided by the Parties to this Protocol.

ARTICLE 7 - TRANSFER OF TECHNOLOGY

Annex 1 Parties shall ensure:

(a) That the best available technologies, practices and processes that control, reduce or prevent anthropogenic emissions of greenhouse gases not controlled by the Montreal Protocol in all relevant sectors, including the industrial, energy, transport, industry, agriculture, forestry and waste management sectors, are expeditiously transferred to developing country Parties to this Protocol;

(b) That every practicable step is taken to support the development and enhancement of the endogenous capacities and technologies of developing country Parties;

(c) That the transfers referred to in subparagraph (a) occur under fair and most favourable conditions.

ARTICLE 8 - MEETING OF THE PARTIES

1. A Meeting of the Parties is hereby established. The Meeting of the Parties shall keep under regular review the implementation of the Protocol and shall make, within its mandate, the decisions necessary to achieve its effective implementation.

To this end, it shall:

(a) Periodically review the commitments of the Parties and the institutional arrangements under the Protocol, in the light of Objective and Principles of the Convention, the experience gained in the Protocol's implementation and the evolution of scientific and technological knowledge;

(b) Adopt targets and timetables referred to in Article 3.1;

(c) Review and revise the commitments of Annex Parties referred to in Article 3.2;

(d) Receive, review and ensure the publication of information submitted to it, including the reports submitted by Parties pursuant to Article 5;

(e) Regularly assess the overall aggregated effect of the steps taken by Annex I Parties in the light of the latest scientific assessments concerning climate change, and of the Protocol's objective, and ensure the publication of such assessments;

(f) At its first Meeting, agree upon and adopt by consensus, rules of procedure and financial rules for itself and for any subsidiary body;

(g) Receive reports from, and if necessary give guidance to the financial mechanism and to subsidiary bodies on matters relating to the implementation of this Protocol;

(h) Seek and utilize, where appropriate, the services and cooperation of, and information provided by, competent international organizations and intergovernmental and non-governmental bodies;

(i) Establish further subsidiary bodies as may be deemed necessary for the implementation of the Protocol;

(j) Make recommendations on any matters necessary for the implementation of this Protocol;

(k) Consider and if approved, adopt proposals for any amendment of or addition to this Protocol or any annex thereto; and

(l) Exercise such other functions as are required for the implementation of this Protocol, including any functions assigned to it by the Conference of the Parties.

2. The Secretariat shall convene the first Meeting of the Parties not later than one year after the date of the entry into force of this Protocol and if feasible, in conjunction with a meeting of the Conference of the Parties. Thereafter ordinary sessions of the Meeting of the Parties shall be held every year in conjunction with sessions of the Conference of the Parties, unless otherwise decided by the Meeting of the Parties.

3. Extraordinary sessions of the Meeting of the Parties shall be held at such other times as may be deemed necessary by the Meeting of the Parties, or at the written request of any Party, provided that, within six months of such a request being communicated to the Parties by the Secretariat, it is supported by at least one third of the Parties.

4. The United Nations, its specialized agencies and the International Atomic Energy Agency, as well as any State not Party to this Protocol, may be represented at any Meeting of the Parties as observers. Any body or agency, whether national or international, governmental or non-governmental, which is qualified in matters covered by the Protocol and which has informed the Secretariat of its wish to be represented at a session of the Meeting of the Parties as an observer, may be so admitted unless at least one third of the Parties present object. The admission and participation of observers shall be subject to the rules of procedure adopted by the Parties at their first Meeting.

ARTICLE 9 - SETTLEMENT OF DISPUTES

In the event of a dispute arising between any two or more Parties concerning the interpretation or application of the Protocol, the Parties shall seek a settlement in accordance with Article 14 of the Convention.

ARTICLE 10 - AMENDMENTS TO THE PROTOCOL

1. Any Party may propose amendments to the Protocol.

2. Amendments to the Protocol shall be adopted at a Meeting of the Parties. The text of any proposed amendment to the Protocol shall be communicated to the Secretariat who shall inform the Parties of the proposed amendment at least 6 months before the meeting at which it is proposed for adoption.

3. The Parties shall make every effort to reach agreement on any proposed amendments to the Protocol by consensus. If all efforts at consensus have been exhausted, and no agreement reached, the amendment shall as a last resort be adopted by a two-thirds majority vote of the Parties present and voting at the meeting. The adopted amendment shall be communicated by the Secretariat to the Depositary, who shall circulate it to all Parties for their acceptance.

4. Instruments of acceptance in respect of an amendment shall be deposited with the Depositary. An amendment adopted in accordance with paragraph 3 above shall enter into force for those Parties having accepted it on the ninetieth day after the date of receipt by the Depositary of an instrument of acceptance by at least two-thirds of the Parties to the Protocol.

5. The amendment shall enter into force for any other Party on the ninetieth day after the date on which that Party deposits with the Depositary its instrument of acceptance of the said amendment.

6. For the purposes of this Article, "Parties present and voting" means Parties present and casting an affirmative or negative vote.

ARTICLE 11 - ADOPTION AND AMENDMENT OF ANNEXES TO THE PROTOCOL

1. The Meeting of the Parties may adopt annexes to this Protocol. Such annexes shall form an integral part thereof and, unless otherwise expressly provided, a reference to the Protocol shall constitute at the same time a reference to any annexes thereto.

2. Annexes to the Protocol shall be proposed and adopted in accordance with the procedure set out in Article 10, paragraphs 2 and 3 above.

3. An annex that has been adopted in accordance with paragraph 2 above shall enter into force for all Parties to the Protocol six months after the date of the communication by the Depositary to such Parties of the adoption of the annex, except for those Parties that have notified the Depositary, in writing, within that period of their non-acceptance of the annex. The annex shall enter into force for Parties which withdraw their notification of non-acceptance on the ninetieth day after the date on which withdrawal of such notification has been received by the Depositary.

4. The proposal, adoption and entry into force of amendments to annexes to the Protocol shall be subject to the same procedure as that for the proposal and adoption of annexes to the Protocol in accordance with paragraphs 2 and 3 above.

5. If the adoption of an annex or an amendment to an annex involves an amendment to the Protocol, that annex or amendment to an annex shall not enter into force until such time as the amendment to the Protocol enters into force.

ARTICLE 12 - RELATIONSHIP BETWEEN THIS PROTOCOL AND THE CONVENTION

Except as otherwise provided in this Protocol, the provisions of the Convention relating to its protocols shall apply to this Protocol.

ARTICLE 13 - RIGHT TO VOTE

1. Each Party to the Protocol shall have one vote, except as provided for in paragraph 2 below.
2. Regional economic integration organisations, in matters within their competence, shall exercise their right to vote with a number of votes equal to the number of their member States which are Parties to the Protocol. Such an organisation shall not exercise its right to vote if any of its member States exercised its right, and vice versa.

ARTICLE 14 - DEPOSITARY

As provided under Article 19 of the Convention, the Secretary-General of the United Nations shall be the Depositary of the Protocol.

ARTICLE 15 - SIGNATURE

The Protocol shall be open for signature by States Members of the United Nations or any of its specialised agencies or that are Parties to the Statute of the International Court of Justice and by regional economic integration organisations at Berlin during the first session of the Conference of the Parties, and thereafter at the United Nations Headquarters in New York from 8 April 1995 to 7 April 1996.

ARTICLE 16 - RATIFICATION, ACCEPTANCE, APPROVAL AND ACCESSION

1. The Protocol shall be subject to ratification, acceptance, approval or accession by States and regional economic integration organisations. It shall be open for accession from the day after the date on which the Protocol is closed for signature. Instruments of ratification, acceptance, approval or accession shall be deposited with the Depositary.
2. Any regional economic integration organisation which becomes a Party to the Protocol without any of its member States being a Party shall be bound by all the obligations under the Protocol. In the case of such organisations, one or more of whose member States is a Party to the Protocol, the organisation and its member States shall decide on their respective responsibilities for the performance of their obligations under the Protocol. In such cases, the organisation and the member States shall not be entitled to exercise rights under the Protocol concurrently.

3. In their instruments of ratification, acceptance, approval or accession, regional economic integration organizations shall declare the extent of their competence with respect to the matters governed by the Protocol. These organisations shall also inform the Depositary, who shall in turn inform the Parties, of any substantial modification in the extent of their competence.

ARTICLE 17 - ENTRY INTO FORCE

1. The Protocol shall enter into force on the ninetieth day after the date of deposit of the thirtieth instrument of ratification, acceptance, approval or accession.

2. For each State or regional economic integration organisation which ratifies, accepts or approves the Protocol or accedes thereto after the deposit of the instrument of ratification, acceptance, approval or accession, the Protocol shall enter into force on the ninetieth day after the date of deposit by such State or regional economic integration organisation of its instrument of ratification, acceptance, approval or accession.

3. For the purposes of paragraphs 1 and 2 above, any instrument deposited by a regional economic integration organisation shall not be counted as additional to those deposited by State members of the organisation.

ARTICLE 18 - RESERVATIONS

No reservations may be made to this Protocol.

ARTICLE 19 - WITHDRAWAL

1. At any time after three years from the date on which the Protocol has entered into force for a Party, that Party may withdraw from the Protocol by giving notice in writing to the Depositary.

2. Any such withdrawal shall take effect upon expiry of one year from the date of receipt by the Depositary of the notification of withdrawal or on such later date as may be specified in the notice of withdrawal.

3. Any Party which withdraws from the Convention shall, pursuant to Article 25 of the Convention, be considered to have withdrawn from this Protocol also.

ARTICLE 20 - AUTHENTIC TEXTS

The original of this Protocol, of which the Arabic, Chinese, English, French, Russian and Spanish texts are equally authentic, shall be deposited with the Secretary-General of the United Nations.

IN WITNESS WHEREOF the undersigned, being duly authorised to that effect, have signed this Protocol.

DONE at Berlin this _____ day of ______ one thousand nine hundred and ninety five.

Appendix IV

Summary for Policymakers

The Regional Impacts of Climate Change: An Assessment of Vulnerability

Edited by Robert T. Watson, Marufu C. Zinyowera, Richard H. Moss

A Special Report of IPCC Working Group II, November 1997

6.4 Australasia Australasia includes Australia, New Zealand and their outlying islands. The region spans the tropics to mid-latitudes and has varied climates and ecosystems, ranging from interior deserts to mountain rainforests. The climate is strongly affected by the oceanic environment and the ENSO phenomenon.

Ecosystems: Some of the region's ecosystems appear to be very vulnerable to climate change, at least in the long term, because alterations to soils, plants and ecosystems are very likely, and there may be increases in fire occurrence and insect outbreaks. Many species will be able to adapt, but in some instances, a reduction of species diversity is highly likely. Any changes will occur in a landscape already fragmented by agricultural and urban development; such changes will add to existing problems such as land degradation, weeds and pest infestations. Impacts on aquatic ecosystems from changes in river flow, flood frequency, and nutrient and sediment inputs are likely to be greatest in the drier parts of the region. Coastal ecosystems are vulnerable to the impacts of sea-level rise and possible changes in local meteorology. Tropical coral reefs, including the Great Barrier Reef, may be able to keep pace with sea-level rise,but will be vulnerable to bleaching and death of corals induced by episodes of higher sea temperatures and other stresses. Measures to facilitate adaptation include better rangeland management; plantings along waterways; and research, monitoring and prediction. Active manipulation of species generally will not be feasible in the region's extensive natural and lightly managed ecosystems.

Hydrology and Water Resources: Vulnerability appears to be potentially high. Any reduction of water availability, especially in Australia's extensive drought-prone areas, would sharpen competition among uses, including agriculture and wetland ecosystem needs. Freshwater supplies on low-lying islands are also vulnerable. More frequent high-rainfall events may enhance groundwater recharge and dam-filling events, but they also may increase the impacts of flooding, landslides and erosion, with flood-prone urban areas being heavily exposed to financial loss. Reduced snowpack and a shorter snow season appear likely, and New Zealand's glaciers are likely to shrink further. Some adaptation options are available, but the cost involved would be high.

Food and Fiber Production: Vulnerability appears to be low, at least in the next few decades (potentially high sensitivity coupled with high adaptability). Agriculture in the region is adaptable, and production increases are likely in some cases. However, there may be a trend toward increased vulnerability in the longer term,especially in warmer and more water-limited parts of Australia, where initial gains for some crops are eroded later as the delayed full effects of climate change (e.g., changes in temperature and precipitation) tend to outweigh the more immediate benefits of increased atmospheric CO_2 concentrations. Impacts will vary widely from district to district and crop to crop. There will be changes in growth and quality of crops and pastures; shifts in the suitability of districts for particular crops; and possibly increased problems with weeds, pests and diseases. Rangeland pastoralism and irrigated agriculture will be especially affected where rainfall changes occur. Changes in food production elsewhere in the world, which affect prices, would have major economic impacts on the region. With regard to forestry, the longer time to maturity results in a relatively large exposure to financial loss from extreme events, fire or any locally rapid change in climate conditions.

Coastal Systems: Parts of the region's coasts and rapidly growing coastal settlements and infrastructure are very vulnerable to any increase in coastal flooding and erosion arising from sea-level rise and meteorological changes. Indigenous coastal and island communities in the Torres Strait and in New Zealand's Pacific island territories are especially vulnerable. Many adaptation options exist, although these measures are not easily implemented on low-lying islands. Moreover, climate change and sea-level rise generally are not well accommodated in current coastal management planning frameworks.

Human Settlements: In addition to hydrological and coastal risks, moderate vulnerability is present from a variety of impacts on air quality, drainage, waste disposal, mining, transport, insurance and tourism. Overall, these effects are likely to be small relative to other economic influences, but they still may represent significant costs for large industries.

Human Health: Some degree of vulnerability is apparent. Indigenous communities and the economically disadvantaged may be more at risk. Increases are expected in heat-stress mor-tality, vector-borne diseases such as dengue, water and sewage related diseases, and urban pollution-related respiratory problems. Though small compared with the total burden of ill health, these impacts have the potential to cause considerable community impact and cost.

Conclusions: Australia's relatively low latitude makes it particularly vulnerable to impacts on its scarce water resources and on crops growing near or above their optimum temperatures, whereas New Zealand's cooler, wetter, mid-latitude location may lead to some benefit through the ready availability of suitable crops and likely increases in agricultural production. In both countries, however, there is a wide range of situations where vulnerability is thought to be moderate to high,particularly in ecosystems, hydrology, coastal zones, human settlements and human health.

INDEX

A

B

Q

R

S

T

U

V

W

Y

Advances in Global Change Research

1. P. Martens and J. Rotmans (eds.): *Climate Change: An Integrated Perspective.* 1999
ISBN 0-7923-5996-8
2. A. Gillespie and W.C.G. Burns (eds.): *Climate Change in the South Pacific: Impacts and Responses in Australia, New Zealand, and Small Island States.* 2000
ISBN 0-7923-6077-X

KLUWER ACADEMIC PUBLISHERS – DORDRECHT / BOSTON / LONDON

MIX
Papier aus verantwortungsvollen Quellen
Paper from responsible sources
FSC® C105338

If you have any concerns about our products,
you can contact us on
ProductSafety@springernature.com

In case Publisher is established outside the EU,
the EU authorized representative is:
Springer Nature Customer Service Center GmbH
Europaplatz 3, 69115 Heidelberg, Germany

Printed by Libri Plureos GmbH
in Hamburg, Germany